*Using Statistical Methods
for Water Quality Management*

WILEY SERIES IN STATISTICS IN PRACTICE

Advisory Editor, MARIAN SCOTT, *University of Glasgow, Scotland, UK*

Founding Editor, VIC BARNETT, *Nottingham Trent University, UK*

Statistics in Practice is an important international series of texts which provide detailed coverage of statistical concepts, methods and worked case studies in specific fields of investigation and study.

With sound motivation and many worked practical examples, the books show in down-to-earth terms how to select and use an appropriate range of statistical techniques in a particular practical field within each title's special topic area.

The books provide statistical support for professionals and research workers across a range of employment fields and research environments. Subject areas covered include medicine and pharmaceutics; industry, finance and commerce; public services; the earth and environmental sciences, and so on.

The books also provide support to students studying statistical courses applied to the above areas. The demand for graduates to be equipped for the work environment has led to such courses becoming increasingly prevalent at universities and colleges.

It is our aim to present judiciously chosen and well-written workbooks to meet everyday practical needs. Feedback of views from readers will be most valuable to monitor the success of this aim.

A complete list of titles in this series appears at the end of the volume.

Using Statistical Methods for Water Quality Management
Issues, Problems and Solutions

Graham B. McBride
National Institute of Water & Atmospheric Research
Hamilton, New Zealand

A JOHN WILEY & SONS, INC., PUBLICATION

Copyright © 2005 by John Wiley & Sons, Inc. All rights reserved.

Published by John Wiley & Sons, Inc., Hoboken, New Jersey.
Published simultaneously in Canada.

No part of this publication may be reproduced, stored in a retrieval system or transmitted in any form or by any means, electronic, mechanical, photocopying, recording, scanning or otherwise, except as permitted under Section 107 or 108 of the 1976 United States Copyright Act, without either the prior written permission of the Publisher, or authorization through payment of the appropriate per-copy fee to the Copyright Clearance Center, Inc., 222 Rosewood Drive, Danvers, MA 01923, (978) 750-8400, fax (978) 750-4744, or on the web at www.copyright.com. Requests to the Publisher for permission should be addressed to the Permissions Department, John Wiley & Sons, Inc., 111 River Street, Hoboken, NJ 07030, (201) 748-6011, fax (201) 748-6008, e-mail: permreq@wiley.com.

Limit of Liability/Disclaimer of Warranty: While the publisher and author have used their best efforts in preparing this book, they make no representation or warranties with respect to the accuracy or completeness of the contents of this book and specifically disclaim any implied warranties of merchantability or fitness for a particular purpose. No warranty may be created or extended by sales representatives or written sales materials. The advice and strategies contained herein may not be suitable for your situation. You should consult with a professional where appropriate. Neither the publisher nor author shall be liable for any loss of profit or any other commercial damages, including but not limited to special, incidental, consequential, or other damages.

For general information on our other products and services please contact our Customer Care Department within the U.S. at 877-762-2974, outside the U.S. at 317-572-3993 or fax 317-572-4002.

Wiley also publishes its books in a variety of electronic formats. Some content that appears in print, however, may not be available in electronic format.

Library of Congress Cataloging-in-Publication Data:

McBride, Graham B., 1948–
  Using statistical methods for water quality management : issues, problems and solutions / Graham B. McBride.
    p. cm.
  Includes bibliographical references and index.
  ISBN 0-471-47016-3 (cloth : alk. paper)
    1. Water quality—Management—Statistical methods. 2. Water quality management—Statistical methods. I. Title.
TD367.M34 2005
628.1'61—dc22   2004059090

10 9 8 7 6 5 4 3 2 1

This book is dedicated to

Robyn
Stephen and Christine
Richard, Chris
Desmond

Contents

List of Figures xv

List of Tables xix

Preface xxi

Part I Issues

1 Introduction 1
 1.1 *Conventions* 1
 1.2 *The essentials* 2
 1.3 *Meeting management's information needs* 4
 1.4 *Water quality observations as random variables* 5
 1.5 *Samples and populations* 6
 1.6 *Special characteristics of water quality data* 7
 1.7 *Data analysis protocols* 9

2 Basic concepts of probability and statistics 11

2.1	Probability rules		13
2.2	Representing data		16
	2.2.1	Types of data	16
	2.2.2	Frequency, bar graph, and histogram	17
	2.2.3	Describing the distribution of probability	18
	2.2.4	Discrete versus continuous data	19
	2.2.5	Summary statistics	20
2.3	Exploratory and graphical methods		22
2.4	Important distributions		25
2.5	Continuous distributions		27
	2.5.1	Normal distribution	27
	2.5.2	Lognormal distribution	30
	2.5.3	Gamma distribution	31
	2.5.4	Beta distribution	32
2.6	Discrete distributions		33
	2.6.1	Binomial distribution	33
	2.6.2	Poisson distribution	36
	2.6.3	Negative binomial distribution	36
	2.6.4	Hypergeometric distribution	36
	2.6.5	Multinomial distribution	36
2.7	Sampling distributions		36
	2.7.1	Student's t-distribution	37
	2.7.2	Chi-square distribution	37
	2.7.3	F-distribution	38
2.8	Statistical tables		38
2.9	Correlation and measures thereof		41
2.10	Statistical models and model parameters		42
2.11	Serial correlation, seasonality, trend, and scale		43
	2.11.1	Effect of serial correlation	44
2.12	Regression		47
	2.12.1	Applications to water quality	48
	2.12.2	Nonparametric regression	49
2.13	Estimating model parameters		49
	2.13.1	Point versus interval estimation	49

		2.13.2	Interval estimates	50
		2.13.3	Bias	51
		2.13.4	Percentiles	51
		Problems		52
		Appendix: Conditional probabilities—The Monty Hall dilemma		54
3	**Intervals**			**57**
	3.1	Confidence intervals		59
		3.1.1	For means	59
		3.1.2	For prediction	66
		3.1.3	For percentiles	68
	3.2	Tolerance intervals		70
	3.3	Credible intervals		71
	3.4	Intervals literature		72
		Problems		72
4	**Hypothesis testing**			**75**
	4.1	The classical approach		76
	4.2	The main steps in the classical approach		76
		4.2.1	What is meant by a decision?	78
	4.3	Commentary on the five steps		79
		4.3.1	Step 1: State the hypothesis to be tested	79
		4.3.2	Step 2: State the acceptable error risk	82
		4.3.3	Step 3: Sampling	83
		4.3.4	Step 4: Apply the decision rule	83
		4.3.5	Step 5: Rejecting, and accepting (maybe!)	86
	4.4	An "evidence-based" approach (for dichotomous data)		93
	4.5	Power analysis		94
	4.6	Multiple comparisons		95
	4.7	Nonparametric tests		98
	4.8	Randomization tests		100
	4.9	Bayesian methods		101

4.10	Likelihood approach	102
4.11	Hypothesis testing literature	103
	4.11.1 The point-null hypothesis testing debate	103
	4.11.2 The p-value culture	104
	Problems	105

5 Detection — 107

5.1	One-sided hypothesis tests	108
	5.1.1 Defining detection probability	108
	5.1.2 What burden of proof?	110
	5.1.3 Decision rules and detection probabilities	111
	5.1.4 Detection regions	113
	5.1.5 Non-normality and unequal variances	114
5.2	Point-null hypothesis tests	115
	5.2.1 Defining detection probability	115
	5.2.2 Power depends on sample size	116
	5.2.3 Decision rules and detection probabilities	118
	5.2.4 Detection regions	119
5.3	Equivalence tests	120
	5.3.1 Defining detection probability	120
	5.3.2 Decision rules and detection probabilities	123
	5.3.3 Equivalence tests versus point-null tests	125
	5.3.4 Detection regions	126
	5.3.5 Equivalence versus the Power Approach	128
	5.3.6 A potted history of equivalence testing	130
5.4	p-value behavior under different hypotheses	131
5.5	What has been detected?	132
5.6	Tests and confidence intervals	132
	5.6.1 A Bayesian alternative	134
5.7	Association or causation?	134
	Problems	136

6 Mathematics and calculation methods — 139

| 6.1 | Introduction | 139 |

6.2	Formulae and calculation methods for important distributions		140
	6.2.1	Normal distribution	140
	6.2.2	Lognormal distribution	141
	6.2.3	Gamma distribution	141
	6.2.4	Beta distribution	142
	6.2.5	Binomial distribution	143
	6.2.6	Poisson distribution	144
	6.2.7	Negative binomial distribution	145
	6.2.8	Hypergeometric distribution	147
	6.2.9	Multinomial distribution	148
	6.2.10	Student's t-distribution	148
	6.2.11	Chi-square distribution	149
	6.2.12	F-distribution	149
	Problems		150
6.3	Calculating confidence and tolerance limits on normal percentiles		152
	6.3.1	Two-sided equi-tailed confidence limits on normal percentiles	152
	6.3.2	One-sided confidence (tolerance) limits on normal percentiles	154
	6.3.3	Two-sided tolerance limits on percentiles	154
	6.3.4	Calculating noncentral t-probabilities	155
	Problems		156
6.4	Development of the detection formulae		157
	6.4.1	One-sided test, unknown variance	157
	6.4.2	One-sided test, known variance	158
	6.4.3	Point-null hypothesis test, unknown variance	158
	6.4.4	Point-null hypothesis test, known variance	159
	6.4.5	Equivalence hypothesis, known variance	159
	6.4.6	Inequivalence hypothesis, known variance	161
	6.4.7	(In)equivalence hypotheses, unknown variance	162

xii CONTENTS

	6.4.8 *Detection regions for equivalence tests and the Power Approach*	*163*
	6.4.9 *Calculating the incomplete noncentral t-probabilities*	*165*
	Problems	*167*

Part II Problems and Solutions

7 Formulating water quality standards   171
7.1	Setting the scene	*171*
	7.1.1 *Explicit and narrative statements*	*172*
7.2	Key questions	*173*
7.3	Enforceability	*174*
7.4	Requirements for enforceability	*175*
	Problems	*176*

8 Percentile standards (and the Reverend Bayes)   179
8.1	Two forms of percentile standards	*179*
8.2	Calculating nonparametric sample percentiles	*180*
	8.2.1 *Comparing the alternatives*	*181*
	8.2.2 *Median versus geometric mean*	*182*
8.3	Percentiles versus maxima	*182*
8.4	Percentile compliance rules	*183*
	8.4.1 *Developing rules using the Classical approach*	*184*
	8.4.2 *Developing rules using the Bayesian approach*	*185*
	8.4.3 *Comparing the Classical and Bayesian results*	*190*
	8.4.4 *Conclusion*	*192*
	Problems	*193*

9 Microbial water quality and human health   195
9.1	The key elements	*196*
	9.1.1 *Hazard identification*	*196*
	9.1.2 *Exposure assessment*	*198*
	9.1.3 *Dose-response*	*201*

	9.1.4	Risk profiling	203
9.2	Risk communication		204
9.3	Some useful mathematical details		206
	9.3.1	Added-zeroes distributions for dichotomous data	206
	9.3.2	Hockey stick distribution	207
	9.3.3	Loglogistic distribution	208
	9.3.4	Dose-response curves	208
	9.3.5	Conditional dose-response relationships	210
	9.3.6	What if the distribution of pathogens is not Poisson?	211
	9.3.7	Comparing conditional and average-dose risks	212
	Problems		213

10 MPNs and microbiology — 215

10.1	MPNs in standard tables	216
10.2	MPN "confidence intervals"	218
10.3	Sampling an MPN distribution	218
10.4	Calculating exact MPNs	222
	10.4.1 Considering each set separately	222
	10.4.2 Combining the sets	223
	Problems	224

11 Trends, impacts, concordance, detection limits — 225

11.1	Detecting and analyzing trends	225
	11.1.1 Trend slope estimator	225
	11.1.2 Seasonal Kendall trend test	226
	11.1.3 Two anomalies	227
	11.1.4 Multi-site trends; a Bayesian approach	228
11.2	Impacts and nominal data	231
11.3	Concordance assessment	232
	11.3.1 Cohen's kappa for dichotomous variables	232
	11.3.2 Continuous variables—Lin's concordance correlation coefficient	236

	11.4 Detection limits	236
	11.4.1 Type A censoring	236
	11.4.2 Type B censoring	237
	11.4.3 Handling "less than" data	238
	Problems	239
	Appendix	240

12 Answers to exercises — 243

References — 271

Author Index — 295

Topic Index — 299

Appendix A Statistical Tables — 305
- A.1 Normal distribution — 306
  - A.1.1 Cumulative unit normal distribution — 306
  - A.1.2 Percentiles of the unit normal distribution — 307
- A.2 t-distribution — 308
  - A.2.1 Cumulative Student's t-distribution — 308
  - A.2.2 Percentiles of Student's t-distribution — 309
- A.3 Binomial distribution — 310
  - A.3.1 Cumulative binomial probabilities — 310
- A.4 Normal confidence and tolerance intervals — 311
  - A.4.1 k factor for two-sided 95% confidence intervals on a normal percentile — 311
  - A.4.2 k factor for one-sided confidence (or tolerance) intervals on a normal percentile — 312
  - A.4.3 k factor for two-sided tolerance intervals on a normal percentile — 313

List of Figures

1.1	Accurate observations are both precise and unbiased.	7
2.1	Bar graph of MPN data.	18
2.2	Cumulative distribution function of MPN data.	19
2.3	Symmetry and skewness.	23
2.4	Lowess fit through data for the Ngakaroa Stream.	24
2.5	Boxplot for somatic coliphage in recreational waters.	25
2.6	Probability density function (pdf) for the unit normal distribution.	28
2.7	Cumulative distribution function (CDF) for the unit normal distribution.	29
2.8	Normal and lognormal probability density functions.	31
2.9	A variety of shapes for the pdf of the two-parameter gamma distribution.	32
2.10	A variety of shapes for the pdf of the two-parameter beta distribution.	33

2.11	Probability mass functions for three common discrete distributions.	35
2.12	Student t-distributions and the unit normal distribution.	37
2.13	Probability density functions for the χ^2- and F-distributions.	38
2.14	Five possibilities for reporting areas under the t-distribution.	39
2.15	Possible linear correlations.	41
2.16	Field Raynes effluent suspended solids data and serial correlation structure.	45
3.1	Geometric mean confidence limits as a function of sample size.	62
3.2	"Error bars."	66
3.3	Prediction intervals and confidence intervals for linear regression.	68
3.4	Confidence limits for the 95%ile as a function of sample size.	69
3.5	Highest density and equi-tails regions.	72
4.1	Critical region for a one-sided test using the unit normal distribution.	84
4.2	Critical region for a two-sided test using the unit normal distribution.	86
4.3	Binomial distribution for $n = 20$ and $\theta = 0.5$.	89
4.4	Negative binomial distribution for $k = 5$ and $\theta = 0.5$.	90
4.5	Binomial distribution for $n = 40$ and $\theta = 0.5$.	92
4.6	ANOVA: inferring differences in means from differences in variances.	96
5.1	Detection curves for two sample sizes for a one-sided inferiority test.	110

LIST OF FIGURES   xvii

5.2	Probabilities of accepting and rejecting complementary one-sided hypotheses.	111
5.3	Detection regions for inferiority and superiority tests.	115
5.4	Detection curve for the point-null hypothesis test.	116
5.5	Effect of increasing sample size on detection curves for a point-null test.	117
5.6	Detection region for the point-null hypothesis.	119
5.7	Detection probability for testing the equivalence hypothesis.	121
5.8	Detection probabilities for testing the equivalence and inequivalence hypotheses (variance unknown).	122
5.9	Detection probabilities for testing the equivalence and inequivalence hypotheses (variance known).	123
5.10	"Two one-sided test" definitions in terms of confidence intervals.	124
5.11	Detection curves for equivalence vs. point-null tests.	127
5.12	Detection regions for testing the inequivalence and equivalence hypotheses.	128
5.13	Detection regions for testing the inequivalence and equivalence hypotheses, compared to the Power Approach.	129
6.1	Unit normal, Student's t, and noncentral t pdfs.	153
8.1	Ratio of 95%ile to maximum for E. coli at 317 sites.	183
8.2	Maximum and 95%ile E. coli at 317 sites.	183
8.3	Four prior densities compared.	188
8.4	"Confidence of Compliance" with a 95%ile: sensitivity to prior distributions.	190
8.5	Confidence of Compliance for five percentiles.	191
9.1	General QRA flow diagram.	197

9.2	Fitting shellfish meal size data.	200
9.3	Virus and bacteria dose-response models.	202
9.4	"Hockey stick" empirical influent virus distribution.	207
9.5	Comparing simple exponential and binomial models.	212
9.6	Comparing beta-Poisson and beta-binomial models.	213
10.1	Binning procedure for MPN data.	221
11.1	Lowess fit through flow-adjusted data for the Ngakaroa Stream.	229
11.2	Robust fill-in for "less than" data.	239

List of Tables

2.1	Most probable number (MPN) data for shellfish-growing waters	17
2.2	Summary of terms for sample data	21
2.3	Summary of terms for populations	21
4.1	Matrix of possible results from a hypothesis test	77
5.1	Definition of population variables	112
5.2	Inferiority and superiority hypotheses	112
5.3	Decision rules and detection probabilities for inferiority and superiority tests	113
5.4	Formulae for inferiority and superiority tests and for point-null tests	114
5.5	Sample size and confidence interval width	117
5.6	Hypotheses for point-null hypothesis tests	118
5.7	Decision rules and detection probabilities for point-null hypothesis tests	118

5.8	Equivalence and inequivalence hypotheses	124
5.9	Decision rules and detection probabilities for equivalence hypothesis tests	125
5.10	Formulae for equivalence tests	126
5.11	Criteria for evidence of association	136
8.1	Permissible exceedances in a proof of safety approach for a 95%ile standard	192
8.2	Permissible exceedances in a proof of hazard approach for a 95%ile standard	193
9.1	Coastal shellfish bioaccumulation factors for two seasons (USA)	200
9.2	Calculated Campylobacter infection rates	204
10.1	Comparing MPN values	217
10.2	Comparing MPN confidence and credible limits.	219
10.3	MPN occurrence probabilities.	220
11.1	Groundwater trends in Canterbury	229
11.2	Presence/absence table	233
11.3	Frequency table for presence/absence data	233
11.4	Chance-correction table for presence/absence data	233
11.5	Suggested agreement descriptors.	234

Preface

When managing water resources we often seek to separate pattern from randomness. For example, is the quality of water in a lake really deteriorating? Is that because the land-use patterns in its watershed are intensifying? Just how sure can we be of a conclusion about that? What are the health risks from swimming in this river or at that beach? Is there an important change in stream benthic communities downstream of mine waste discharges? These are issues that statistical methods can shed light on. I hope this book will help water resources managers and scientists to formulate, implement and interpret better and more appropriate methods to address such matters.

To do so, I have not sought to repeat material readily available in other texts.[1] Rather than being a "how to" compendium of the many procedures to be found in them, I have focused on material not generally available elsewhere, both in general (such as in Bayesian modes of inference) and in particular (such as percentile standards, water-related human health risk modeling, and MPN methodology for microbiological enumerations).

This book's origins lie in an invitation from my colleague, Dr Bryan Manly, to present a paper to the Joint Statistical Meetings held in 2000 at Indianapolis. The title of the paper (as suggested by him) was "Statistical Methods Helping and Hindering Environmental Science and Management." It drew on experience I had gained over some years in using statistical methods to address practical questions that commonly

---

[1] Green (1979 [127]), Berthouex & Brown (1980 [23]), Snedecor & Cochran (1980 [291]), Sokal & Rohlf (1981 [292]), Iman & Conover (1983 [161]), Gilbert (1987 [118]), Krebs (1989 [176]), Ward *et al.* (1980 [338]), Gibbons (1994 [117]), Zar (1996 [349]), Manly (2001 [198]), Millard & Neerchal (2001 [226]), Helsel & Hirsch (2002 [147]), Townend (2002 [319]), Bolstad (2004 [28]).

arise in managing water resources. For example: Just what does the analyst mean when stating that a result is "statistically significant"? How should one devise and document percentile standards for effluent discharges, drinking water, or environmental standards? How many samples should be required in drinking water standards in order to provide a satisfactory level of assurance that the waters' quality sufficiently protects public health? When and how can one take account of what we think we already know when framing compliance rules or when analyzing experiments?

The Indianapolis paper, with its somewhat provocative title, was subsequently published (McBride 2002 [206]), and it resulted in an invitation to write this book. In doing so, I have attempted to give "lines of approach" (rather than "answers") to a range of issues that arise in water management (and in water science), such as those given above. I say "lines of approach" very deliberately because I have so often been asked, "What is the statistically correct way of analyzing these data?" In fact, there is seldom (if ever) *one* correct way, either for analysis of data or for design of sampling programs. We make judgments, not rules (Stewart-Oaten 1995 [300]). There may be a number of alternative approaches, each with its merits and drawbacks, and awareness of these can only be beneficial. There are of course many more inappropriate and incorrect lines of approach. So a function of this book is to clarify the appropriateness of such methods for various issues. By doing so, I hope to alert scientists and water resources managers to the wide variety of fruitful statistical methods that can be used. For example, it can come as a surprise that statistical methods do not just deal with numbers; incorporation of nominal data can greatly enhance their utility (e.g., using water color, wind direction, and octants of cloud cover in models of near-shore contamination by fecal bacteria).

In planning this book, I had first thought that it would be wise to introduce statistical concepts gradually, by way of a series of practical examples. As an instance of that, one could introduce Bayesian approaches to data analysis when considering the framing of percentile compliance rules. However, on reflection and discussion with colleagues, it became apparent that some general and pervasive issues needed first to be discussed in their own right. Accordingly, the book is in two parts. The first, *Issues*, consisting of six chapters, presents the material I consider to be important to be understood when contemplating using statistical methods for water resources management. It includes a number of practical examples. The five chapters in the second part, *Problems and Solutions*, draw on material given in the first part. They cover a range of topics that often arise but are not covered in most texts (formulating environmental standards, using percentile standards, microbiological water quality, "most probable numbers" (MPNs) and human health, and a catch-all chapter including material on trends, impacts, concordance and detection limits). Each chapter contains a number of set problems, for which full "answers" are give in the final (12th) chapter. Some of these problems are more difficult than others; some are discursive, in which case "answers" are longer than is usual in a statistics text.

So what are those "general and pervasive issues"? Some are details—but important details (especially concerning correct usage of standard errors and explaining "error bars"). Others are substantial. They have to do with identifying management's information needs, the role of various forms of hypothesis testing, and the types of intervals that may be used to account for uncertainty (confidence intervals, tolerance intervals, and credible intervals). To set the scene for these issues, Chapter 1 discusses the use of statistical populations and samples in water management. Chapter 2 presents basic concepts of probability and statistics, including different modes of statistical inference, especially frequentist versus Bayesian (throughout the book I present the case for using both). Chapter 3 discusses and contrasts confidence intervals, tolerance intervals and credible intervals. Chapters 4 and 5 then cover general and specific issues to do with statistical hypothesis testing—these tests are widely used in water management, and not always appropriately. These chapters address questions such as: When are "one-sided" tests appropriate? What can we actually infer from a test of a single "null" hypothesis? When should we use nonparametric approaches? Is there a corresponding Bayesian test? What is the role of equivalence testing? How do we give effect to the precautionary approach? We also introduce, in Chapter 5, the "detection probability" in the context of one-sided and two-sided tests. Detailed (though not rigorous) mathematical arguments and calculation routines are presented in Chapter 6—finer mathematical details are admirably covered in papers and other texts (e.g., Ferguson 1967 [90], Lehmann 1986 [183], Freund 1992 [101], Lee 1997 [179], Gelman *et al.* 2000 [110], Casella & Berger 2002 [45], Wellek 2003 [340]). This chapter can be skipped without losing too much of the main information presented.

A number of these issues are seldom addressed in much detail in applied statistical texts. For example, most workers actually interpret confidence intervals in a Bayesian manner, yet the Bayesian view of probability and associated modes of inference are seldom presented. It is important for environmental professionals to grasp such matters and their consequences. I became aware of this in 1989 when, while browsing in the basement of the Colorado State University library, I stumbled across the 1970 text by Morrison & Henkel *The Significance Test Controversy* ([232]). My response was one of surprise: "What controversy?" I read avidly, from a text arising not from environmental science, but from psychology (from where so many statistical advances have emanated). Many colleagues involved in water science and management (in my home country and abroad) have had the same response when confronted by the notion that statistical methodology, especially hypothesis testing, is accompanied (to this day) in the statistical literature by a degree of controversy. Some statisticians have even advocated that tests be abandoned altogether [e.g., in chapters of a recent book, edited by Harlow, Muliak, and Steiger (1997 [136]): *What if there were no significance tests?*]. But, in water management at least, we cannot abandon tests—comparisons, hence tests, are often required (between sites, between treatment levels, with standards,...).

Throughout I have made use of footnotes, giving detail that readers can pass over. More substantial technical details are contained in Appendices to some chapters. These too can be passed over without compromising understanding of the text. References are contained in a single list at the book's end, along with an author index and a subject index.

The book's text was prepared using LATEX, using the versatile WinEdt software. Most of the graphics have been prepared using Kaleidagraph™. A lot of the text was prepared listening to Miles Davis' *In a Silent Way*.

Acknowledgements Firstly, I thank Julian Ellis (Water Research Centre, England), whose "NS 29" deserves to be more widely known and read (see reference [84]), Robert Ward and Jim Loftis at Colorado State University, and Bryan Manly at West Inc., Wyoming (formerly at Otago University, New Zealand). Some of the material herein has been derived from Short Courses that Jim, Robert, Bryan and I have taught together. Other Fort Collins colleagues (Tom Sanders at CSU and Roy Irwin at the US National Park Service) have been a stimulation, as have many colleagues at NIWA (the National Institute of Water and Atmospheric Research)—Russell Cole, Bryce Cooper, Rob Collins, Rob Davies-Colley, Chris Francis, Mal Green, Judi Hewitt, Ian Jowett, Charles Pearson, Rick Pridmore, Kit Rutherford, Mike Scarsbrook, Rebecca Stott, James Sukias, Simon Thrush and Bob Wilcock. Another NIWA colleague, Mark Meleason, made a comprehensive and enthusiastic review of draft chapters as did also Alec Zwart (University of Waikato, Department of Statistics, Hamilton, New Zealand) and David Smith (Bureau of Water Supply, New York City Department of Environmental Protection). Murray Jorgensen, Bill Bolstad, James Curran, and Judith McWhirter (University of Waikato) have provided insights and feedback, as have Michael Taylor and Alexander Kouzminov (Ministry of Health, New Zealand). I also benefitted from discussions of many issues with other colleagues, both in New Zealand (Stan Abbott, Trevor Atkins, Andrew Ball, Noel Burns, James Court, Chris Hatton, Rob Lake, Gillian Lewis, Megan Linwood, Dorothy McCoubrey, John Maxted, Chris Nokes, Terry Ryan, Chris Skelly, Brett Stansfield, Desmond Till, Cliff Tipler, Peter van der Logt and Bill Vant), and abroad (Steve Hrudey, Dennis Helsel, Chuck Haas, Joe Germano, Sander Greenland, David Parkhurst and Samantha Rizak).

My employer (NIWA) has been very supportive. Funding for some of the work presented here has come from the New Zealand Foundation for Research, Science and Technology, and also from the New Zealand Ministry for the Environment, the Ministry of Health, and the Ministry of Research, Science and Technology. John Moore (Christchurch City Council) made coastal outfall human health risk analysis results available (for Chapter 9). Chris Hatton and Grant Barnes (Auckland Regional Council) made the Ngakaroa Stream data available (for Chapters 2 and 11). Bruce Apperley (Gisborne District Council, New Zealand) and Carl Hanson (Environment Canterbury, Christchurch, New Zealand) provided data cited in Chapter 11. Joke Baars (NIWA's librarian) has been unfailing in her industry, getting many obscure books and articles. Alison Bartley (NIWA) typed endless citations. I thank Susanne Steitz, Lisa Van Horn, Heather Bergman and Stephen Quigley at Wiley, and especially Amy Hendrickson at TeXnology Inc., for their unfailing help in producing the text.

Finally, I thank my wife Robyn for her constant support and encouragement.

GRAHAM B. MCBRIDE

Hamilton, New Zealand

Part I
Issues

'Far better an approximate answer to the right question, which is often vague, than the exact answer to the wrong question, which can always be made precise.'
—John Tukey (1962 [321])

'It's all in the mind, you know.'
—Wallace Greenslade and Spike Milligan, *The Goon Show* (1952–1960)

1
Introduction

1.1 CONVENTIONS

Key statistical terms are written in **bold type** when they first appear. Italicized upper case Roman letters are used to denote data and the statistics derived from them, making it easier to distinguish them from text. For example X_i refers to the ith numerical **datum** in a **sample**, and $\bar{X}$ is their **arithmetic mean**—usually just stated as the **mean**.[1] Data do not have to be numbers, they can be **categories** (such as water color), but in that case a mean is not defined. Italicized lowercase Greek letters are used for **population parameters** such as the **true mean**, which is **estimated** by $\bar{X}$ (so the true mean is denoted by μ). Samples are **drawn** from populations, and the parameters are a succinct way to summarize the shape of the populations. Lowercase italicized Roman letters (such as x and y) refer to the population **random variables** from which samples may be drawn. Particular values of those random variables are written in uppercase letters (e.g., X and Y). X_p represents the $100p$th percentile of the x distribution (so $X_{0.95}$ is the 95th percentile—hereafter written as "95%ile"). Overbars always denote a mean; for example, the difference between the means of two **independent** samples drawn from x and y is written as $\bar{X} - \bar{Y}$, whereas if those samples were **paired** we would be contemplating the mean difference $\overline{X - Y}$.

Like any discipline, statistics has its own language conventions. While most can be explained as we go along, those to do with the words **sample** and **error**, the symbol p, and even the word **statistics** need some explanation (though the context should make the meaning of such phrases clear).

---

[1] Other types of means may be used, especially the **geometric mean**.

2 INTRODUCTION

To an environmental professional a sample is a volume or mass of material taken from the environment—for example, a container of stream water for subsequent phosphorus analysis in a laboratory, or a reading of pH from a probe. But to a statistician, a sample is a collection of results or **observations**—for example, phosphorus concentrations in a set of water "samples." So to a statistician one **sample** contains **many data** and the **sample size** is the number of data in the sample—not the volume of the container. Similarly, an environmental professional would regard an **error** as just that—a mistake. But to a statistician, **sampling error** is the natural **variability** inherent among data taken from a **population** and is therefore always present and to be accounted for. If there are errors in measurement (nothing in this world is perfect), these are referred to as **measurement error**.

Then we have a **standard error**. This is not a recommended way to make a mistake! It is another measure of variability—not of the data, but of a parameter estimated from the data, such as the mean.[2] Next, because of conventions in wide use, we are forced to use the symbol *p* ambiguously (it can mean *p*-**value** (an **exceedance probability**) or **proportion**. Uppercase *P* generally denotes a **cumulative probability**, expressed as a percentage (e.g., the *P*th percentile).

Finally, the term "**statistics**" itself can have two meanings. The singular **statistic** is a number that characterizes some feature of a set of data—for example, the sample arithmetic mean as a measure of the **central tendency** of data. The plural **statistics** can also refer to the methods that use such numbers to make estimates and comparisons. In this text we generally use the term **statistic** for the former meaning and **statistical methods** for the latter.

1.2 THE ESSENTIALS

The measurements we make are taken from samples that are only a tiny proportion of the water body that we want information about. So any **inferences** drawn from the data are **uncertain** to some degree. Statistical methods are a means of handling this uncertainty, both in designing sampling programs and in obtaining useful information from the results of sampling.

Classical (**frequentist**) statistical methods allow us to draw inferences from sample data with, under certain assumptions, a known degree of uncertainty. Furthermore, they can permit different data analysts to follow the same procedures, thus reaching the same conclusions from a given set of data.[3] Statistical methods also provide a means of specifying how monitoring data should be analyzed before they are collected, thus ensuring that the information needs of management will be met from a planned monitoring program.

---

[2] If that parameter is the mean the standard error is in fact the **standard deviation** *of the mean* of the data, *not* of the data themselves. Importantly, it tends to get smaller as the number of data increases. More on that later (Section 2.13.1).

[3] This is not the case for the **Bayesian** methods that we shall meet later—although these can have some definite advantages, as we shall see.

Awareness of the need for the application of statistics in water quality management is growing, yet the application of statistics is still often not the general rule. An influential handbook in the UK (Ellis 1989 [84]) states:

> Generally speaking, the (water) industry seems much better served by publications and expertise dealing with the practical physical and chemical sides of sampling and analysis than it is equipped with advice on the statistical aspects of sample programme design.

That situation has been improving, especially with the appearance of texts such as those of Gilbert (1987 [118]) or Helsel & Hirsch (2002 [147]) Nevertheless, we begin with three pleas (given in the light of responses one sometimes hears in water management forums):

- The taunt "You can prove anything with statistics" is 100% *wrong*. In fact, you *can't* prove anything with statistics. What you *can* do is quantify and account for uncertainty and seek to distinguish pattern from randomness. That means one can design data collection programs and interpret results in an efficient and timely way. If statistical methods are not considered when designing and analyzing sampling programs, chances are that effort will be wasted, and inappropriate conclusions may be drawn.

- We cannot establish cause and effect with statistical methods *alone*, not even in **designed experiments** and certainly not in the **observational studies** that most environmental studies employ. A rise in dissolved oxygen concentrations ("DO") in the Mataura River (South Island, New Zealand) in the 1970s is correlated with a decrease in oxygen demand in the Waikato River in the North Island (they both had large-scale waste treatment improvements installed at about the same time), but improved effluent treatment in a North Island watershed was not the cause of improving DO in a South Island river. They are **associated** but **causality** should not be inferred. The safe option is to only infer causality if extra information makes this credible, even when notable correlation has been detected.

- When confronted by unwelcome results, one cannot stop people using the (in)famous remark variously attributed to Disraeli and Mark Twain. Disraeli's use of it has been thought to have been in response to the release of accurate numbers of war casualties: "there are three kinds of lies: lies, damned lies, and statistics." The best counter to this charge is the competent and honest use of statistical methods. Of special note is the need to explain the results of an analysis clearly. To merely say that a result is "statistically significant" is not meaningful to many (and may not even be clear to the claimant).

Always keep in mind the real, practical questions to which answers are needed:

- How do I phrase defensible water quality standards?

- What are the average water quality conditions in an area?

- What are the extreme water quality conditions in an area?
- Are effluent and receiving water standards being met?
- How different are the aquatic communities or water quality between regions (e.g., upstream and downstream of an outfall), and why are they different?
- Is an observed trend in water quality a cause for concern?

1.3 MEETING MANAGEMENT'S INFORMATION NEEDS

Water quality sampling is most often carried out in support of water quality management; some is for scientific enquiry. **Routine monitoring** is often a check on the effectiveness of a pollution control scheme—evaluating compliance with standards or long-term trends in quality. An **intensive survey** is often an attempt to gain a better understanding of the chemical, biological, and physical processes operating in a particular body of water. This information is useful for designing pollution control schemes and (when repeated over time) for assessing their effectiveness.

Statistic methods are useful in designing both types of sampling programs and in interpreting their results. But to apply statistics effectively, it is wise to attempt to enunciate the information needs of water quality management in statistical terms before monitoring begins. For example:

- Determine annual mean concentrations of certain water quality variables with specified precision.
- Identify long-term **trends** in water quality variables with a stated **probability** (i.e., "**power**") when trends exceed a specified magnitude over a specified time scale.
- Verify whether a standard has been met at least a critical proportion of the time, with specified **level of confidence**.
- Determine, with specified **precision**, the fraction of a population of lakes that exceed a certain level of acidity.

Water quality management implementation has often not stated an explicit need for a statistical view. An exception is water quality microbiological standards for recreational water requiring that a median or geometric mean of samples should not exceed some stated limit, as is common (e.g., in the USA, Australia, and New Zealand). Increasingly, we see water quality managers accepting a more statistical view of monitoring and wanting regulations to be expressed in more statistical terms. For example:

- A specified number of ground water samples are collected before site operation begins or "up-gradient" of the facility. These are termed background samples. Each year thereafter a given number of detection monitoring samples are taken.

Statistical tests (at a specified confidence level) are used to determine whether the mean of the detection monitoring samples is greater than that of the background samples. If so, a change in water quality due to a facility's operations might have occurred, and some action is taken—initially additional monitoring.

- Some water quality management regimes and effluent standards, and even receiving water standards, may require that a critical concentration limit should not be exceeded for at least 90% or 95% of a compliance assessment period—these are **percentile standards**).

This shift in emphasis away from simple data collection toward the adoption of statistical information objectives is expected to continue.

Information on water quality conditions that is generated by a monitoring program needs to be conveyed to users in a timely manner. So there is also a need for careful documentation of the monitoring system design, including its data analysis component (Ward *et al.* 1990 [338]). There is also a need for the allocation of staff time and other resources specifically for routine data analysis. Otherwise, data may pile up without any useful information being generated—the "data-rich – information-poor" syndrome (Ward *et al.* 1986 [336]).

Finally, three distinct approaches may be taken regarding the "burden of proof" to be adopted in management (e.g., in assessing compliance with wastewater effluent standards, compliance with drinking water standards, or in assessing environmental impacts). These are:

Permissive in which data have to be "convincing" before a breach of compliance or detection of an environmental impact is inferred (sometimes called the "benefit of doubt" approach, Ellis 1989 [84], or "proof of hazard").

Precautionary in which data have to be "convincing" before it is concluded that there has *not* been a breach or an impact (sometimes called the "fail safe"approach, Ellis 1989 [84], or "proof of safety").

Face value in which no allowance is made for sampling error, and data are taken—at face value—to represent the population from which they have been taken.

The appropriate burden of proof is not an issue that an environmental professional or statistician can resolve alone. Matters of policy, costs and fairness are involved, and it is wise to seek to clarify the issue with the water management agencies as much as possible. Further discussion, with respect to setting standards and assessing compliance with them is given in Chapters 7 and 8.

1.4 WATER QUALITY OBSERVATIONS AS RANDOM VARIABLES

The water quality measurements we make reflect a myriad of processes in the environment. It is well known how some of these processes operate (e.g., dilution, in-stream reaeration), and they can be quantified. Other processes are not so well understood.

6 INTRODUCTION

So whenever measurements are made, say of an effluent BOD_5 or a stream pH, some part of their variation has to be attributed to **chance** (i.e., variation we cannot explain).

Statistical methods offer a means of doing this in an objective way, rather than the all-too-common confusion of conflicting subjective opinions. They are capable of distinguishing between **randomness** (**noise**, random variability) and **pattern** (e.g., **seasonality**, trend) by repeatable procedures.

Because of the role of chance in measurements, statistical methods call for a water quality variable to be viewed as a **random variable**, or as a **stochastic process**. The distinction is that whereas a random variable does not imply some natural ordering of the results, a stochastic variable does (e.g., a time series of data at a particular site, or a set of samples down a river at the same time). In either case the value that a water quality variable may take has at least some element of randomness in it, and this needs to be recognized (Ward & Loftis 1983 [335]).

It would be tidier perhaps to use just one term, but we're stuck with the fact that the common literature uses both, with "stochastic" being much used in hydrological science. For practical purposes it is acceptable to use either, so we'll use the term **random variable**.

The major point is the idea of **randomness**. If you want to make an estimate of some water quality variable, or compare one water quality "population" with another, **random samples** may be needed at some stage (in space and/or in time). Indeed, some advocate that sampling should *always* be **random** (e.g., Cotter 1985 [60]). At first sight this is alarming: How do you fit random temporal sampling into a normal workday? It would make work scheduling much more difficult than it is now—and the reaction of the laboratory manager can be imagined. Random spatial sampling may also invoke difficulties of site access.

Fortunately, some form of **systematic sampling** is often an acceptable substitute for random sampling, as we shall see—especially for trend assessments. It is even possible—though it sounds like a contradiction in terms—to adopt **systematic random sampling**.

1.5 SAMPLES AND POPULATIONS

Samples are taken from **populations**. "Population" is a statistical term meaning all the possible measurements that could be made (but never are). Of course, it is occasionally possible to measure the whole population, but this is pretty much confined to industrial processes where a finite number of production units are made and all are tested. In environmental studies, most variables can be measured at an effectively infinite number of locations and times (e.g., any chemical in a stream) and it is not possible to measure the whole population. Even if only a finite number of values exist in the population (e.g., lead content of adult shellfish in a sandbank), it is never permissible sample them all (to do so would destroy the population).

So we end up taking just a few physical samples from a population. As an example, consider a volume of 1 liter taken once a month from a stream with an average monthly flow of $10 \, m^3 \, s^{-1}$. Of the volume of water flowing past the sampling site in that month, only a proportion of 4×10^{-11} has been taken. Or, consider taking 10 sediment or biomass samples from a sandflat with a surface area of $10,000$ m^2, using a sampler with diameter 10 cm. The proportion of sandflat sampled is 8 in a million.

When we do such sampling, we want our data to be as **accurate** as possible, given the usual constraints of money, time, and the need to avoid destructive sampling. To be accurate, data must be both **precise** and **unbiased**, as shown in Figure 1.1.

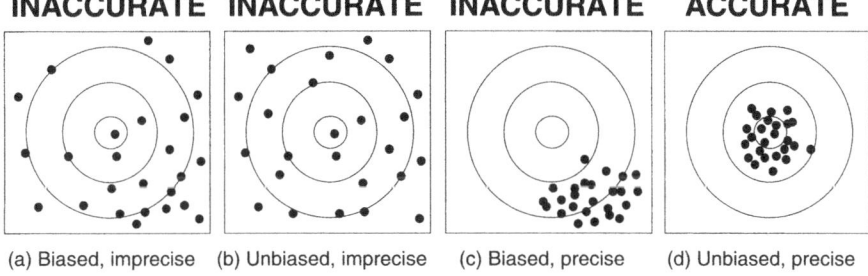

Fig. 1.1 Accurate observations are both precise and unbiased.

Random sampling removes **sampling bias**, but bias can still result from systematic **measurement errors**. Note that precise results can be inaccurate [as in (c) in Figure 1.1]. Imprecision results from high variability. This variability can be in the population and/or in the measurement technique.

If a sample is representative of the population from which it is drawn, then the small sample may be used to draw general conclusions about the entire population. This is called **statistical inference**. Representative samples are independent, and are usually also random samples. We shall talk more about this later (Section 2.11). For now, consider the following example. Imagine that we want to describe average water quality conditions at a particular point in a river using 100 water quality observations. If we collected all 100 of them in a single week, they would certainly not be representative of the population we were trying to describe. A representative sample would consist of observations spaced randomly (or in some cases uniformly) over the entire year or over several years if we want to describe long-term average conditions.

1.6 SPECIAL CHARACTERISTICS OF WATER QUALITY DATA

Consider these typical features of water quality data:

- Censored observations:

    - Nondetects (measurements less than a detection limit, or failure to detect the presence of even one unit).
    - Greater-thans.

- Changing sampling methods, analysis methods, detection limits.
- Missing values.
- Ambiguously recorded values:
  - using the detection limit or zero for the value of nondetects.
  - keeping the same nickname (e.g., "NO3") while changing the analysis method.[4]
  - defining the units of reporting—for example, omitting the "measured as" information (e.g., is that "NO3-N" or just "NO3"?).
  - is "NH3" free ammonia, total ammonia, ammonium?

One can even have both "less thans" and "greater thans" in the same data record—for example, for oxygen demand or microbiological constituents, where there are both data below detection limit and others that are "too numerous to count."[5]

These characteristics can make data analysis fraught with traps for the unwary. It is highly preferable to minimize the occurrence of these features when data are collected and stored rather than to try to deal with them at the data analysis stage. By doing this, more and better information will be obtained. A key element of this is the utilization of appropriate database software, for data storage and documentation. The lack thereof has been one of the main reasons for the "data-rich but information-poor" syndrome in water quality monitoring (Ward *et al.* 1986 [336]). Such software cannot be quickly put together, precisely because water quality data does have these special features. In particular, each datum needs considerable associated documentation to record the place, time, and method of sampling and laboratory analysis. Such documentation is essential if data are to retain value.

Ward *et al.* (1990 [338]) give six guiding principles for design of archive software:

- Data must be able to be stored and retrieved unambiguously.
- Software must be portable.
- Software must be easy to use.
- There must be protection against willful or accidental damage.
- Outward data transfer must also be unambiguous.
- There must be flexible enquiry and reporting features.

It is easier to achieve these principles with newer software, especially spreadsheets. In particular, "less-than" and "greater-than" data can now be properly documented (though we still do find many datasets where the "<" and ">" qualifiers have been dislocated from the numbers, especially in relational database systems).

---

[4]Note that the "3" in "NO3" is not a subscript, as it would normally be. This is because standard (ASCII) computer text (e.g., in a database or spreadsheet package) often does not use subscripts.

[5]These are often coded as "TNTC". Note too that some microbiological assays need to be interpreted with considerable care if reported as less than a limit. Parkhurst & Stern (1998 [254]) give special statistical procedures for the detection of oocysts of *Cryposporidium*.

Example: microbiological data If a record of microbiological concentrations contains numerous *identical* maximum values, one should immediately be suspicious that at least some of these data should be *greater than* that value. For example, the enumeration of fecal coliforms by multiple tube fermentation methods in a "dilution series" can result in all, or all but one, of the tubes showing a positive result. For a common dilution series,[6] if all but one tube is positive, the result is often reported as 1600 MPN per 100 mL.[7] If *all* tubes are positive, the result should be reported as ">1600 MPN per 100 mL." But if ">" has become dislocated from the "1600" figure, an ambiguous result occurs.[8] This is surprisingly common, especially in older datasets.

1.7 DATA ANALYSIS PROTOCOLS

A Data Analysis Protocol (DAP) is a prespecified approach to analyzing data from a certain sampling program or type of sampling program. This has many advantages (Ward *et al.* 1988 [337], Ward 1996 [334]):

- It removes the uncertainty in information generated that is associated with selection of statistical methods. All data analysts should get the same results using the same data and same protocol.[9]

- It confines any arguments over selection of methods to the network design stage, before the data are collected.

- Once data collection has begun, attention is shifted toward obtaining better quality data and away from the search for better statistical methods. (Of course, protocols are not cast in concrete; they can be changed or supplemented with additional analyses when there is good reason.)

- It provides a closer connection between network design and eventual data analysis by allowing the assumptions of the statistical methods and their sample size requirements to be considered in the design.

- It facilitates computerization of data analysis and permits routine analysis to be performed by people who are not statisticians.

Within the protocol, data are checked for certain characteristics, such as the fraction of values recorded as nondetects and the applicability of the normal distribution assumption. The results of this checking are used to determine which statistical tests for compliance will be selected.

Such protocols can be incorporated into software (e.g., IDT 1998 [160]).

---

[6] A $3 \times 3 \times 3$ setup of tubes each containing 100 mL, 10 mL, and 1 mL of sample.
[7] "MPN" is the *most probable number*, discussed further in Chapter 10.
[8] Sometimes the "<" sign is actually reported alongside the detection limit in a laboratory record, but subsequent data processing puts them into separate columns making their eventual dislocation almost inevitable.
[9] This is not quite true if Bayesian methods are to be used, as we shall see.

2
Basic concepts of probability and statistics

There is always **uncertainty** in sampling, whether we recognize it or not. With **statistical methods** we can make **estimates** and **comparisons**, but there is always uncertainty in our estimates and the **risk of error** when making comparisons. Such uncertainty and error risks must be accounted for in designing a sampling program, or in interpreting its data. This means that it may be necessary to compute the **probability** of the error occurring; probability deals with uncertainty.

Probability is a term capable of more than one interpretation. In many texts, one encounters only the **objective**, **frequency** interpretation. This identifies the probability of occurrence of a particular **event** (e.g., obtaining "heads" on a coin toss) as the proportion of the time that it would occur **in the long run**—meaning that were we to sample the same population a "large" number of times, the probability of the event occurring on any occasion would be "very close" to the proportion of times that it would actually happen in an infinite number of such **trials**. On each sampling we would be performing a **chance experiment**, and in each of the trials in these experiments the event must be equally likely to occur. But "equally likely" appears to be synonymous with "equally probable," in which case we have a circular argument. This poses a difficulty: Complete objectivity is hugely attractive to scientists, but this circularity seems to make it unattainable.

Note too that, in practice, it is common to refer to the probability of the event occurring from *one* trial (e.g., the probability of obtaining "heads" on the next toss of a coin). This is generally forced upon us, because we seldom can (or do) perform a long series of chance experiments. But a strict application of the above definition doesn't allow us to attach a probability to the outcome of a single trial, only to a collection of them. To assign a probability to a single coin toss we must *assume* that the coin is in fact unbiased. But the only way to fully check that assumption (i.e., assess the lack of bias) is to toss the coin many times; hence the strictly objective view is that one cannot talk of the probability of a single event occurring, such as "rain tomorrow," or even of "heads" on the next toss of a coin (von Mises 1964 [331]; see also Barnett 1999 [11]).

Statistical methods built on this objective approach are known as **frequentist**, sometimes also called **classical**.[1] They *are* objective in the sense that different data analysts will arrive at the same numerical results (if not their interpretation) when examining the same data.

The main (and earlier) alternative to the objective approach to probability is explicitly **subjective**, in which case probability necessarily incorporates some measure of belief or opinion (or, occasionally, knowledge).[2] Statistical methods built on this view are **Bayesian** (the Reverend Thomas Bayes, $\sim$1702–1761, was the first to formally enunciate this approach).[3] A Bayesian probability is sometimes called **inverse probability**. These methods are subjective in the sense that different analysts examining the same dataset may not arrive at the same numerical results. This is essentially because calculations of these probabilities require the updating of **prior probabilities**, and these can differ between analysts. These prior probabilities will have at least some influence on the final result—a **posterior probability**. A difficulty is therefore invoked, because they can incorporate "mere opinion." This has been a matter of some controversy among statisticians (Dennis 1996 [71])—"feelings" versus "frequencies" (Poole 1988 [260]). Nevertheless, Bayesian methods have been increasingly used in environmental sciences (Ellison 2004 [85]).

Both views have their advantages and strong advocates (as reviewed by Barnett 1999 [11]), and it is this author's view that both can be used to advantage, depending on the circumstances.[4] Both are presented in this text. The frequentist view may be most fruitful when planning a sampling program, because a large number of possible outcomes will need to be considered. This is sometimes called **before-trial betting** (Edwards 2000 [78][5]). But once the data are at hand—Edwards's **after-trial evalua-**

---

[1] But note that "classical" implies antiquity, whereas frequentist approaches are relatively new (one of the first applications was by father and son Benjamin and Charles Peirce in a US Federal Court case in 1867, concerning allegations of forgery of a signature on a will; see Menand (2001 [220]). For the most part they only entered the mainstream of statistical thought in the 20th century.

[2] A further approach is **fiducial**, introduced by the celebrated geneticist and statistician R. A. Fisher in the 1930s (Fisher 1930 [92]—see also Fisher 1959 [94]). This can be seen as some kind of a bridge between frequentist and Bayesian approaches, but has been held to be not well-defined and has yet to find much favor (Barnett 1999 [11])—but it may yet do so (Nelder 1999 [238]). We do not pursue it here.

[3] See Barnard (1958 [10]) or Press (2003 [264]) for a detailed presentation of Bayes' essay and its impact. Bayes was both an English Nonconformist minister and a mathematician. His essay was presented to the Royal Society by his friend Richard Price, in 1763. It was published in the Society's Philosophical Transactions in the same year, two years after Bayes' death.

[4] The authoritative *Advanced Theory of Statistics* by Stuart *et al.* (1999, Chapter 26 [304]) concludes that neither approach can claim superiority, as they each have to borrow ideas off the other to set up and justify a comprehensive theory. The authors of an influential text on statistical inference state their belief that "...there is no one best way to do statistics; some problems are best solved with classical statistics and some are best solved with Bayesian statistics" (Casella & Berger 2002, p. 437 [45]).

[5] Edwards attributes these useful terms to Hacking (see Hacking 1965 [133]).

tion—the frequency interpretation can be more problematic. This is particularly the case when interpreting **confidence intervals**, and the subjective view becomes more appropriate. In that case Bayesian methods may be more appropriate and useful.

In either case it is important that an **axiomatic** approach is used. Here probabilities are defined as mathematical entities and are required to behave according to defined rules. Either the objective or subjective approaches can be used in a manner consistent with these rules (there are a number of axiomatic systems that need not concern us here).[6] Accordingly, we do not need to separate these two types of probabilities in presenting probability rules.

Finally we note that while probability deals with uncertainty, it does not necessarily follow that it deals with **evidence**. The best measure of evidence may be a term related to probability, known as **likelihood**. Likelihood is not probability; essentially it provides a means of comparing the strength of support for alternative models or hypotheses.[7] This is discussed further in Section 4.10.

These issues will all be further highlighted and explained as we proceed through this text.[8]

2.1 PROBABILITY RULES

In the axioms of probability we have these important rules, for events A, B,....[9]

- Range of probability for any event A

$$0 \leq \Pr(A) \leq 1. \tag{2.1}$$

- A is contained in B

$$\Pr(A) \leq \Pr(B). \tag{2.2}$$

- For mutually exclusive events A and B

$$\Pr(A) = 1 - \Pr(B). \tag{2.3}$$

- For non-mutually exclusive events A and B

$$\Pr(A \text{ or } B) = \Pr(A) + \Pr(B) - \Pr(A \text{ and } B), \tag{2.4}$$

or

$$\Pr(A \text{ and } B) = \Pr(A) + \Pr(B) - \Pr(A \text{ or } B). \tag{2.5}$$

---

[6]These include Kolmogorov, Savage, Rényi—see Press (2003 [264]).
[7]A good rule of thumb (van Belle 2002 [327]) is to use frequentist, Bayesian or likelihood approaches as appropriate, in an "eclectic and ecumenical approach to inference." Nevertheless, one eminent statistician (Lindley 2000 [187]) holds that statistical inference is based on probability alone, not on the likelihood, *and* that such probabilities must be personal (i.e., Bayesian).
[8]See also the recent text on such topics, Taper & Lele (2004 [306]).
[9]In these rules "and" corresponds to the union symbol $\bigcup$ and "or" corresponds to the intersection symbol $\bigcap$.

14 BASIC CONCEPTS OF PROBABILITY AND STATISTICS

- Distributive rules for three events A, B, and C
$$\Pr[(A \text{ or } B) \text{ and } (A \text{ or } C)] = \Pr[(A) \text{ or } (B \text{ and } C)], \qquad (2.6)$$
or
$$\Pr[(A \text{ and } B) \text{ or } (A \text{ and } C)] = \Pr[(A) \text{ and } (B \text{ or } C)]. \qquad (2.7)$$

- Event B occurs, given that event A has occurred[10]
$$\Pr(B|A) = \frac{\Pr(A \text{ and } B)}{\Pr(A)}. \qquad (2.8)$$

- Rules defining independent events
$$\Pr(A \text{ and } B) = \Pr(A)\Pr(B), \qquad (2.9)$$
and
$$\Pr(B|A) = \Pr(B). \qquad (2.10)$$

- Multiplication rule
$$\Pr(A \text{ and } B) = \Pr(A)\Pr(B|A). \qquad (2.11)$$

- Rule of total probability for multiple mutually exclusive events
$$\Pr(A) = \sum_{i=1}^{k} \Pr(B_i)\Pr(A|B_i), \qquad (2.12)$$
where k is the number of events of type B.

- Bayes' rule for a single event[11]
$$\Pr(B|A) = \frac{\Pr(B)\Pr(A|B)}{\Pr(A)}. \qquad (2.13)$$

- Bayes' rule for multiple mutually exclusive events
$$\Pr(B_j|A) = \frac{\Pr(B_j)\Pr(A|B_j)}{\sum_{i=1}^{k}\Pr(B_i)\Pr(A|B_i)}, \qquad (2.14)$$
where k is the number of events of type B.

Care is needed when interpreting conditional probabilities,[12] especially because they lie at the heart of Bayesian analyses. This is dramatically demonstrated in the celebrated "Monty Hall dilemma," presented in this chapter's Appendix.[13] Note also that $\Pr(A \text{ and } B)$ and the conditional probability $\Pr(A|B)$ both refer to the probability that A and B occur, yet they are different numbers. Problems 2.1 and 2.2 address these issues.

---

[10]"$\Pr(B|A)$" is the probability that B occurs, given that A has occurred. It is a **conditional probability**.
[11]There is no controversy about this rule *per se*; controversy arises only when the events are hypotheses (rather than data), and the rule is used to manipulate probabilities about hypotheses.
[12]According to Casella & Berger (2002 [45]) they can be "particularly slippery entities."
[13]Casella & Berger (2002 [45]) present another intriguing puzzle to do with conditional probabilities, the *Three Prisoners* puzzle.

Example: estuarine sedimentation probabilities   A hydrodynamic and sediment transport modeling study of sedimentation in an estuary has shown that a combination of "strong westerly" winds (event A) and "greater than moderate" rainfall (event B) produces ecologically damaging deposition in a sensitive shoreline region. This estuary is subject to increased sediment loads due to urbanization of the surrounding catchment, and the exposed earth is more easily eroded by the rain and strong winds that drive the flow within the estuary. Meteorological data show that the probability of strong westerly winds occurring on any given day is $\Pr(A) = 0.1$, and the probability of greater than moderate rainfall on any given day is $\Pr(B) = 0.2$. If these events were independent, then by applying the first independence rule [Equation (2.9)] the probability of ecologically damaging deposition occurring on any given day would be $\Pr(A)\Pr(B) = 0.1 \times 0.2 = 0.02$. However, such independence is unlikely—both rainfall and wind are derived from meteorological forcing and are therefore unlikely to be independent. (Tides, on the other hand, are derived from astronomical forcing and are likely to be independent of winds and rain—disregarding any possibility of stormtides.) If further examination of meteorological data shows that that the probability of at least moderate rain given strong westerly winds is $\Pr(B|A) = 0.5$, then the probability of deposition on any given day [using Equation (2.11)] is $\Pr(A \text{ and } B) = \Pr(A)\Pr(B|A) = 0.1 \times 0.5 = 0.05$. This increase in calculated probability (from 0.02 to 0.05) may be very significant for management of the estuarine environment and associated land-use planning.

Example: Bayes' rule   Continuing the estuarine deposition example, let's imagine that the above calculations are contained in evidence placed before a land-use planning tribunal. The tribunal must consider what terms and conditions can and should be placed on development of land in the estuary's watershed. Furthermore, another party to the hearing (a land developer) tables evidence claiming that wind–rain–erosion is very unlikely to cause any significant deposition in the estuary. This party claims that whatever erosion occurs will be minor and will be the result of natural processes. Who and what should the tribunal believe? They could be advised to proceed as follows. Let's use **H** to denote the hypothesis that deposition will result from erosion and also **K** to denote the land developer's contention.[14] These hypotheses are mutually exclusive. Past experience indicates that **H** is rather more likely than **K**. So before proceeding to obtain further evidence, the tribunal assigns probabilities to the two hypotheses as $\Pr(\mathbf{H}) = 0.8$ and $\Pr(\mathbf{K}) = 0.2$; that is, we have prior probabilities in ratio 4 to 1. Further evidence is then tabled that in a similar estuary such erosion and undesirable deposition did actually occur (under similar conditions of wind and rain). Let's call this evidence E. The tribunal now needs to know how this evidence changes the prior probability of **H**; it wants to know the conditional probability $\Pr(\mathbf{H}|E)$. Bayes' rule [Equation (2.14)] is the vehicle by which this can be done—with $k = 2$, and with A playing the role of E, B_1 playing the role of **H** and B_2 playing the role of **K**. To use, it we need extra information: the conditional probability that the observed deposition

---

[14] We use the convention that when the event referes to a hypothesis, rather than data, that event is in bold non-italics characters.

16   BASIC CONCEPTS OF PROBABILITY AND STATISTICS

(in the other, similar estuary) could have occurred under each of the two hypotheses. Call these probabilities $\Pr(E|\mathbf{H})$ and $\Pr(E|\mathbf{K})$. It solicits expert opinion which says the former is more likely than the latter, with $\Pr(E|\mathbf{H}) = 0.7$ and $\Pr(E|\mathbf{K}) = 0.05$.[15] We can now calculate

$$\begin{aligned}\Pr(\mathbf{H}|E) &= \frac{\Pr(\mathbf{H})\Pr(E|\mathbf{H})}{\Pr(\mathbf{H})\Pr(E|\mathbf{H}) + \Pr(\mathbf{K})\Pr(E|\mathbf{K})} \\ &= \frac{0.8 \times 0.7}{0.8 \times 0.7 + 0.2 \times 0.05} = 0.982.\end{aligned}$$

The new evidence has considerably increased the weight that the tribunal would give to the original evidence, of undesirable deposition arising from wind and rain causing erosion of denuded land. That is because the prior probability (0.8) has been raised to the higher posterior probability (0.982). Had the new opinion been the converse of the above—$\Pr(E|\mathbf{H}) = 0.05$ and $\Pr(E|\mathbf{K}) = 0.7$—the posterior probability would be 0.222, a considerable reduction from the prior probability.

2.2   REPRESENTING DATA

2.2.1   Types of data

There are two main categories of data: **continuous** and **discrete**. As examples, data on salinity, chemical concentration, and aquatic organism size are continuous and thus include an effectively infinite number of values (in theory, if not in their actual measurement). In contrast, MPN microbiological enumeration data are discrete, having only a (usually small) finite number of possible values (MPNs—most probable numbers—are discussed in detail in Chapter 10).

Some further distinctions should be made. All continuous data and some discrete data (such as MPNs) are measured on the **interval scale**. On this scale, intervals between two data values have a clear meaning and are always numbers. Ratios of such numbers have a clear interpretation only if there is an absolute zero, in which "0" is the bottom of the scale. On such scales, twice 10 units is 20 units and so we talk of these data being on the **ratio scale**. Some data are reported on scales that do not have such an absolute zero (e.g., temperature, pH), and ratios are therefore problematic. Still further we have **circular scale** data, such as time of day, wind octant, and compass direction.

Discrete data can fall into two further categories, not on the interval scale, in which intervals do not have a clear meaning. First is the **ordinal scale**, where an order—a ranking—is implied. For example, observations of water clarity can use a scale such as "clear, slightly turbid, turbid, muddy." Nondetect data ("less thans") and "greater thans" are reported on this scale (though they are not necessarily measured on it). Interval scale data can be reduced to this scale so that nondetect data and "greater

---

[15]Note that $\Pr(E|\mathbf{H})$ and $\Pr(E|\mathbf{K})$ do not have to sum to one—they are not mutually exclusive events. In fact they refer to the *same* event, but with different conditionals.

thans" can be included in an analysis. Finally we have **nominal scale** data measuring attributes of some variable. These can be **dichotomous**, such as presence/absence of a microorganism, or they may be **polychotomous** such as water hue (blue, blue-green, green-yellow, ...).

> Statistical methods can be applied not just to discrete interval scale data, such as MPNs, but also to discrete ordinal and nominal scale data. This is of great importance for water quality management, because such data are very common—and useful.

Nevertheless we concentrate for now on discrete ordinal data and on continuous data (an example use of polychotomous data is given later, in Section 11.2).

2.2.2 Frequency, bar graph, and histogram

We shall concentrate on discrete ordinal data for a start. Such data can be summarized in a **frequency table**. For example, consider the MPN data in Table 2.1, where we have denoted the sample size by n (in the table, $n = 30$). The table shows how frequencies and relative frequencies are defined and calculated. In particular, the sum of the frequencies should equal the sample size, and the sum of the relative frequencies should be 100% (note that checking this property is compromised by excessive rounding of the individual relative frequencies). The cumulative frequency for a given MPN value is calculated merely by summing all frequencies up to that value.

Table 2.1 Most probable number (MPN) data for shellfish-growing waters[a]

MPN (/100 mL)	Frequency	Relative frequency (frequency/n) (%)	Cumulative frequency	Cumulative relative frequency (%)
2.9	8	26.7	8	26.7
3.6	10	33.3	18	60.0
9.1	6	20.0	24	80.0
23	3	10.0	27	90.0
43	2	6.7	29	96.7
460	1	3.3	30	100.0

[a] Data taken from United States Food and Drug Administration (1989). NSSP shellfish sanitation program manual of operations Part I: sanitation of shellfish growing areas, 1989 revision. Shellfish Sanitation Branch, Washington, D.C. (example in Appendix, p. APF-4).

Plotting the absolute or relative frequency versus MPN gives us a **bar graph**. This is a neat pictorial way to show the frequency of all observations—much easier to assimilate than a table (e.g., compare Table 2.1 with Figure 2.1).

2.2.3 Describing the distribution of probability

The bar chart in Figure 2.1 gives a picture of the **frequency distribution** of the data. It shows the three main features of distributions:

- central tendency (i.e., the most common data values)
- spread (variability around the central values)
- skewness (the degree of asymmetry)

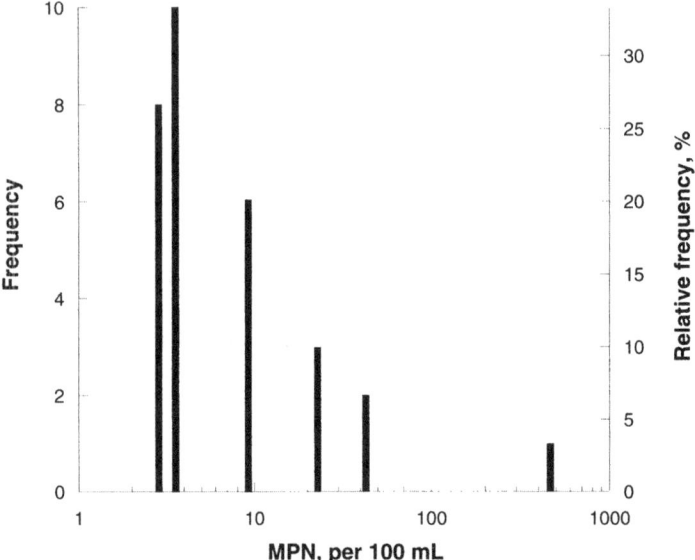

Fig. 2.1 Bar graph of MPN data in Table 2.1.

As more and more **random** observations are taken,[16] the graph comes closer to the frequency distribution *of the population*. If we take the graph as an approximation to the true population graph (called the **probability mass function**, or **"pmf"**), we can use it to calculate the probability of a particular MPN occurring—after all, objective probability is defined as the long-run relative frequency. This is best seen using relative frequencies: If we took the above bar graph as the probability distribution, then we could say that the probability of getting MPN = 23 (per 100 mL) is Pr(MPN = 23) = 10%.

But say our interest lies not so much in the probability of a particular value occurring, but rather in the probability of getting *up to* a given value (as in the case

---

[16]Strictly, we can't claim that observations are truly random, because to examine the notion we use tests of *non*-randomness (there are no tests of randomness *per se*). The failure of a suite of tests to find a pattern does not guarantee the absence of a pattern—a future, better, test may in fact do so.

of environmental bacteriological standards). Here we plot cumulative frequencies, rather than just the frequencies as in the bar graph. The resulting graph is called the **Cumulative distribution function (CDF)**, or **Empirical distribution function**, as shown in Figure 2.2.

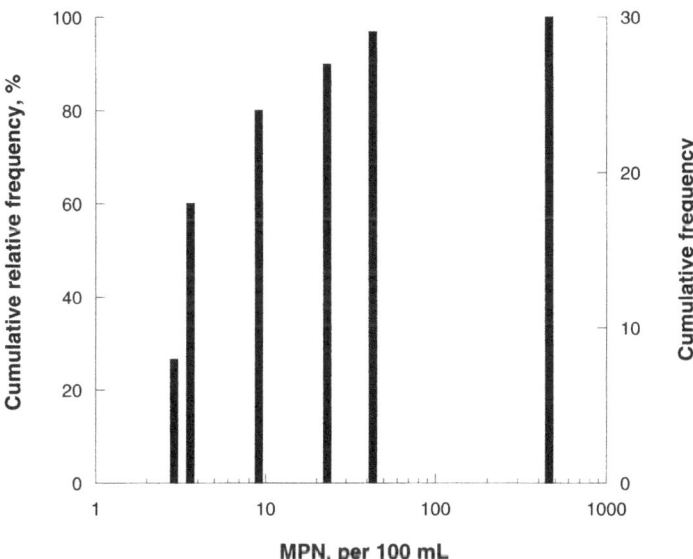

Fig. 2.2 Cumulative distribution function of MPN data in Table 2.1.

As more and more random data are collected, the sample CDF comes closer to the population CDF. This is the key to computing the exact probability that a given value will not be exceeded (so we obtain the **nonexceedance probability**). For example, to obtain the probability of the MPN being no greater than 23 per 100 mL, the graph gives $Pr(MPN \leq 23) = 90\%$. Note that important features of the distribution's shape are not as apparent on the cumulative graph as they are on its noncumulative counterpart (the bar graph). The **skewness** of the MPN data (typical of aquatic microbiological data) so clearly evident in the bar graph is not nearly as obvious in the sample CDF.

2.2.4 Discrete versus continuous data

So far we have considered discrete ordinal data. With two exceptions, plotting data frequencies and population distributions of continuous data is entirely equivalent to this case. Here we give a brief discussion of this topic; more detail appears later (in Section 2.4).

The first difference is that plotting sample frequencies versus values of the variable measured gives a histogram (in which adjacent bars do touch each other), not a bar graph (in which adjacent bars are separated). The number and width of the bars is up to you (or your software)—because you must divide the range of the data into a set of groups (e.g., between 0 and 2, between 2 and 4, ...) and then count up how many data lie within each group. While older texts may advocate that data should be plotted with relatively few groups, to avoid long and tedious plotting of every datum by hand, we now have computers to do all that. So it is a good idea to experiment with the width of the groups; information is lost with excessive grouping.

The other major difference lies in the use of the **probability density function** (**pdf**—we will discuss such a function in some detail in Section 2.5.1). As the number of random samples increases, the bars of the histogram can be made ever narrower so that the histogram takes on the smooth shape of the pdf. So it is tempting to think that for many samples the pmf becomes the pdf. But whereas the pmf height displays probabilities, the height of the pdf does not. This is a fundamental distinction. The probability at x for the pdf is effectively zero (even though a continuous variable must take *a* value, there are an infinite number of values that it may take). Probabilities for continuous variables are only defined in terms of **areas under the pdf**, not for a particular value of the pdf. So we can get probabilities attached to **ranges** of continuous data, but not to a datum. To illustrate, let's say that we want to know the probability that the variable x is less than a particular value X. This probability is just the value of the population CDF at $x = X$, which is the same as the area under the pdf to the left of X.

This means that the CDF is the integral of the pdf (or, equivalently, the pdf is the derivative of the CDF). So why introduce this mathematical artefact at all? There are many good reasons. One is that noted at the end of the last section—cumulative graphs do not show the distribution shape nearly as well as their noncumulative counterparts. Another is the fact that probability tables for particular distributions (such as those in Appendix A) are usually stated in terms of areas under the pdf.

The various ways in which discrete and continuous data and population variables may be presented are given in Tables 2.2 and 2.3.

2.2.5 Summary statistics

We often need to summarize data or populations by a few numbers. These are **summary statistics**. For populations, these are called **parameters** of the distribution, and summary statistics of samples are used to **estimate** (approximate) them. The most important sample statistics are as follows.

For central tendency

Mean $\bar{X} = \frac{1}{n}\sum_{i=1}^{n} X_i$. The mean of random samples is always an unbiased estimate of the true mean (μ), whatever the distribution data are drawn from.

Mode The most frequently occurring value.

Median The middle value—half the data are smaller, half are larger. If n is even use the mean of the two middle values. The median of random samples is always an unbiased estimator of the population median, regardless of the distribution that samples are drawn from.

Table 2.2 Summary of terms for sample data

Data type	Frequency	Cumulative frequency
Discrete	Bar graph	Sample cumulative distribution function (sample CDF)
	Sample frequency distribution	Empirical distribution function (EDF)—but only for raw data
	Frequency polygon (midpoints of bar graph bars joined by straight lines)	Cumulative frequency polygon
Continuous	Histogram	Sample cumulative distribution function (sample CDF)
	Sample frequency distribution	EDF (if no data grouping)
	Frequency polygon	Cumulative frequency polygon
		Ogive (using the upper class boundary)

Table 2.3 Summary of terms for populations

Data type	Occurrence probability	Cumulative probability
Discrete	Probability mass function	Cumulative distribution function (CDF)
		Distribution function (DF)
Continuous	Vanishingly small[a]	Cumulative distribution function (CDF)
		Distribution function (DF)

[a] *Not* the probability density function; the height of the pdf is a density, not a probability. A probability for a continuous population variable is an area under the pdf.

Geometric mean The nth root of n samples: $\bar{X}_g = \sqrt[n]{X_1 X_2 \ldots X_n}$. It is less than the sample mean. It is zero if any datum is zero, regardless of how large the others are.[17] If all data are positive, it can also be calculated as the antilog of the mean of the logarithms of the data. It is used in some bacteriological standards, where it is often (inappropriately) called the "mean log" or "log mean." The sample geometric mean may be greater than or less than the sample median

[17] Therefore, data should always be scrutinized for zeroes before calculating geometric means. Such zeroes commonly arise in manipulated datasets containing "less thans."

(by as much as a factor of 3), an issue of some importance in setting percentile standards (as discussed in Section 8.2.2).

For spread

Range largest value to smallest value.

Interquartile range (Median of the top half of the data) − (Median of the bottom half of the data). Denoted by **IQR**.

Variance $S^2 = \frac{1}{n-1}\sum_{i=1}^{n}(X_i - \bar{X})^2$. Using $n-1$ **degrees of freedom** in the divisor maintains S^2 as an unbiased estimate of the true variance (in effect, one degree of freedom has already been used up in calculating $\bar{X}$ as an estimate of the true mean).

Standard deviation S, the positive square root of the variance. It has the same units as the central tendency statistics, which can be a great advantage. Because it is the square root of an unbiased quantity (S^2) it is actually a slightly biased estimator of the true value (σ), especially for small samples (Zar 1996, p. 37 [349]). This bias is usually ignored.

Coefficient of variation CV = standard deviation divided by the mean ($S/\bar{X}$), dimensionless.

IQR/median A dimensionless **nonparametric** measure.

For asymmetry

Coefficient of skewnness $g_1 = \frac{m_3}{m_2^{3/2}}$ (sometimes written as γ_1 or as $\sqrt{\beta_1}$), where the 2nd and 3rd moments are defined as $m_2 = \frac{1}{n}\sum_{i=1}^{n}(X_i - \bar{X})^2$ and $m_3 = \frac{1}{n}\sum_{i=1}^{n}(X_i - \bar{X})^3$.

In summary, we can depict all this as follows (Figure 2.3 for typical—somewhat idealized—distribution shapes).

There are other summary statistics that indicate distribution shape (e.g., kurtosis, which measures the degree of peakiness). Published formulae for skewness and kurtosis coefficients can differ slightly (the simpler versions don't remove a small bias at low n).

2.3 EXPLORATORY AND GRAPHICAL METHODS

Before any formal calculations of means, variances, or other model parameters, a visual inspection of the data is always in order. One should be wary of statistical conclusions that are not in keeping with general, visual appraisals. A very influential

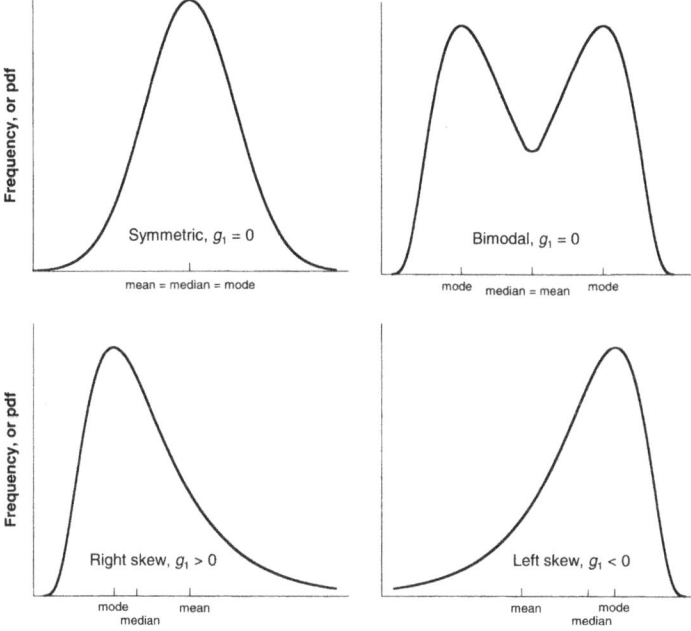

Fig. 2.3 Symmetry and skewness.

statistician (Tukey) has been a strong advocate of this point of view; always look for visual clues about patterns in your data before employing formal procedures.

A **time series plot** of the data is always a good starting point. From such a plot, one can assess the general level of temporal variability, completeness of record, presence of extreme values, seasonal variation, and, to a certain extent, **serial correlation**. These plots can be examined for any obvious changes (e.g., a step-change coinciding with a change in wastewater treatment plant—or a change in laboratory analytical method (see Section 11.1.3). Only impose a straight line on such plots if there appears to be a linear trend (as opposed to curvilinear or step trends). Sometimes one can use a **smoother** to reveal patterns in data, as shown in Figure 2.4.

The next step is to get an idea of the data distribution and how it changes over time. **Histograms** and **bar graphs** are excellent for this.

A **stem-and-leaf diagram** is a close relative of a histogram or bar graph that is very useful at the exploratory data analysis stage. The construction of a stem-and-leaf plot is very simple and easily illustrated by an example.

Suppose we have the following data: 58, 59, 32, 21, 50, 42, 43, 79, 22, 86, 74, 15, 98, 86, 65, 15, 93, 27, 73, 57. We first construct a stem—one node for each interval of ten is logical in this example—as on the left-hand column. And then we add the leaves, one for each data point, as in the right-hand column.

The stem-and-leaf plot is really a histogram or bar graph turned on its side, with the added feature of displaying data values directly.

24 BASIC CONCEPTS OF PROBABILITY AND STATISTICS

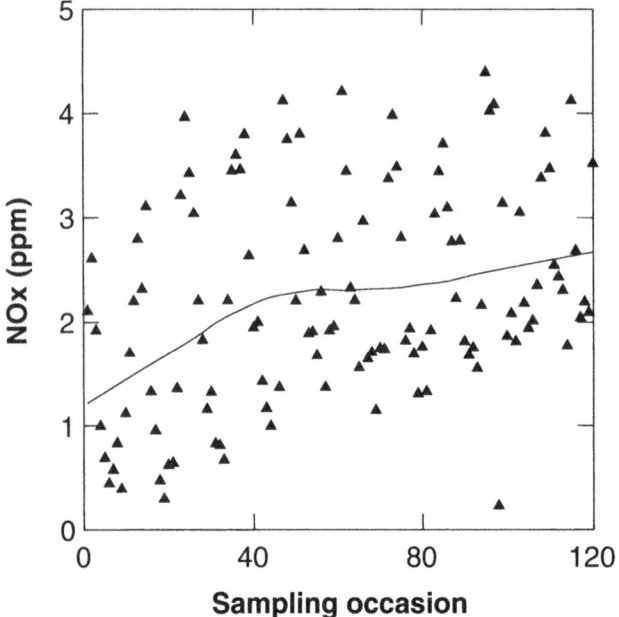

Fig. 2.4 Lowess fit for Ngakaroa Stream data, prepared using Systat (SPSS 1998 [294]).

Another type of plot that is extremely useful is the **box-and-whiskers** plot (sometimes called just **boxplot**). Many different versions of box-and-whisker plots are in use (e.g., McGill *et al.* 1978 [218]), and for that reason it is wise to always state conventions used in a particular boxplot. The plot usually portrays the median, interquartile range (middle 50% of the data), and the extremes for each class interval. Class intervals could logically correspond to years or to seasons to indicate how the distribution changes over time (i.e., to indicate **seasonality** or trend). An example is shown in Figure 2.5 for somatic coliphage concentrations in recreational water distributed over five New Zealand watershed types. The units are plaque-forming units ("pfu") per 100 mL of water. Data are plotted on log scale (plotting on natural scale

```
1 ‖ 55
2 ‖ 127
3 ‖ 2
4 ‖ 23
5 ‖ 8907
6 ‖ 5
7 ‖ 943
8 ‖ 66
9 ‖ 83
```

Stem and leaf data display

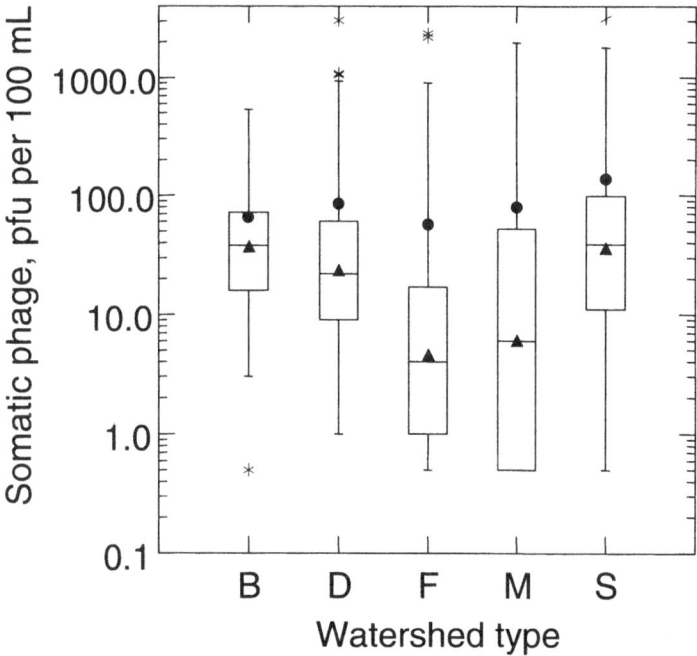

Fig. 2.5 Boxplot for somatic coliphage in recreational waters (McBride *et al.* 2002 [211]), prepared using Systat (SPSS 1998 [294]). The box covers the interquartile range (IQR); its bottom is the 25%ile and its top is the 75%ile. The line within it is the median (50%ile); solid circles are means; solid triangles are geometric means. Whiskers extend from the box to the furthest datum inside the inner fence (defined as $1.5 \times IQR$ beyond the box). Data between the inner and outer fence (defined as $3 \times IQR$ beyond the box) are outliers and are plotted as starbursts. Any data beyond the outer fence are extreme outliers and are plotted as circles (though none appear in this figure). Note that the DataDesk software (Velleman 1997 [330]) generally uses the same conventions but reverses the role of starbursts and circles.

fails to demonstrate the observed pattern over the watershed types). Means (solid circles) and geometric means (solid triangles) have been added.

2.4 IMPORTANT DISTRIBUTIONS

A number of distributions are in common use—some more than others. They are used to characterize the distribution of water quality variables and to make estimates and inferences. Some apply to continuous variables, some to discrete variables. Mathematical details and calculation methods are described in Chapter 6. Here we introduce them and describe their main features:

Distributions for continuous random variables

The normal distribution is used for symmetric continuous data and also—in the form of the unit normal distribution—as the distribution of a **test statistic**. (The unit normal distribution has zero mean and unit variance.)

The lognormal distribution is used for right-skewed continuous data, as is common with water quality variables (especially microorganisms).

The gamma distribution is also used for right-skewed continuous data (it too is always right-skewed). For technical reasons it is preferred over the lognormal distribution to describe skewed data in Bayesian analyses.

The beta distribution can take a variety of shapes, even though it has only two parameters. It is often used as a **prior distribution** in Bayesian analyses, as we shall see in Chapter 8.

Other distributions may also be used from time to time, especially in **quantitative risk assessment**. For example, the loglogistic distribution is used in Chapter 9.1.2 to describe the distribution of shellfish meal size in a national nutrition survey.

Distributions for discrete random variables

The binomial distribution is used for discrete dichotomous data, where each sampling event can result in one of only two outcomes (e.g., breach versus comply). This includes ordinal data (such as ranks), and nominal data (categories, such as exceed/not exceed, or turbid/clear). The distribution of such outcomes in random samples is *always* binomial, provided the probability of success is the same for each trial. This is a most important result. That is, one does not assume that they follow the binomial distribution, they *automatically do* follow that distribution. This distribution is **underdispersed**—its variance is always less than its mean.

The Poisson distribution is used for randomly distributed entities (e.g., bacteria colonies on an incubation plate or in a water body). Its variance always equals it mean, reflecting the fact that it mimics **randomness**.

The negative binomial distribution is used for right-skewed discrete data (e.g., MPN enumerations, stream invertebrates). It is **overdispersed** in that its variance is always greater than its mean.

The hypergeometric distribution is used in place of the binomial distribution when sampling is from a finite population and is **without replacement**.

The multinomial distribution is used for discrete polychotomous data, where each sampling event can have more than two outcomes. It is used in performing calculations of a Most probable number (MPN) (discussed in Chapter 10).

Sampling distributions Finally we have three **sampling distributions**, in addition to the unit normal distribution. They specify the distribution of a test statistic and are used in the construction of various types of intervals (as discussed in Chapter 3).

Student's t-distribution is used in place of the unit normal distribution when the population variance is unknown and the sample size is "small" (typically less than 30).

The chi-square distribution is used when making inferences about variances.

The F-distribution is used when making inferences about the *ratios* of variances, as in **Analysis of variance** problems.

Mathematical details of these distributions are given in Chapter 6; summary information and discussion of their applicability is presented below.

2.5 CONTINUOUS DISTRIBUTIONS

2.5.1 Normal distribution

This distribution's pdf is the familiar "bell-shaped curve".[18] It is symmetric and needs only two parameters to specify it—the mean (μ) and variance (σ^2). It can also be specified in terms of summary statistics, using the sample mean ($\bar{X}$) as an estimate of the population value (μ) and the sample standard deviation (S) as an estimate of the population standard deviation (σ).

When doing statistical calculations we often need tables of values of the normal distribution. The book of tables required to list a separate table for each possible value of the mean and standard deviation would be gigantic. This problem is completely avoided by using the (faintly troubling) **normal deviate** (this may go by a number of other names: **normal score, standardized score, standardized mean**), generally denoted as Z (occasionally as K) and defined as

$$Z = \frac{X - \mu}{\sigma} \quad \text{or} \quad Z_i = \frac{X_i - \mu}{\sigma}, \quad (2.15)$$

depending on whether we are referring to a particular value of the random variable x (i.e., X) or to a datum (X_i).[19]

Recalling that both μ and σ have the same units as X or of our data ($X_i, i = 1, \ldots, n$), we see that units cancel in this formula, so it is dimensionless. All we will need is the distribution of Z, which is known as the unit normal distribution (or standard normal distribution—an example follows), and so we'll need only one table. The probability density function of this distribution is shown in Figure 2.6.

---

[18] It was thought at one time that all measurements fitted a normal distribution (Menand 2001, chapter 8 [220]).
[19] Z can also be defined for the sample mean, with $\bar{X}$ replacing X, in which case σ is replaced by the "standard error."

28 BASIC CONCEPTS OF PROBABILITY AND STATISTICS

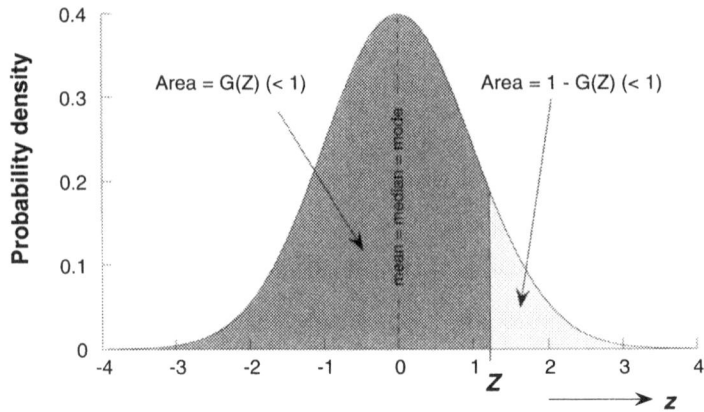

Fig. 2.6 Probability density function (pdf) for the unit normal distribution.

The mean of this distribution is zero (as are the median and mode) and its standard deviation is 1 (hence the term "unit normal distribution"). The total area under the pdf is one, and so the sum of the areas of the shaded and clear areas under the curve is one. The clear region is the **area in the tail** of the normal distribution. This is the **upper tail**, also called the **right tail**. The shaded area (G) to the left of Z is what we referred to earlier as the area under the pdf, sometimes called the **area under the normal curve**. The notation $G(Z)$ emphasizes that G is a function of Z only. The Z value is the **critical value** of the distribution, meaning that the area to the left of it is G (and the area to the right of it is $1 - G$). Tables of these values are called **proportions of the normal curve** or the **cumulative unit normal distribution**. Critical values may also be called **percentiles, quantiles**, or **percentage points**. Tables of the cumulative unit normal distribution and its percentiles are given at the end of this book (in Appendices A.1.1 and A.1.2).

In using this distribution to make probability calculations, we only ever need to look up *one* of these standard tables and do some simple one-line calculations. For example to calculate a percentile we

- read the standard normal table

- transform back from the normal deviate to the original measurement scale; that is, by simple algebra we can transform the Z-equation [Equation (2.15)] back to the following X-equation:

$$X = \mu + Z\sigma \qquad (2.16)$$

Example Take the MPN data (Table 2.1). Say we wish to calculate the 90th percentile (i.e., "90%ile") of the MPN distribution. So $G = 0.9$. Let's assume that the MPN distribution is normal (we will come back to this assumption in Section 2.5.2).

From Table A.1.2 we read that $Z_{G=0.9} = 1.2816$ (in Figure 2.6, when $G = 0.9$ then $Z = 1.2816$). If we further assume that the sample mean and standard deviation (24.29 and 83.02 per 100 mL) are the true values for the population, then, from Equation (2.16), $X_{0.9} = \mu + Z_{0.9}(\sigma) = 24.29 + 83.02 \times 1.2816 = 130.7$ per 100 mL. This is the calculated 90th percentile of the MPN distribution (assumed to be normal).

For completeness, Figure 2.7 shows the cumulative distribution function (CDF) for the unit normal distribution. Note that it gives much less visual evidence about the shape of the distribution than does the pdf. However, the *height* of the curve does give the probability (i.e., G) of getting up to the value Z which is also given by the area under the pdf curve.

Always be aware that the height of the pdf is *not* a probability.

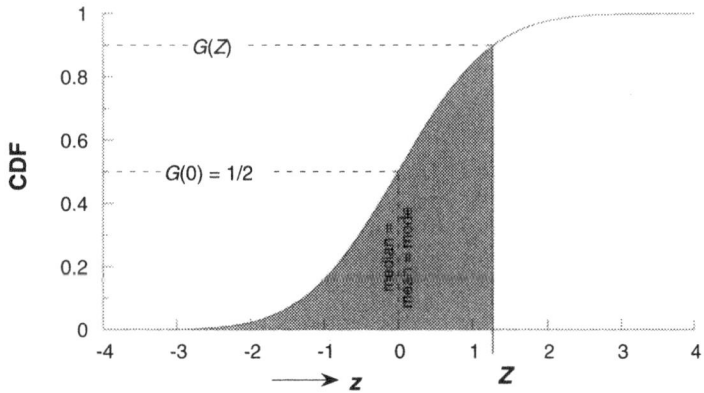

Fig. 2.7 Cumulative distribution function (CDF) for the unit normal distribution.

Some commonly used critical values of the unit normal distribution (as can be read from Table A.1.2) are:

G	0.500	0.800	0.900	0.950	0.975	0.990	0.995
Z	0.0000	0.8416	1.2816	1.6449	1.9600	2.3263	2.5758

Using Table A.1.1, we can show that the area under a normal distribution between the mean plus or minus one standard deviation is 0.6826 and that the area under that distribution between the mean plus or minus two standard deviations is 0.9544 (see Problem 2.9). This uses the important result for any symmetric continuous distribution, that, for a positive Z value

$$G(Z) = 1 - G(-Z), \qquad (2.17)$$

and so the area between Z and $-Z$ is

$$G(Z) - G(-Z) = 2G(Z) - 1, \qquad (2.18)$$

the total tail area (less than $-Z$ and greater than Z) is

$$1 - [G(Z) - G(-Z)] = 2[1 - G(Z)] = 2G(-Z), \qquad (2.19)$$

and, by differentiating either equation, we obtain

$$Z_G = -Z_{1-G}. \qquad (2.20)$$

These three results hold true for cumulative areas of *any* symmetrical distribution.

2.5.2 Lognormal distribution

The normal percentile calculation in the previous section is a simple procedure, and it is often used unthinkingly as though it is always valid. It isn't. The main problem with it is that we have assumed that the distribution of x is *normal*. When we find, as in this example, that the sample standard deviation greatly exceeds the mean, this is a signal that the data may be very skewed (you could easily see that by looking at boxplots or a histogram of the data). This indicates that the data have been drawn from a distribution that is distinctly non-normal (it is strongly skewed to the right) and so the procedure is not appropriate. Such skewness is very common for aquatic microbiological data.

A common way out of this is to observe that if we were to take logarithms of the MPN data, the histogram would become much more symmetrical. Indeed, it does (try it). So we could apply the normal calculation method to the logarithms of the data, and then having got the result just antilog it to get our answer. This is a particular example of the use of **transformations**.

This implies that our data—before we take logarithms—could have been drawn from the **lognormal distribution**, which can also be characterized by just two parameters, but is right-skewed. In fact it is always right skewed (g_1, its coefficient of skewness, is always positive, as shown in Section 6.2.2). It is asymmetric and needs only two parameters to specify it—the mean (μ_y) and variance (σ_y^2) of the logarithms of the transformed variable $y = \ln(x)$. It too can be specified in terms of summary statistics, namely, the mean and variance of the logarithms of the data. Sometimes the standard deviation is specified, rather than the variance, so it can be important to check.

This distribution has the advantages of (a) mimicking the right-skew so often shown by water quality variables and (b) disallowing negative values that are allowed in the normal distribution. It can also be described by three parameters, to accommodate a shift along the x-axis without changing the distribution's shape. This is shown on Figure 2.8, in which the distribution is contrasted with the unit normal distribution. Note that whereas a lognormal population's median and geometric mean are identical, their sample counterparts are not. Furthermore, the sample geometric mean is a biased estimator of its population value,[20] whereas the sample median is an unbiased estimate of its true value. This is one of the reasons that some water quality microbiological standards use the sample median, rather than the sample geometric mean. Lognormal CDF calculations are simply derived from the normal distribution's CDF, using logarithms (see Problem 6.1).

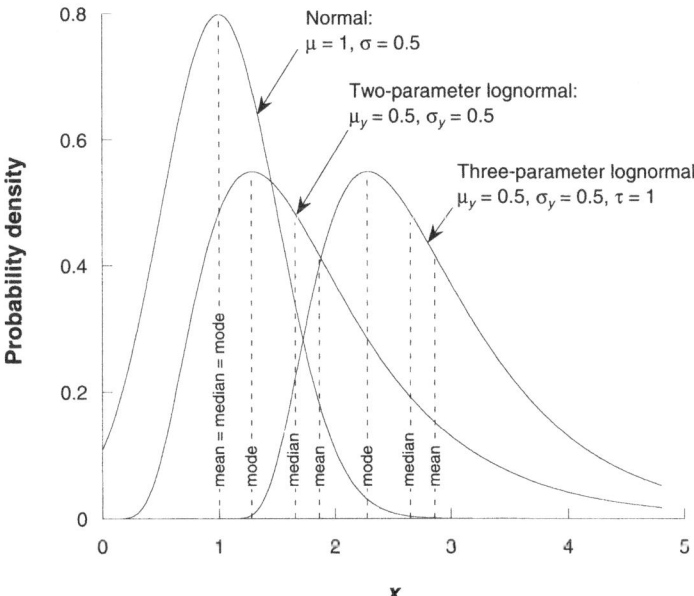

Fig. 2.8 Normal and lognormal probability density functions. Lognormal pdfs are defined in terms of the mean and standard deviations of the logarithms of the x-axis (μ_y and σ_y), plus an optional shift parameter (τ). The modes, medians, and means of these distributions can be calculated from the formulae in Section 6.2.2; that is, mode = $\exp(\mu_y - \sigma_y^2) + \tau$, median (= geometric mean) = $\exp(\mu_y) + \tau$, mean = $\exp(\mu_y + \frac{1}{2}\sigma_y^2) + \tau$.

Gilbert (1987 [118]) has an excellent discussion of this distribution and of the use of transformation methods in general. The use of the lognormal distribution has diminished in water quality statistics because of the advantages of **nonparametric methods**, in which no distribution assumption is required (e.g., Chapter 8 applies these methods to percentile standards).

2.5.3 Gamma distribution

---

[20]The bias factor is $\exp[\sigma_y^2/(2n)]$ (Gilbert 1987 [118]). The estimated geometric mean should be divided by this factor. Note that the bias decreases as the sample size (n) increases. For a "small" sample size ($n = 5$) and large standard deviation of the logarithms (e.g., $\sigma_y = 0.7$, as is commonly reported for faecal indicator bacteria at marine bathing beaches; see McBride *et al.* 1998 [210]), the factor is 1.05. It will generally be smaller than this, so the bias is seldom serious.

This is a more versatile two-parameter (α and β) right-skewed distribution than the lognormal.[21] This is seen from the variety of shapes it can assume shown in Figure 2.9. Two special cases are the **exponential distribution** ($\alpha = 1$) and the **chi-square distribution** (for even values of α and for $\beta = 2$). It too disallows negative values, but can extend over all positive values of its abscissa. Nevertheless, the lognormal tends to be more convenient to use in classical analyses with the gamma being used more commonly in Bayesian analyses.

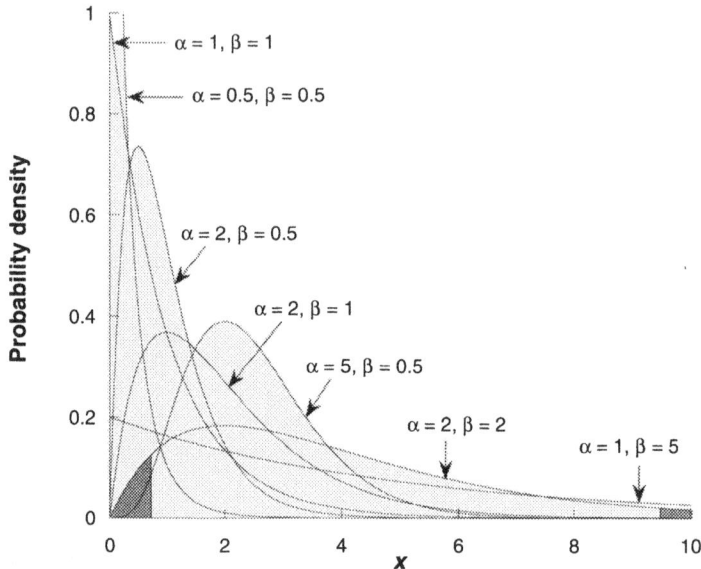

Fig. 2.9 A variety of shapes for the pdf of the two-parameter gamma distribution, $\Gamma(\alpha, \beta)$. The inside edges of the shaded areas (at $x = 0.711$ and 9.488) denote the 5%ile and 95%ile of the $\Gamma(2, 2)$ distribution, which is also the chi-square distribution with $f = 2\alpha = 4$ degrees of freedom (i.e., χ_2^2, see Section 6.2.11). Each shaded area encompasses an area of 0.05 (the right-hand area extends to $+\infty$).

2.5.4 Beta distribution

This distribution is also characterized by two parameters (α and β) and is used in Bayesian analyses. It is also used to derive the "beta-Poisson" dose-response model widely used in quantitative health human health risk analysis (Chapter 9). It is defined over the interval [0, 1] and is extraordinarily versatile, as shown on Figure 2.10. It can be left-skewed or right-skewed. The heavy line in the first quadrant in that figure (for

---

[21]This α is not the right-tail area of a distribution, as was the case on Figure 2.6. Such double usage of symbols is quite common, as noted in Section 1.1. So it is always a good idea to check the context in which such a parameter appears, to be sure of its meaning.

$\alpha = \beta = \frac{1}{2}$) describes Jeffreys' prior—we will be using this prior when considering "Confidence of Compliance" (Chapter 8).

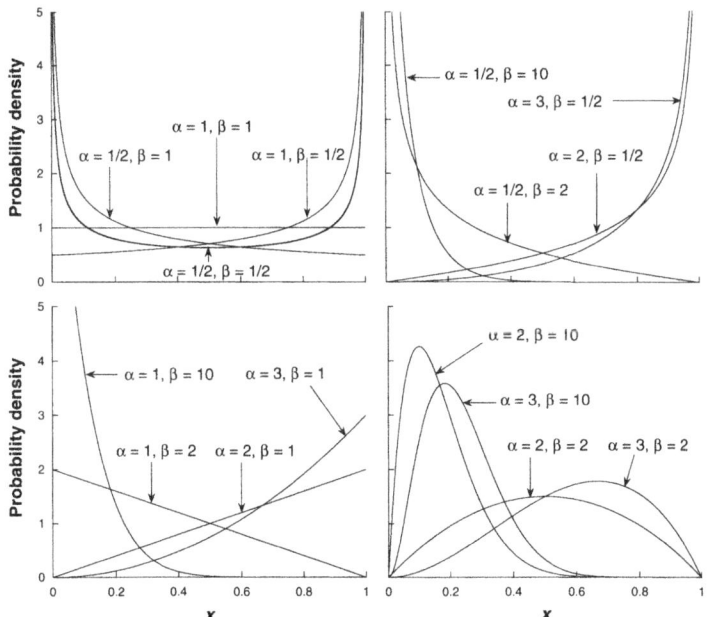

Fig. 2.10 A variety of shapes for the pdf of the two-parameter beta distribution.

2.6 DISCRETE DISTRIBUTIONS

2.6.1 Binomial distribution

The importance of this two-parameter discrete distribution is often overlooked. It applies when a collection of random samples can have one of two outcomes—for example, pass or fail, exceed or not exceed, turbid or clear. Then, the distribution of those outcomes always follows the binomial distribution, regardless of the distribution that the data happen to follow. The probability mass function is $b(e; n, \theta)$, representing the probability of getting e successes (e.g., pass, exceed,...) in n random trials, each with a probability θ of success.[22]

For example, say you are interested in testing whether an effluent concentration has exceeded a certain limit for more than 5% of a year (the approach used in English water law). Let's suppose that the effluent was in fact borderline for compliance with that limit; that is, it was exceeded for exactly 5% of the year. Then the probability of any individual random sample exceeding the limit is $\theta = 0.05$. Let's say that 20

---

[22] In this text the exponential number e (the base of natural logarithms) is not italicized, so that it is distinguishable from the parameter e.

samples are taken to assess compliance[23] and that a breach of the effluent standard will be inferred if there is more than 1 exceedance of the limit in those 20 samples (noting that 1:20 is 5%). One should ask, What then is the risk of falsely inferring that a breach has occurred? Say 2 or more out of the 20 samples exceeded the limit, and so a breach of standard would be inferred. Tables of the binomial distribution can be used to calculate the risk, as follows.

- The probability of an individual random sample exceeding the limit is $\theta = \Pr(\text{exceed limit}) = 0.05$

- $\Pr(2 \text{ or more exceedances of the limit}) = 1 - \Pr(0 \text{ or } 1 \text{ exceedances})$—using Equation (2.3). For samples sizes up to to about 20 the latter probability can read be directly from tables of the **cumulative binomial distribution**.[24] The cumulative probability of getting up to e successes in n random trials, each with a probability p of success, is $B(e; n, \theta) = \sum_{i=0}^{e} b(i; n, \theta)$. For example, in Iman & Conover (1983 [161]) there are rows in the table (using our nomenclature) for $n = 20$ that read:

n	e	$\theta =$ .05	.10	.15
20	0	.3585	.1216	.0388
	1	.7358	.3917	.1756
	2	.9245	.6769	.4049

where e is the number of exceedances of the limit with a probability (θ) of occurrence in any one sample.[25] So $\Pr(0 \text{ or } 1 \text{ exceedances}) = 0.7358$. The required probability is therefore: $\Pr(2 \text{ or more exceedances}) = 1 - 0.7358 = 0.2642 \approx 26\%$.

Many people are surprised to find such a high risk. A discharger would feel entitled to protest that such a risk is unfair. Of course, the risk is less if the effluent was not borderline, but that may not mollify the discharger's concerns.

Table A.3.1 gives cumulative binomial probabilities corresponding to common compliance assessment periods for daily sampling—weekly, fortnightly, and four-weekly. Section 6.2.5 gives a number of useful formulae for this distribution. Of particular note is its versatility, and its role in Bayesian analyses with its continuous cousin (the beta distribution), as we shall see in Chapter 8. It can be left-skewed or right-skewed, and it is always underdispersed (its variance is always less than its mean). Its pmf can be written in terms of the mean, as $b(e; n, \mu) = \binom{n}{e} n^{-n} \mu^e (n-\mu)^{n-e}$. Some examples for a range of means are shown on Figure 2.11, along with the following two related distributions, Poisson and negative binomial. In some circumstances the

---
[23] A compliance assessment rule may state that 20 samples are to be taken in a given period; see Chapter 7.
[24] For larger sample sizes, one can obtain these probabilities from spreadsheet packages, or by programming simple-yet-accurate methods, such as that given in Section 6.2.5 for calculating binomial and cumulative binomial probabilities.
[25] Iman & Conover use x where we have used e.

binomial distribution can be approximated by the unit normal distribution (details are presented in Section 6.2.5).

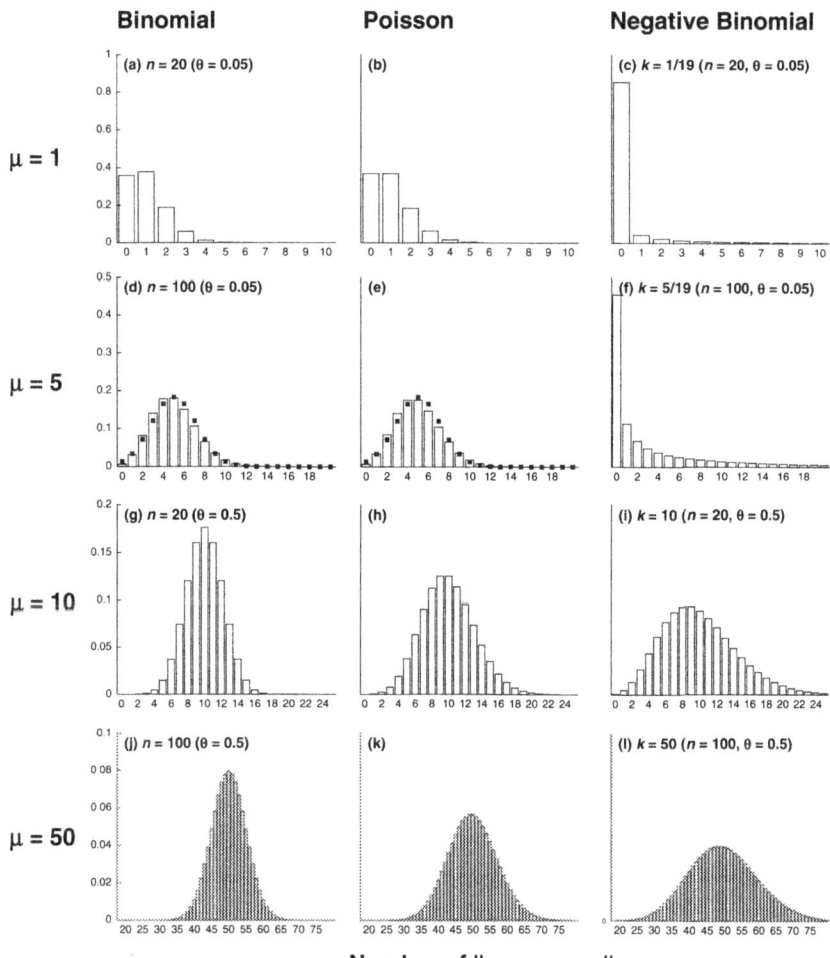

Fig. 2.11 Probability mass functions for three common discrete distributions for four values of their mean μ: n is the number of random trials, θ is the probability of "success" on an individual trial, and k is negative binomial dispersion parameter. The relationships between these parameters and μ for the binomial and negative binomial distributions are given in Sections 6.2.5 and 6.2.7. (The dots on parts (d) and (e) display the normal approximation to the Binomial and Poisson distributions, also presented in Section 6.2.5.)

2.6.2 Poisson distribution

In situations where there are a large number of random trials (n is large) of our random variable (x), the calculation of binomial probabilities can be cumbersome. Under easily satisfied conditions, x follows the computationally simple Poisson distribution, with mean and variance both equal to μ. The Poisson pmf is simply $p(e; \mu) = \mu^e \exp^{-\mu}/e!$, where $e! = e(e-1)(e-2)\ldots(2)(1)$ is the factorial function. It is often used to describe the distribution of microorganisms in water. Like the binomial distribution, it can in some circumstances be approximated by the unit normal distribution (details are presented in Section 6.2.6). It is always right-skewed.

2.6.3 Negative binomial distribution

This is also related to the binomial distribution but is more skewed (always to the right). It is often used to describe contagious distributions, for example, for the distribution of *Cryptosporidium* oocysts in water bodies (e.g., Gale 1998 [104]). It is always overdispersed (its variance is always greater than its mean, as shown in Section 6.2.7). Its pmf can be written in terms of that mean: $b_s^*(s; k, \mu) = \binom{s+k-1}{s}\left(\frac{k}{\mu+k}\right)^k\left(\frac{\mu}{\mu+k}\right)^s$, where s is the number of "successes" and k is the overdispersion index. This index is not constrained to be an integer. The smaller the value of k, the more dispersed the distribution is, as can be seen in Figure 2.11.

2.6.4 Hypergeometric distribution

The binomial distribution may be rather inaccurate when taking random dichotomous samples from finite populations, without replacement. The hypergeometric distribution must then be used. It too may be left-skewed or right-skewed.

2.6.5 Multinomial distribution

This is a generalization of the binomial distribution for polychotomous data, when sampling effectively infinite populations. We will use it in Chapter 10 when developing exact MPN formula for a set of three or more dilution series (as is common).

2.7 SAMPLING DISTRIBUTIONS

We have already seen that when we know what distribution our data follow (lognormal and binomial in the examples given) we may need to consult statistical tables to obtain the required probabilities or critical values—or we may rely on software to do that. This *may* be necessary in **estimation**, which is the first major branch of statistics, particularly when we calculate confidence intervals. It is *always* necessary in the second branch, **hypothesis testing** (where quantities are compared). Hypothesis testing uses results from mathematical statistics that show what distribution the test statistic follows when the tested hypothesis is assumed to be true (these are known as sampling distributions). Three common examples follow.

2.7.1 Student's *t*-distribution

The standardized mean of data drawn from a normal distribution with *unknown* variance follows Student's t-distribution. This is a generalization of the unit normal distribution. Like the unit normal, it is symmetrical about zero, but it is wider and less tall. For large **degrees of freedom** the two become indistinguishable, see Figure 2.12.[26]

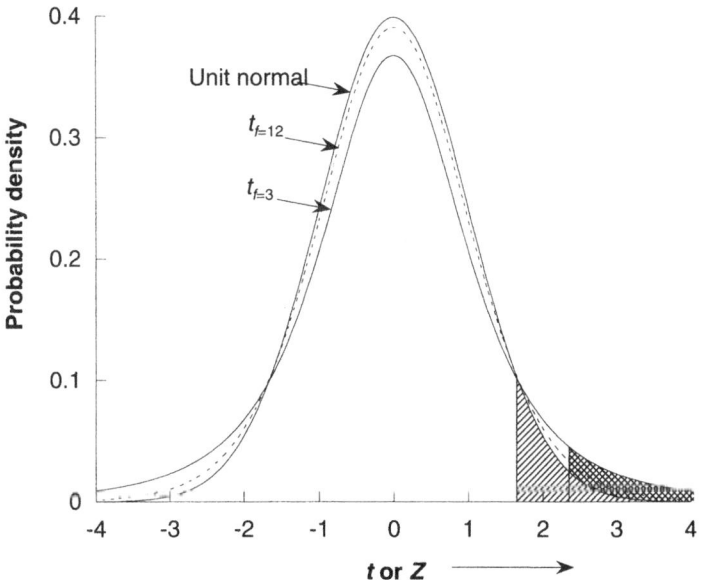

Fig. 2.12 Student t-distributions for 3 and 12 degrees of freedom, and the unit normal distribution. The inside edges of the shaded areas denote the 95%ile of the unit normal distribution (at $Z = 1.6449$) and of the t_3-distribution (at $t = 2.3534$). Each shaded area, extending to $+\infty$, encompasses an area of 0.05.

2.7.2 Chi-square distribution

The standardized variance of data drawn from a normal distribution, defined as $\frac{fS^2}{\sigma^2}$, follows the "chi-square" distribution with f degrees of freedom. This distribution is commonly abbreviated as χ_f^2 (χ is the Greek letter chi), as depicted on Figure 2.13. It forms the basis of many "chi-square" procedures. It is a special case of the gamma distribution, i.e., $\chi_f^2 = \Gamma(\frac{f}{2}, 2)$ (see Chapter 6.2.3).

---

[26] Degrees of freedom is typically $n - 1$ or $n - 2$, where n is the sample size (i.e., the number of data). The actual choice depends upon the problem being addressed. Degrees of freedom is sometimes denoted by ν, df or f. We use f.

38 BASIC CONCEPTS OF PROBABILITY AND STATISTICS

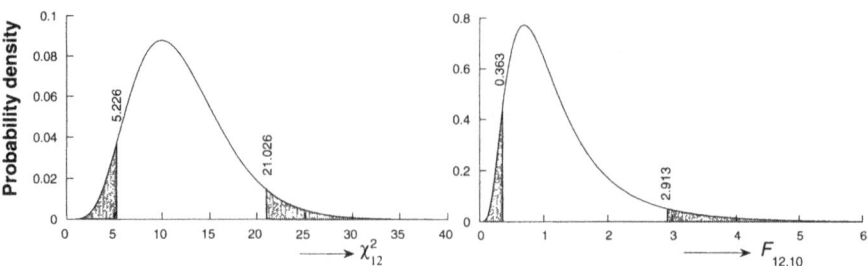

Fig. 2.13 Probability density functions for the χ^2-distribution (12 degrees of freedom) and F-distribution with 12 (numerator) and 10 (denominator) degrees of freedom. The shaded regions each contain an area of 0.05.

2.7.3 F-distribution

The standardized ratio of variances drawn from two normal distributions follows the F-distribution (the "F" honors one of the great 20th century statisticians, Fisher (Sir Ronald Alymer), 1890–1962. It is also depicted on Figure 2.13.

Note that the χ^2 and F distributions are always positive. That is because they relate to variances and variance ratios, which are squared quantities and are therefore positive. The F-distribution has both numerator and denominator degrees of freedom because it relates to *ratios* of variances, which is why it is so prominent in **Analysis of variance** (see Section 4.6).

2.8 STATISTICAL TABLES

It is essential to be able to read statistical tables of these distributions correctly. Note that no two texts follow the same convention in their tables of distributions. There are two issues to be aware of.

Which area under the pdf does the table refer to? Tables may be defined in terms of areas to the right of a critical value or to the left. This has consequences for nomenclature. As an example, consider percentiles of the t-distribution with f degrees of freedom. If the table of the distribution is based on an area α to the right of the percentile, that percentile would be written as $t_{\alpha,f}$ (Bowker & Lieberman 1972 [31], Freund 1992 [101]). Other authors refer to the area cut off to the left—the cumulative area—and so the same quantity would be written as $t_{1-\alpha,f}$ (e.g., Gilbert 1987 [118]). Still others refer to the the areas in each tail = $\frac{\alpha}{2}$ or to the right of the y-axis (Snedecor & Cochran 1980 [291]). These possibilities are displayed on Figure 2.14.

> **In this text we only use cumulative areas—because that is consistent with the concept of cumulative probability.**

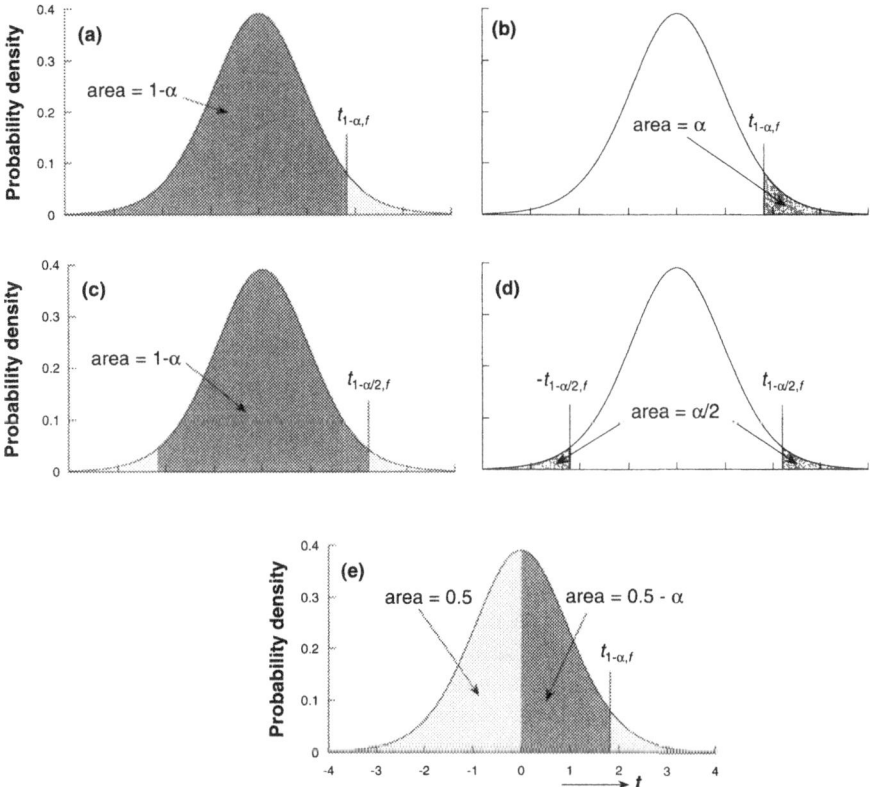

Fig. 2.14 Five possibilities for reporting areas under the t-distribution ($f = 12$, $\alpha = 0.05$). (Part (e) is more common for the unit normal distribution.)

For example, the table in Gilbert (1987 [118]) gives $t_{0.95,12} = 1.782$ while the same quantity in Bowker & Lieberman (1972 [31]) is given as $t_{0.05,12}$. Perhaps the most helpful table is that of Zar (1984 [348]). The following is an excerpt, where "$\alpha(2)$" means two-tailed and "$\alpha(1)$" means one-tailed, with the total tail area $= \alpha$ in either case. This makes it clear whether you're dealing with one-tailed or two-tailed distributions. (Some problems require us to consider both tails of these distributions.) And we can see on the graphs on Figure 2.14 that the one-tailed critical value (for $\alpha = 0.05$, $f = 12$) used is 1.782 and the two-tailed value used is 2.179.

ν	$\alpha(2)$: $\alpha(1)$:	0.50 0.25	0.20 0.10	0.10 0.05	0.05 0.025	0.02 0.01	0.01 0.005
12		0.695	1.356	1.782	2.179	2.681	3.055
20		0.687	1.325	1.725	2.086	2.528	2.845
∞		0.6745	1.2816	1.6449	1.9600	2.3263	2.5758

Some of the numbers on the bottom line of Zar's unit normal table should be familiar. Remember that for large degrees of freedom the t-distribution becomes the normal distribution? Then if $t = +1.9600$, the right-tail area is 0.025 and so if $t = \pm 1.9600$, the combined area of both the left and right tails is 0.05. So you can use this to check your interpretation of whatever table you're presented with.

You may have to interpolate to get what you want. Tables of continuous distributions come in one of two forms. The first presents values of areas under the pdf for a number of critical values. For example, the unit normal distribution table given by Gilbert (1987 [118]), and in many other texts, includes the following cumulative normal distribution for values of G corresponding to critical values (Z) of the unit normal distribution (using our nomenclature):

Z	.00	$\cdots$	.08	.09
1.2	.8849	$\cdots$	.8997	.9015
1.3	.9032	$\cdots$	.9162	.9177

You can obtain the area under the normal curve for a given critical value from such tables. For example, if $Z = 1.28$, the value of G is 0.8997. Most t, χ^2, and F tables are of this type. But say your question inverts this: You want the critical value *for a given area under the normal curve.* You have to interpolate in the above table to find an approximate value for this. For example, to find the value Z corresponding to an area to the left of $0.9 (Z_{G=0.9})$, we must interpolate between $Z = 1.28$ and 1.29.[27] A few texts do give the inverted table. For example, Iman & Conover (1983 [161]) have the following for the unit normal distribution, giving the exact answer we want:

G	.000	$\cdots$	.008	.009
.89	1.2265	$\cdots$	1.2702	1.2759
.90	1.2816	$\cdots$	1.3285	1.3346

Of course, the need to use tables has diminished in recent years, with the widespread availability of statistical computer software. But there *will* be occasions on which you'll need to know how a table is to be (or has been) read.

Statistical tables in this book Tables for cumulative areas and percentiles of the unit normal distribution and Student's t-distribution are given in the Appendices at the end of this book (Appendices A.1.1 – A.2.2). Appendix A.3.1 lists cumulative binomial probabilities for daily sampling for one, two, and four weeks. The remaining tables in that Appendix are discussed in Chapter 3.

---

[27] Using linear interpolation we obtain the critical value as $1.28 + \frac{(1.29-1.28)(0.9-0.8997)}{0.9015-0.8997} = 1.2817$ (the accurate answer is 1.2816).

2.9 CORRELATION AND MEASURES THEREOF

Correlation is usually defined as a measure of the **linear association** between random variables. It is most often measured by the familiar **Pearson's correlation coefficient**, r, defined as

$$r = \frac{\sum_{i=1}^{n}\left[(X_i - \bar{X})(Y_i - \bar{Y})\right]}{\sqrt{\sum_{i=1}^{n}(X_i - \bar{X})^2 \sum_{i=1}^{n}(Y_i - \bar{Y})^2}}. \quad (2.21)$$

The correlation coefficient lies in the interval $[-1, 1]$, with a value of zero implying no correlation. A positive correlation between x and y implies that y tends to increase as x increases, and vice versa (see Figure 2.15). Pearson's r is a widely accepted measure of the strength of a *linear* relationship—especially in linear regression. However, r is a **random variable**, and if X and Y are not normally distributed, then the distribution of r is not known. Thus it may not be possible to test the significance of a given value of r.[28]

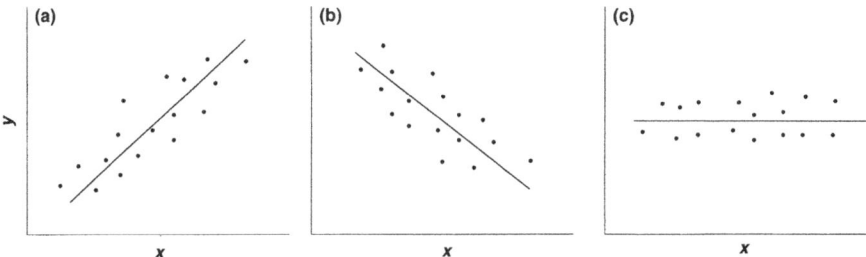

Fig. 2.15 Possible linear correlations: (a) positive, (b) negative, (c) undefined.

There are other measures of correlation that are based on the ranks of data rather than the actual values. Since they are not influenced by the underlying distribution, they are **nonparametric** and can be used to test the correlation in non-normal data.

Rank correlation measures the strength of linear dependence between the **ranks** of data, but it does not imply that the association between the random variables from which the data are drawn is necessarily linear. In other words, general **monotonic** relationships are measured. One of these nonparametric measures is **Spearman's rho**. Another is **Kendall's tau**. Both are discussed in Conover (1980 [56]) and Zar (1996 [349]).[29] Rho (often denoted as r_s) and tau will have different numerical values for given datasets (rho tends to be the larger). However, tests based on either would generally lead to the same conclusion. Both measures are used frequently in water quality studies, and there is no particular reason to prefer one over the other. Spearman's rho is calculated simply as Pearson's r computed on the ranks of the

---

[28]The meaning of "significance" will be discussed when we look at hypothesis tests, in Chapter 4.
[29]Various approximate tables of "critical values" of Spearman's coefficient appear in texts. Of these the table by Zar (1996 [349]) is the most accurate.

data with ties replaced by the average rank (indeed it is sometimes called the rank correlation coefficient). Thus software which can compute r and can rank a set of observations can also compute rho. Most software includes a Spearman's rho option these days.

Kendall's tau is the basis for the **Mann–Kendall test** and the seasonal Kendall for trend that we will discuss later (in Chapter 11). While requiring a greater calculation effort, it has the advantage over Spearman's rho of approaching the unit normal distribution more quickly. This is why Kendall's tau, and not Spearman's rho, is used in trend tests.[30] In brief, tau is a measure of **concordance** between pairs of data. For example, the two pairs of data—day = 7 with a COD = 52, and day = 9 with a COD = 85—are concordant because COD increases as day increases; discordant data pairs would be (day 7, COD = 52), and (day 9, COD = 45). Kendall's tau is the proportion of concordant or discordant pairs among all possible pairs of data. We discuss this statistic more fully later on.

2.10 STATISTICAL MODELS AND MODEL PARAMETERS

In all applications of statistical methods, some **model** is assumed. If the model accurately reflects the real situation, then our statistical conclusions are likely to reflect the real situation as well. Unfortunately, statistical models are often inappropriate, and conclusions drawn from them do not apply to the real world. We have already seen an example of this, when calculating the MPN percentiles, in Sections 2.5.1 and 2.5.2.

In the classic textbook model a set of observations are drawn:

- at random
- from a normally distributed population
- from a population with constant (usually unknown) mean and variance

Estimation of the model parameters (mean and variance) is then straightforward—as are tests for changes in these parameters over time or space. However, this simple classic model is not generally appropriate in water quality for several reasons:

> **The appropriate underlying distribution is not usually normal.** It may be lognormal, a commonly encountered distribution in water quality. Or, perhaps most often, no one particular analytical form of the distribution would seem to be appropriate.
>
> **The observations are not usually drawn at random.** Typically, they are evenly spaced over time. Information may be carried over from one observation to the next. Such data, having redundant information, are said to be

---

[30] Also, Kendall's tau can be generalized to a partial correlation coefficient, whereas Spearman's rho can't (Legendre & Legendre 1998 [181]).

serially correlated. Evenly spaced data would be very weakly serially correlated if they were spaced far enough apart, but they would still not be random. For example, sampling always at the same time of day or the same day of the week could provide a distorted picture of water quality because of predictable cyclic variation, as described below.

The random variable's mean and variance are not constant over time.
These parameters, as well as other statistical characteristics of water quality random variables, may change over time in cyclic fashion (e.g., in diurnal, weekly, or seasonal cycles). Many of these cycles are well understood, such as the diurnal cycle of dissolved oxygen in a stream or lake, seasonal water temperature variations, or weekly variations in rivers influenced by treatment plant discharges or reservoir releases. In these cases, sampling times should be planned to provide the information sought. If one always samples at the same point in a cycle, a true picture of average conditions will not be obtained. However, if one is interested only in the worst-case conditions, then sampling can be timed accordingly. Other patterns of variation are less well-understood, and this may call for a random-sampling approach.

Our general model for water quality observations should therefore include (i) **nonnormality**, (ii) **seasonality**, (iii) **autocorrelation** (also called **serial correlation** when considering temporal variations, see Section 2.11), and (iv) **trend**. To address issue (i) we might use a distribution other than normal for our assumed model, or we might not assume any particular form for the distribution, leading to a **nonparametric** analysis. To account for the remaining issues it is helpful to write an analytical expression for the general model being used (better still, for a number of candidate models). Although it might not be really necessary to do this for routine data analysis, one should always be very clear about what model is being assumed (explicitly or implicitly). A general statistical model for water quality might look like the following

$$y(t) = \mu(t) + b(t) + \epsilon(t), \tag{2.22}$$

where t is the time of observation; $y(t)$ is the value to be observed; $\mu(t)$ is the seasonal mean component; $b(t)$ is the deterministic trend component; $\epsilon(t)$ is the random noise ("**residuals**") component, serially correlated to at least a small degree. This model could be expanded to consider more than one water-quality variable and/or more than one location at a time. In so doing, we can consider correlation between water quality variables or correlation over space.

Even though we may not write down an algebraic expression for our assumed model, it is very important to *write down the major assumptions*, such as normality, independence of residuals, and seasonality.

2.11 SERIAL CORRELATION, SEASONALITY, TREND, AND SCALE

There is only so much information to be had about long-term patterns from sampling in any one time period. Some information gets carried over from one observation to

44 BASIC CONCEPTS OF PROBABILITY AND STATISTICS

the next in fixed-interval sampling. This is serial correlation. The closer observations are spaced, the more information is carried over. Samples that are truly random are not serially correlated. But when invoking the notion of random sampling, we must carefully consider the temporal and spatial scales of our sampling. For example, if all our observations are made in one year, they will not be random with regard to estimating a longer-term mean—but they could be random with regard to estimating the mean for that specific year. So the meaning of correlated versus independent observations depends on the nature of our assumed model, since our model, be it explicit or not, assumes some time and spatial scale of interest.

Serial correlation can be most clearly characterized using the **correlogram**. This plots the lag-k serial correlation coefficient r_k against k (where k is an integer). This coefficient is usually taken as the Pearson's correlation coefficient (r).[31] At each lag it is computed for the two random variables $X(t)$ and $X(t-k)$.[32] The correlogram of a raw dataset will show the effects of seasonality and trend as well as serial correlation. If seasonal means are appropriately subtracted from each data point and trend components are removed as well, the resulting correlogram will reveal the correlation structure of the residuals, unaffected by seasonality or trend.

For example, take the suspended solids effluent data shown in Figure 2.16(a) (based on the "Field Raynes" data of van Dijk & Ellis 1995 [328]). In Figure 2.16(b) we see their correlogram, showing strong seasonal periodicity, and a downward drift. On removing the seasonality we get Figure 2.16(c), where the downward drift remains. Finally, on detrending the deseasonalized data we get the "residuals" shown in Figure 2.16(d), in which the serial correlation coefficients show no particular pattern.

These figures show how one can discern structure in data using correlograms, and that the effect of removing both seasonality and trend over the period of record can result in residuals exhibiting little systematic pattern and with "small" serial correlation coefficients. Furthermore, the coefficients in Figure 2.16(d) are mostly contained within the **confidence limits**, and this is consistent with treating the residuals as random (the confidence limits used to assess randomness are $\pm Z_{1-\frac{\alpha}{2}}/\sqrt{n}$, where n is the sample size.)[33] Therefore Equation (2.22) could be used as a statistical model, with the residuals being described by the noise term $\epsilon(t)$.

2.11.1 Effect of serial correlation

In using a model of the general form of Equation (2.22) it must be noted that we are *not* concerned about serial correlation in the raw data we use for model fitting.

---

[31] We could also use Spearman's rho if we want.

[32] For example, to calculate r_k from a series of data $X_i : i = 1, \ldots, 100$, put X_i in cells A1 to A100 in a spreadsheet, and put a copy into cells B1 to B100. Then push the B column down one space, by inserting a blank cell into B1, and do a linear regression of all the 99 nonblank data pairs (i.e., for rows 2 to 100). The correlation coefficient of this regression is r_1. This "push-down" process is repeated to obtain r_2 from the data pairs in rows 3 to 100, and so on, leading to the correlogram. The maximum lag taken is often recommended to be $k_{\max} = n/4$.

[33] Wider intervals are used for fitting models; that is, $\pm Z_{1-\frac{\alpha}{2}} \sqrt{\left(1 + 2\sum_{i=1}^{k-1} r_i^2\right)/n}$.

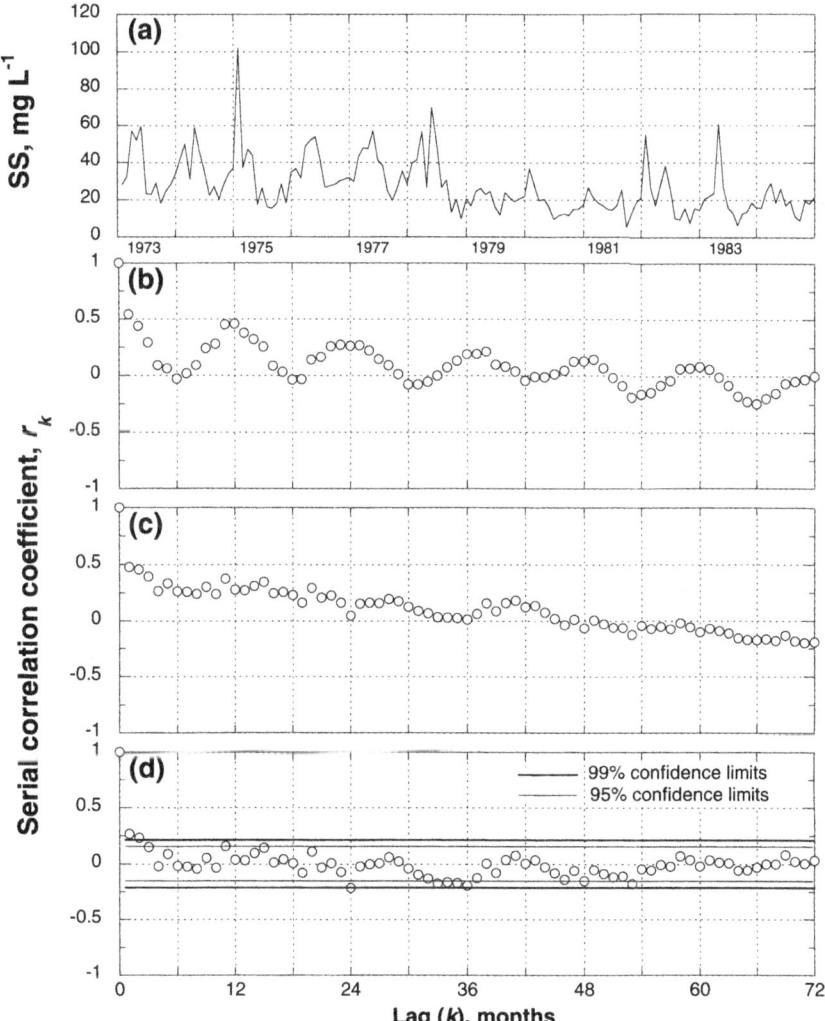

Fig. 2.16 Field Raynes monthly effluent suspended solids data ($n = 144$) and serial correlation structure: (a) time series of data; (b) correlogram of raw data for 72 lags; (c) correlogram of data with seasonality removed (the mean monthly SS value over the 12 years has been subtracted from each monthly value); (d) correlogram of data with seasonality *and* trend removed (the correlogram is for the residuals about a linear regression fitted to the deseasonalized data).

Autocorrelation *in the residuals* is what must be considered. And as Figure 2.16 suggests, removing trend and seasonality (in either order) greatly reduces the amount of serial correlation from that present in the raw data to that exhibited by the residuals. In water quality studies it is therefore wise to remove seasonality and trend if at all possible, using models that contain these terms.[34] If we then end up with "white noise," as on Figure 2.16(d), we may safely ignore serial correlation—even though the raw data exhibited large serial correlation coefficients (Reckhow *et al.* 1993 [269]).

However, issues do arise if a model's residuals still exhibit strong serial correlation. It is generally held that such correlation always inflates the standard error of those residuals, by reducing the number of "effectively independent" samples present (a standard error is a data standard deviation divided by a square root of the sample size, as we shall see in Section 2.13.1). But this depends on our scale of interest— whether estimates are being made for the longterm or just for the period of record. For long-term scales the residuals' standard error is indeed inflated, but over the period of record it is actually deflated (Müskens & Kateman 1978 [236], Ellis 1989 [84], Loftis *et al.* 1991 [190]). This reflects the following intuitive understanding. For serially correlated data, we know more about the interval over which the data are collected and less about the long term than we would for an independent series. So to estimate *long-term parameters of a population*, such as long-term means, we need to collect more data for a serially correlated series than for an independent series to achieve the same level of precision—that is, the same confidence interval width. But, if the data are serially or spatially correlated, we need *fewer* sample points to estimate parameters *for a specific period or region for which we have data*.[35] Note that there can be a "chicken or egg problem" here—we need many data to estimate the **autocorrelation function**, via the correlogram, in order to see how few data we need to make our estimates. Cutting correlograms off at lag $n/2$, rather than the commonly recommended $n/4$ (as in Figure 2.16), helps to alleviate this issue, especially when abetted by considerations of residuals' autocorrelation functions for other similar water bodies.

As a consequence, it is safe to ignore serial correlation in performing trend analyses or calculating means within periods, so long as the analysis is restricted to the period of record and you use a model that assumes that observations are independent.[36] This is because we would then be assuming a larger residuals' standard error than is actually the case. The calculation of the standard error in this situation is rather complicated (Müskens & Kateman 1978 [236]), again requiring knowledge of the correlation structure. Kateman & Müskens (1978 [172]) give an example application to river water quality studies. So it is probably best to be conservative and assume independent residuals in the calculations.

In summary:

- In estimation or testing for long-term means or trends (i.e., extrapolating beyond our data), we need more data if residuals are serially correlated.

[34] Note that in removing seasonality it is generally better to subtract the seasonal average from each seasonal datum, as was done for the data on Figure 2.16, rather than fitting an explicit form, such as a sinusoid. In the latter case the residuals may turn out to be strongly serially correlated if the data follow a seasonal pattern more complicated than that shown by a sinusoid (Ellis 1989 [84]).
[35] This is the concept behind **kriging**, the geostatistical interpolation technique.
[36] Of course, the best approach in this situation is to use a method that explicitly accounts for serial correlation, but few are available (Legendre 1993 [180]).

- In estimating or testing specific-period means or trends (within the scale of our data), we need fewer data if residuals are serially correlated. In this case it is safe to ignore serial correlation in a model that assumes independence in the residuals.

- The distinction between serial correlation and trend is scale-dependent.

These issues are further addressed in Section 11.1.3.

2.12 REGRESSION

Whereas correlation measures the *strength* of a relationship, regression *defines* it. Simple linear regression is one of the most common statistical estimation procedures. It applies to models of the form

$$y = bx + a + \epsilon, \tag{2.23}$$

where a and b are the model's coefficients and ϵ is a random noise term whose **expectation** (i.e., mean) is zero. If the slope $b \neq 0$, then the random variables x and y are correlated. For predicting the value of y given a value of x, the random noise term is set always equal to zero. Then, the above expression provides the mean value of y for a given value of x, which we can write as $\hat{Y} = bX + a$, where the circumflex (^) denotes an estimated quantity.

The parameters, a and b, are estimated using a set of data points (X_i, Y_i) and standard regression equations from any basic statistics text or, more likely, using statistical software. Even spreadsheets do regression. These equations are developed by minimizing the sum of squared vertical distances of all the Y data from the straight line—the method of **Ordinary Least Squares (OLS)**. In other words, it is assumed that the X_i data are measured *without error*; only the Y_i values are **noisy**. If the x-axis is time, that is certainly appropriate. But if x represents a river's rate of flow, there may be a problem of **errors-in-variables**, in which case your estimates of a and b may be biased. This problem is often not addressed in environmental studies, nor (explicitly) in many software packages, but techniques for greatly reducing the bias are available, requiring only a little more effort than standard regression.[37]

The regression output will also produce a value of Pearson's r (or, more usually, r^2), indicating the strength of the linear relationship between x and y. The value of r^2 indicates how much of the variance in the data is explained by the regression model.

Regression can also be extended to multivariate models of the form

$$y = b_1 x_1 + b_2 x_2 + \cdots + b_n x_n + a + \epsilon, \tag{2.24}$$

again using standard statistical packages.

---

[37] Reckhow & Chapra (1983 [268]) describe a rather simple errors-in-variables technique, in which one first performs a regression of Y on X (i.e., $\hat{Y} = a + bX$). Then, a regression of X on Y is performed (i.e., $\hat{X} = c + dY$). The GM regression is $\hat{Y} = u + gX$, where $g = \pm\sqrt{a/c}$ and $u = \bar{Y} - g\bar{X}$. This is known as "Ricker's Geometric Mean regression." See Problem 2.14 for an example application.

More versatile approaches to regression use "Generalized Linear Models" (McCullagh & Nelder 1995 [215]). These use alternative distributions for the model's random component and also "link functions". For example, say we have binary response data so that y in Equation (2.24) is a zero–one dichotomous discrete variable, such as effect/no effect. The OLS model is not appropriate here, because the dependent variable (y) is restricted to be between zero and one. Instead we can use a Generalized Linear Model by forming a new dependent variable using the **logit function**, $\ln\left(\frac{\theta}{1-\theta}\right)$, where θ is the mean of y when this new model is used. This is the link function and is the logarithm of the **odds ratio**, the probability of presence divided by the probability of absence. The logit ranges from $-\infty$ to $+\infty$, as required by a linear model. Also, we note that a normal distribution of the random error term is inappropriate; it should be a binomial distribution. So we obtain a special form of the Generalized Linear Model, the **logistic regression** equation

$$\ln\left(\frac{\theta}{1-\theta}\right) = b_1 x_1 + b_2 x_2 + \cdots + b_n x_n + a + \epsilon', \qquad (2.25)$$

where ϵ' follows the binomial distribution, with parameter θ. The output from this model is θ, ranging between zero and one. This model has been used in epidemiological studies of the impact of water quality on the health of bathers (e.g., Wymer & Dufour 2002 [346]). See Problem 2.15 for a discussion of this model's properties.

2.12.1 Applications to water quality

Linear regression—both simple and multiple—has many applications in water quality assessment, including

1. Describing relationships between water quality variables, using one or more variables to predict another.

2. Describing relationships between water quality and flow, using flow as a predictive variable.

3. Describing a general increase or decrease in a water quality variable over time as a linear trend and testing its importance.

In (1) and (2) especially, the relationship may be nonlinear—in particular, it may be exponential. In this case the values of the random variables may be replaced by their logarithms before the regression is performed. The model is then

$$\log_{10}(y) = b \log_{10}(x) + a + \epsilon, \qquad (2.26)$$

or, equivalently,

$$y = x^b (10^a)(10^\epsilon). \qquad (2.27)$$

A straight line on log–log graph paper would suggest this type of relationship.[38] Note that this model contains a multiplicative, rather than an additive, error term.

2.12.2 Nonparametric regression

The usual linear regression estimators for a and b are based on least-squares theory, and their calculation does not invoke any assumptions about distributions. In that sense, they are nonparametric. However, to test the significance of the parameters or to obtain interval estimates for b or a from the regression, the noise term must be normally distributed. For many situations, the log transform mentioned above may be used to achieve normality.

Another alternative is to use Spearman's rho rather than Pearson's r to test the significance of the relationship between x and y following the usual linear regression. However, Spearman's rho will gauge the strength of a general **monotonic relationship**, rather than a strictly linear relationship.

A **completely nonparametric regression** can be obtained by replacing the data (for both x and y) with their ranks and performing the usual linear regression. The resulting linear equation between the ranks of the Ys and the ranks of the Xs provides a nonlinear monotonic relationship between Y and X (Conover 1980 [56]).

Prediction equations for y given x are accomplished by:

- Establishing a rank for the desired value of Y by interpolating between the observed X_i values.

- Using the regression equation to predict a rank for the desired Y: Rank(Y) = $b \times$ Rank(X_i) + a.

- Interpolating between the appropriate observed values of Y.

This type of regression relationship cannot be used to extrapolate beyond the range of observations—and one should always be careful when using methods that do allow it (i.e., in parametric regression).

The **Sen slope estimator** (Gilbert 1987 [118]) is a nonparametric estimate of slope for a set of points $(X_i, Y_i, i = 1$ to $n)$ and may be obtained by computing the slope between each pair of points and then using the **median slope** as the estimator.

2.13 ESTIMATING MODEL PARAMETERS

2.13.1 Point versus interval estimation

Our statistical models may have a large number of parameters—means, variances, slopes, and so on—for which we can never know the "true" (population) value. We **estimate** these parameters from a limited number of observations (the sample). The

---

[38] That is why logarithms to base 10 are generally adopted in this modeling—there is no such thing as log–log paper for natural logarithms (to the base e).

50 BASIC CONCEPTS OF PROBABILITY AND STATISTICS

sample mean ($\bar{X}$) is an example of a **point estimate** of the true mean (μ) of a random variable. We refer to $\bar{X}$ as an **estimator** of μ. Likewise, the true variance (σ^2) is estimated by the sample variance S^2, for which formulae were given earlier.

Precision of the estimate But how good is our estimate of a given parameter in a particular situation? If we repeat the estimation with a new dataset of n observations, we will get a different answer every time. But as n gets larger, the variability in the repeated estimates tends to decrease. This variability is called the **precision of the estimate**.[39] Precision may be quantified by the **variance of the estimate** or, more commonly, by its square root, the **standard error of the estimate**. For example, the standard error of the mean is defined as

$$S_e = \frac{S}{\sqrt{n}}. \quad (2.28)$$

It is worth noting that S_e is in fact a standard deviation, but of the estimate *not* of the data.

Smaller values of S_e imply greater precision of the estimate. Note that the standard error is *not a measure of data variability*: It measures precision of our estimate of a parameter, such as a mean, with the data we have. This is such a common misconception that it is worth elaborating. Say from $n = 25$ total phosphorus concentration data we compute a standard deviation of $S = 19.2$ mg L^{-1}. The standard error is $S_e = 19.2/\sqrt{25} = 3.84$ mg L^{-1}. But if we had collected $n = 100$ samples and obtained a similar standard deviation (say $S = 18.6$ mg L^{-1}), the standard error would be $S_e = 1.86$ mg L^{-1}. This shows that although the variability of the two datasets is hardly different (i.e., their standard deviations are very close), one standard error is about twice the other. Accordingly, the often-heard claims that small "error bars" on data bar graphs and histograms demonstrate low data variability are not justified when those bars are in fact standard errors.

> Standard errors are used when we want to think about precision of an estimate, not about the variability of our data.

2.13.2 Interval estimates

Perhaps a clearer way to express precision is an interval estimate or **confidence interval** for the estimate of the parameter. Such intervals may be either one-sided or two-sided. They will contain the true value of the estimated parameter for a specified fraction of the time. This specified fraction is referred to as the confidence level. The confidence level can be anything between zero and one, but 90%, 95%, and 99% levels are most commonly used.

To elaborate on just what a confidence interval implies, we would expect a 95% confidence interval for the mean of a given random variable to contain the true value of that mean for 95% of the time, on average, if the sampling and interval calculation

---

[39]In Section 1.5 we described precision *of measurements*. This is not the same as precision *of estimates*.

procedure were to be repeated many times.[40] As the sample size (n) increases, the precision of our point estimates increases, so the variance of the estimate and the associated confidence interval width decreases.

2.13.3 Bias

If the mean of repeated applications of an estimator is the true value of the parameter, the estimator is said to be **unbiased**. On the other hand, biased estimates tend to be high or low, on average, compared to the true value. Dividing by $n - 1$ rather than n in the equation for sample variance produces an unbiased estimate of the true variance (essentially because one degree of freedom has been used up when estimating the mean—and the sample mean must be known before the sample variance can be calculated).

Bias can also be introduced by a sampling strategy—for example, by collecting samples at the same time of day or same day of the week when daily or weekly cycles are present. That sampling strategy will fail to detect the cyclic variations. This is a type of bias, called **aliasing**.[41]

Accuracy is the difference between an estimated value and the true value. In most real estimation problems, accuracy is rather difficult to assess since we don't know the true value. Thus precision and unbiasedness are the usual criteria by which we evaluate estimating methods or estimators.

Random sampling is typically assumed when estimating model parameters but is rarely achieved. Fixed interval (say monthly) sampling is more common and is appropriate for analyzing trends.

Fixed interval sampling is not usually a problem *if one is careful to define what is actually estimated* in view of underlying cycles that are present. For example, if dissolved oxygen (DO) is sampled at 10:00 a.m. every day for a year, one would have a very precise estimate of mean DO at teatime (10:00 a.m.) for that year. But this would be a biased or aliased estimate of 24-hour average DO for that year.

2.13.4 Percentiles

Water quality management is increasingly using percentile standards, in response to dissatisfaction with standards based only on means or on maxima. For example, consider an effluent standard for a wastewater treatment plant stipulating merely that the daily median load of biochemical oxygen demand (BOD_5) should not exceed 300 kg. Citizens can be rightly concerned that such a standard says nothing about how much above 300 kg the effluent may discharge for up to half the days (though it does tell the treatment plant operator what is expected on average). The effect of those exceedances could be environmentally significant. Conversely, some permits have been issued stating merely that the maximum daily (BOD_5) load should not

---

[40]This is the result of taking the classical view of statistics. But "to be repeated many times" raises some serious practical issues—experiments and surveys simply can't be repeated many times. Maybe there is an alternative interval? We discuss these matters further in Chapter 3.

[41]Aliasing is sometimes also known as "fold-over" or "mixing."

exceed, say, 800 kg. This too is unsatisfactory, because it ignores the possibility of an exceedance of that limit through legitimate causes (e.g., power shutdown or laboratory measurement error). Commonly, when brief breaches of the maximum do occur, there is no follow-up action, because the incident was minor in its effect. That seems unsatisfactory. A way around that is to set the maximum to a very large figure, but that satisfies nobody.

For these reasons effluent standards are increasingly being set using a combination of requirements on average conditions (using arithmetic means, geometric means, or medians) *and* on high percentiles (typically the 95%ile) (see Problem 6.2). Furthermore, these are best expressed as *percentiles of time*, not as *percentiles of samples*. Some detailed statistical issues arising from this are treated in some depth in Chapter 8, but as a general matter one needs to be aware that *there are several ways to calculate percentiles*. Many software packages fail to advise the user just what formulae they use.

Environmental water quality standards (e.g., for recreational activities) are increasingly using 95%iles (e.g., WHO 2003 [342]), as are drinking-water standards (e.g., MoH 2000 [229]). When consenting wastewater discharges, also using limits expressed as percentiles, percentile mixing calculations are required. Monte Carlo methods (Section 4.8) can be used for simple cases (e.g., a single discharge into river). Alternative approximating methods may be used where there are computational limitations—for example, in costal regions when water movement is predicted using computational hydrodynamic mixing models (Murdoch 2001 [235]). Maximum standards may still be retained however, especially in a "surveillance" mode, where the health of recreational water users need to be protected (see Chapter 7).

Problems

2.1 What is the distinction between the probabilities $\Pr(A|B)$ and $\Pr(A \text{ and } B)$? Give an example.

2.2 What can you say about the relative values of the probabilities $\Pr(A|B)$ and $\Pr(A \text{ and } B)$?

2.3 What is the difference between independent events and mutually exclusive events?

2.4 Continuing the estuarine deposition example in Section 2.1, let's consider that the urbanization pressure is not expected for another 20 years, and that climate scientists advise that there is a 70% chance that there will be a climate shift in that period such as to modify the probabilities to $\Pr(A) = 0.15$, $\Pr(B) = 0.25$ and $\Pr(B|A) = 0.6$. What is the daily probability of deposition in 20 years time, if that is when urbanization will occur?

2.5 In the estuary deposition example in Section 2.1, what would be the impact on the posterior probability had the tribunal taken an impartial stance regarding the merits of the two competing hypotheses, before hearing further evidence?

2.6 For three events A, B, and C, where A is sampled from one population and B and C are both sampled from another separate population, show that $\Pr[(A \text{ and } B) \text{ or } (A \text{ and } C)] = \Pr(A)[\Pr(B) + \Pr(C)]$. (This result is used in the answer to Problem 4.6.)

2.7 Why is precision of measurements different from precision of an estimate?

2.8 What is the total area under the "sampling distributions" (unit normal, t, χ^2 and F), and why is this?

2.9 Show that the area under a normal distribution between the mean plus or minus one standard deviation is 0.6826 and that the area under that distribution between the mean plus or minus two standard deviations is 0.9544. Show that these results hold for *any* normal distribution.

2.10 Show that the sample geometric mean can be calculated as the antilog of the mean of the sample logarithms. Apply this to the MPN data given on Table 2.1. What are the implications if one of the data values is zero?

2.11 In Section 2.5.2 we noted that one of the reasons that water quality microbiological standards may use the sample median rather than the sample geometric mean is that the sample geometric mean is always a biased estimator of the true geometric mean. Can you think of another reason?

2.12 Does the sample geometric mean overestimate or underestimate the geometric mean of a lognormal distribution?.

2.13 We have the following two groups of data for the concentration of the bacterium *E. coli* per 100 mL of water on each of 11 separate days: site 1—100, 82, 1198, 336, 5, 1071, 156, 31, 1120, 156, and 5 (average = 387.27); site 2—50, 20, 146, 738, 98, 1401, 98, 6488, 970, 98, and 50 (average = 923.36). The between-site averages for each day are therefore: 75, 51, 672, 537, 51.5, 1236, 127, 3259.5, 1045, 127, and 27.5 (average 655.32). But these averages are not representative of the central tendency of the data (as a bar graph of results clearly shows), so we could think of using the sample geometric means by taking averages of the logarithms of the data and then exponentiating the result. It is simplest to do this for the between-site averages. In doing so we obtain the mean of the natural logarithms of these data as 5.4682 and so the sample geometric mean is $e^{5.4682} = 237.03$. (We get the same result if we take logarithms to base 10, giving an average value 2.3748, and then computing the sample geometric mean as $10^{2.3748} = 237.03$.) Is this the correct geometric mean?

2.14 Forty one pairs of data (X and Y) (in parts per billion) are obtained for two methods of estimating chlorophyll *a*, with $\bar{X} = 2.07146$ and $\bar{Y} = 2.81805$ and the following regressions: $Y = 0.86947X + 1.0170$ and $X = 0.94901Y - 0.60288$. Which regression equation should be used? What are its main features?

2.15 Discuss the meaning of the coefficients in the logistic model [Equation (2.25)].

54 BASIC CONCEPTS OF PROBABILITY AND STATISTICS

Appendix: Conditional probabilities—The Monty Hall dilemma

Monty Hall was the host of the TV game show *Let's Make a Deal*, in which a contestant comes on the stage and is presented with three doors. Monty explains that behind one door is a prize (e.g., a car) but that there is a goat behind each of the other two doors. The contestant is to open a door and will win whatever is behind it.

At Monty's invitation, the contestant chooses a door, but doesn't open it. Monty then says that he will open one of the other doors to reveal one of the goats, and he does so. The contestant has seen *Let's Make a Deal* many times, and on all those occasions Monty has opened a door to reveal a goat.[42] He then invites the contestant to open their chosen door, or to choose the other (unopened) door. The question is: Should the contestant switch to that other door?

The answer is yes—in doing so the contestant would win two times out of three. This answer often clashes violently with intuition, even among very numerate professionals (Hoffman 1998 [154]). Such responses claim that the probability is $\frac{1}{2}$ of either unopened door concealing the prize; all that happened when Monty opened a door was to change the probability for each unopened door from $\frac{1}{3}$ to $\frac{1}{2}$, so switching doors makes no difference to the chance of winning the prize. This reasoning ignores the subtle effects of conditional probabilities.

**Probabilities of events occurring can change
when new information is made available.**

The proof runs as follows.[43] Let B_1 denote the event that the prize is behind door number 1, let B_2 denote the event that the prize is behind door number 2, and similarly for B_3. Then, before Monty opens a door, the contestant faces the prior probabilities $\Pr(B_1) = \Pr(B_2) = \Pr(B_3) = \frac{1}{3}$, meaning that each door is equally likely to conceal the prize. Now suppose the contestant chooses door number 1 and Monty then opens door number 2 to reveal a goat. Let's use A to denote the event that Monty opens door number 2 and consider the long-run probability that he will have opened that door (he won't have done so if the prize happened to be behind it—because he said he would open a door to reveal one of the goats, so he knows where the prize is). This is a "total probability", comprising three separate components, because we need to consider the three possible locations of the prize. If the prize is behind door number 1 he knows that he may choose either door number 2 or door number 3, because either conceals a goat. So $\Pr(A|B_1) = \frac{1}{2}$. If the prize is behind door number 2 he cannot choose it (and must choose door number 3). So $\Pr(A|B_2) = 0$. On the other hand if the prize is behind door number 3 he is certain to choose door number 2, so that $\Pr(A|B_3) = 1$. Then by the "Rule of total probability for multiple mutually exclusive events" [Equation (2.12)] we have

---

[42] He didn't always open a door—we're just keeping things simple here.
[43] A similar proof can be found in Gill (2002 [119]).

$$\begin{aligned}
\Pr(A) &= \Pr(B_1)\Pr(A|B_1) + \Pr(B_2)\Pr(A|B_2) + \Pr(B_3)\Pr(A|B_3) \\
&= \left(\frac{1}{3} \times \frac{1}{2}\right) + \left(\frac{1}{3} \times 0\right) + \left(\frac{1}{3} \times 1\right) = \frac{1}{2}
\end{aligned}$$

Given that the contestant has chosen the first door, intuition suggests that the probability that Monty chooses the second one should be $\frac{1}{2}$, as indeed it is.

Finally, by Bayes' rule [Equation (2.13)], the probability that the prize is behind door number 1, given that Monty opens door number 2, is

$$\Pr(B_1|A) = \frac{\Pr(B_1)\Pr(A|B_1)}{\Pr(A)} = \frac{(1/3)(1/2)}{(1/2)} = \frac{1}{3}$$

The same result holds for any other labeling of the doors. So, using Equation (2.3), the probability of the prize being behind door number 3 is $\Pr(B_3|A) = 1 - \Pr(B_1|A) = \frac{2}{3}$. Therefore, by switching doors the contestant's chance of winning the prize would be doubled.

The key to resolving this dilemma is to recognize that in opening a door, Monty knows which door conceals the prize. That information destroys the prior equality of probabilities between the unopened doors.

> **This analysis shows that probabilities can vary between individuals, depending on what information they have.**

For example, if more contestants arrive on the stage just after Monty had opened the second door, and they know nothing of the prior conversation between Monty and the first contestant, their probability for each unopened door is $\frac{1}{2}$.

Finally, note that for the first contestant the probability of the prize being behind the door originally chosen is *preserved* by Monty's actions. For example, were there to have been four doors on the stage the prior probability of the chosen door is $\frac{1}{4}$, and this remains unaltered after Monty opens another door to reveal a goat, and so the contestant should switch doors—although the increase in the chances of winning are less than the doubling that occurs when only three doors are on the stage.[44]

If none of this convinces you, imagine that there were 100 doors and Monty opened 98 of them, all revealing goats. Would you switch then? Or a million doors, and he opened 999,998 of them!

---

[44]The long-run probability that Monty opens door number 2 (of the four doors) is $\Pr(A) = \left(\frac{1}{4} \times \frac{1}{3}\right) + \left(\frac{1}{4} \times 0\right) + \left(\frac{1}{4} \times \frac{1}{2}\right) + \left(\frac{1}{4} \times \frac{1}{2}\right) = \frac{1}{3}$. Therefore the probability that the prize is behind door number 1 is $\Pr(B_1|A) = \frac{(1/4)(1/3)}{(1/3)} = \frac{1}{4}$. So there's a $\frac{3}{4}$ chance that the prize is behind one of the two other unopened doors, and therefore a $\frac{3}{8}$ chance that it lies behind the third door (and the same chance that it lies behind the last door). By changing to one of these two doors the contestant's chances of winning the prize increase by 150% (from $\frac{1}{4}$ to $\frac{3}{8}$).

3
Intervals

When we make a point estimate of a parameter, we obtain an indication of a possible value for the water quality variable of interest. But multiple estimates can seldom be expected to return exactly the same value. Because of this variability in estimates, our interest may lie more in stating a region—an interval—in which the true value of the parameter probably lies. In particular, we often want to have some measure of precision of an estimate we may make (e.g., of a mean, a median, or a percentile). In another situation we may want to know the proportion of a population's distribution that lies between certain limits, especially if we want to assess a shift in its spread or location. There are three intervals that may be used in such circumstances, depending on the questions addressed.

Confidence intervals are ranges in which the parameter may lie most of the time, in repetitive sampling.

Tolerance intervals are ranges covering a stated proportion of the population most of the time, in repetitive sampling.

A credible interval is a range in which the parameter probably lies.

The first two of these intervals are frequentist, in that they strictly only have meaning under repetitive sampling (i.e., in the "long run"). The last is Bayesian, in that the probability statement made relates to the particular sample at hand. Bayesian tolerance intervals can also be defined, but are seldom featured in the literature (Guttman 1970 [129]), and will not be discussed here. We discuss these three types of intervals in turn. These intervals also have subdivisions: Confidence intervals may actually be of two types (confidence intervals or prediction intervals); tolerance intervals may also be of two types (β-content and β-expectation). Any interval may also be one-sided or two-sided, and two-sided intervals may be "equi-tailed" or "highest density." These divisions are discussed also.

Interpreting confidence intervals   The bulk of this chapter concerns confidence intervals, because these constitute most of the literature on intervals. But in focusing on them a fundamental point must be made concerning their interpretation. A common (mis)interpretation of a confidence interval is that the true value of the variable probably lies between the interval's limits.[1] For example, if the 95% confidence interval's lower and upper limits for mean growing season lake chlorophyll *a* concentration are 2 and 17 parts per billion, such a statement would say "I am 95% sure that lakes' mean chlorophyll *a* concentration in the growing season is between 2 and 17 parts per billion." This interpretation is not correct. All one can say is that were the lakes to be sampled in the growing season many times over, and were a 95% confidence interval to be calculated for the growing-season for each of those times, then 95% of the intervals so calculated, each with different limits, would contain the true mean value. In the frequentist interpretation of probability, one cannot attach a probability to a particular interval. The quoted interpretation is invalid.

However, this is often the interpretation that investigators actually wish to make, and so it may be held that the finer points of interpretation are of no concern (and are just plain awkward). But they are of concern, for the following reason. Let's say that a Bayesian credible interval calculation gives the same upper and lower limits as those calculated using the usual frequentist confidence interval formula. In that case, taking the Bayesian view of probability, the above interpretation *is* valid. So we should ask; Under what circumstances would the limits of the confidence interval and the credible interval be the same? Sidestepping the mathematics, the answer is: When the prior distribution is "vague" or "diffuse," stating that all values of chlorophyll *a* are equally likely.[2] In such cases the formulae for the calculation of the interval's limits are identical. But this prior distribution postulates that negative concentrations are as likely in the population as are positive values. It also postulates that "large" positive values (say above 500 parts per billion, up to infinity) are *more likely* than small values (between 0 and 500 parts per billion).[3] Yet lake studies have never reported values

---

[1]This is quite widespread. For example, a popular biostatistics text states the following: "What it tells us is that we can ... say that we are 95% confident that the stated interval encompasses μ" (Zar 1996, p. 101 [349]); later, on the same page, the correct interpretation is given. A manual for popular software (Mathsoft 1997 [203]) says: "With probability $1 - \alpha$, the mean μ lies within the *confidence interval* (L, U)." Even an advisory booklet on using confidence intervals states "... there is a 95% chance that the indicated range included the 'population' difference in mean blood pressure levels" (Gardner & Altman 1989 [108]). While this is followed by the correct interpretation (that were a series of identical studies carried out repeatedly on the same population then in the long run 95% of such intervals would contain the true mean), the statement quoted above is given as its practical interpretation. The versatile DataDesk software's guide states the correct interpretation (Velleman 1997, p. 18/3 [330]).
[2]This is often presented in Bayesian literature as an "uninformative" prior distribution. But any statement of a prior distribution contains *some* information (Lindley 1971 [186], Stuart *et al.* 1999, p. 446 [304]), so that term is not used in this text.
[3]Problem 3.1 invites further discussion on this point.

above 500 parts per billion.[4] Accordingly, a narrower prior distribution is appropriate (e.g., between 0 and 500 parts per billion), setting it to zero outside that range. The resulting credible interval will not be as wide as the confidence interval, particularly for low sample sizes.

Other writers have addressed this issue as follows:

> It is interesting that most researchers are taught statistics from a classical perspective, yet confidence intervals are often interpreted in a Bayesian sense. When the Bayesian interpretation is adopted, the analyst should realize that this implies a subjective interpretation for probability, and this should be specified in the analysis ... the prior probability distribution must be stipulated if the Bayesian interpretation for confidence intervals is adopted.... (Reckhow & Chapra 1983 [268]).
>
> The statement that such-and-such is a 95% confidence interval for μ seems objective. But what does it say? It may be imagined that a 95% confidence interval corresponds to a 0.95 probability that the unknown parameter lies in the confidence range. But in the classical approach, μ is not a random variable, and so has no probability. Nevertheless, statisticians regularly say that one can be '95% confident' that the parameter lies in the confidence interval. They never say why (Howson & Urbach 1991 [157]).

It appears that the only simple solution to the issue of interpreting a single confidence interval is to make the Bayesian inference, realizing that a diffuse prior has been invoked.[5] We revisit such matters in Chapter 10 when discussing uncertainty in water microbiological MPN assay results.

3.1 CONFIDENCE INTERVALS

We first discuss confidence intervals for means, both because these are commonly reported and because they serve to illustrate some important features common to all confidence intervals, such as dependence on sample size, equi-tailed versus asymmetric tails, two-sided versus one-sided intervals, and the relationship between confidence intervals and hypothesis tests. First, note that it is conventional to refer to "$100(1-\alpha)\%$" confidence intervals, where α is a small number ($<\frac{1}{2}$). So the phrase "95% confidence interval" means that $\alpha = 0.05$.

3.1.1 For means

Single normal mean The two-sided confidence interval for the mean of a normal population is calculated from

---

[4]There are strong phenomenological grounds for never expecting such high values, because of nutrient limitation and phytoplankton self-shading.

[5]In cases where confidence intervals are calculated many times, such as in a Monte Carlo analysis discussed in Section 4.8 and Chapter 9, the frequentist interpretation of confidence intervals may be very appropriate, precisely because those intervals may be calculated many times.

60 INTERVALS

$$\bar{X} \pm t_{1-\frac{\alpha}{2},f} \left(\frac{S}{\sqrt{n}} \right), \tag{3.1}$$

where S is the sample standard deviation, $f = n - 1$ is the degrees of freedom (n is the sample size, i.e., the number of samples), and $t_{1-\frac{\alpha}{2},f}$ is the percentile of the t-distribution cutting off an area of $\frac{\alpha}{2}$ in the *right* tail, with $f = n - 1$ degrees of freedom [and also, from Equation (2.20), $-t_{1-\frac{\alpha}{2},f}(= t_{\frac{\alpha}{2},f})$ is the value of the t-distribution cutting off an area of $\frac{\alpha}{2}$ in the *left* tail, with $f = n - 1$ degrees of freedom—see Figure 2.14]. The term in parentheses ($\frac{S}{\sqrt{n}}$) is the standard error of the mean (S_e), already discussed in Section 2.13.1.[6] Accordingly we can write the equation as $\bar{X} \pm t_{1-\frac{\alpha}{2},f}(S_e)$. The confidence interval width is $2t_{1-\frac{\alpha}{2},f}(S_e)$; the half-width is $t_{1-\frac{\alpha}{2},f}(S_e)$. Indeed, confidence interval half-widths are generally expressed as the product of a percentile (critical value) of a sampling distribution multiplied by the relevant standard error.

Small sample sizes For small n (e.g., $n < 30$), Equation (3.1) requires that the data be drawn from a normal distribution, or at least "close to" a normal population. However, for n greater than about 30 the **central limit theorem** says that the means of samples drawn from a non-normal distribution tend to be distributed normally and the standard error of that distribution is the "standard error of the mean."[7] Therefore the above estimator is also valid for non-normal distributions if more than about 30 observations are available.[8] Furthermore, for $n > 30$ the critical values of the normal and t-distributions are very similar, and so Z percentiles could be used in place of t percentiles. (While this may have been convenient in the past, there seems little point in doing so nowadays, with the widespread availability of software and statistical tables.)

Non-normal means, small sample size In this case it is often best to select a statistic other than the mean for which to compute an interval. As noted in Section 2.2.5, the mean is not a good index of central location for skewed data. So if central location is the interest, it would be better to compute an interval for a more appropriate statistic, such as the median or geometric mean. Indeed, if the data are drawn from a lognormal population, the equivalent confidence interval on the population's geometric mean or median (μ_g) of a two-parameter lognormal distribution is

$$\frac{\bar{X}_g}{e^{t_{1-\alpha/2,f}S_y/\sqrt{n}}} \leq \mu_g \leq \bar{X}_g e^{t_{1-\alpha/2,f}S_y/\sqrt{n}}, \tag{3.2}$$

---

[6]The interval is based on the result that $100\alpha\%$ of all possible sample means from a population with mean μ will yield T-values that are larger than $t_{1-\frac{\alpha}{2},f}$ or smaller than $-t_{1-\frac{\alpha}{2},f}(= t_{\frac{\alpha}{2},f})$, where $T = |\bar{X} - \mu|/S_e$.

[7]The standard error of the mean is a standard deviation *of the distribution of the sample means*; it should not be (but often is) confused with the standard deviation of our data—see Section 2.13.1.

[8]Various authors have used $n = 10, 30,$ or 100 as the value at which the normal distribution approximates the t-distribution; 30 is a reasonable compromise (Iman & Conover 1983 [161]).

where $\bar{X}_g$ is the sample geometric mean and S_y is the standard deviation of the natural logarithms of the data. Equation (3.2) is easily derived from Equation (3.1), see Problem 3.2. Note that this equation ignores the bias in $\bar{X}_g$ as an estimate of μ_g, so it is only approximate for "small" sample sizes if the true geometric standard deviation is "large" (see footnote 20 on page 31). Intervals calculated from Equation (3.2) are asymmetric (about the sample geometric mean $\bar{X}_g$), whereas the interval in Equation (3.1) is symmetric (about the sample mean $\bar{X}$). Note that if logarithms to base 10 are taken, then "e" in Equation (3.2) must be replaced by "10." Indeed, the base of logarithms that have been used is a potential source of confusion, especially when it is noted that "LOG" functions in various calculators and computer software may denote either natural logarithms or logarithms to base 10.

> **Always check that logarithms and antilogarithms are being referred to the same base.**

Note too that in some cases the mean *is* of interest regardless of skew—for example, when computing the total amount of a contaminant in sediments. In that case, if samples have been drawn from a right-skewed lognormal distribution, Land's procedure may be used to calculate confidence limits on the arithmetic mean (see Gilbert 1987, p. 169 [118]).

Finally, if data do not appear to have been drawn from normal or lognormal distributions, one can use nonparametric confidence intervals. These are wider than their parametric cousins, and they will be wider for the median rather than for the mean (see Problem 3.4). Details are given in many texts (e.g., Iman & Conover 1983 [161], Zar 1996 [349]).

Example Microbiological standards for marine bathing waters may require that a *minimum* of 5 samples be taken in a bathing season, with the geometric mean or median of their enterococci concentrations being compared to an acceptable limit, such as 35 enterococci per 100 mL (e.g., USEPA 1986 [324]). A question often asked is, what is a *desirable* number of samples to take? Some light can be shed on this by examining the way in which confidence limits for these microorganisms change with sample size. The lognormal distribution is usually found to be appropriate for such data, so Equation (3.2) would be appropriate. To keep things simple, let's assume that the sample median is 15 enterococci per 100 mL and that the standard deviation of the logarithms (to base 10) is 0.7 (a common value for marine waters), whatever the sample size.[9] Then we may use Equation (3.2), with "e" replaced by "10," to prepare Figure 3.1. The curvature of the confidence limits is such that there are dramatic increases in precision of the estimate of the true geometric mean as the sample size is increased above the minimum of 5. The figure suggests that at about 20 samples, there is a rapidly diminishing return of information for the increased effort in sampling.

---

[9] Of course, this won't be the case—sample medians will differ between sets of observations. But we lose no generality by making this simplification.

62 INTERVALS

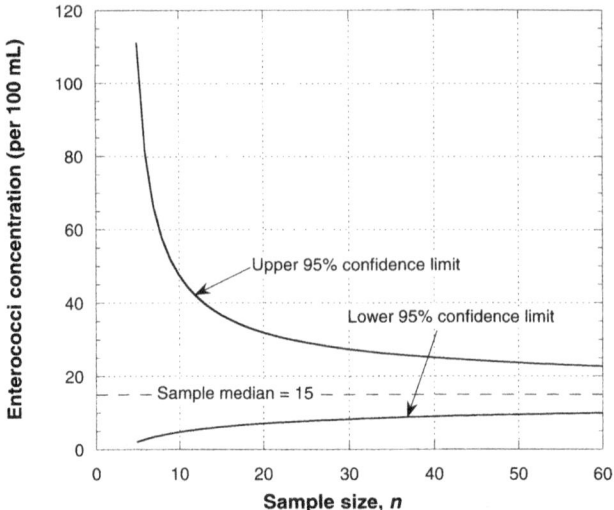

Fig. 3.1 Geometric mean confidence limits as a function of sample size.

Equi-tailed confidence intervals The use of $\frac{\alpha}{2}$, not just α, in Equations (3.1) and (3.2) arises because these formulae are for two-sided intervals: They have both a finite lower bound and a finite upper bound. Also, the area in each tail of the distribution of the mean (or median or geometric mean) is $\frac{\alpha}{2}$ (e.g., 0.025 for a 95% confidence interval), because the confidence interval is "equi-tailed." But we can also consider 95% confidence intervals that are not equi-tailed, because all that is required is that the total tail area is 5%. Little use is made of such intervals in applications of *frequentist* statistical methods, and we do not pursue them here. However, in computing *Bayesian* credible intervals, such an approach is common and can be highly advantageous (because it minimizes the interval's width), as we shall see in Section 3.3.

One-sided confidence intervals As already noted, the use of $\frac{\alpha}{2}$ instead of α in Equations (3.1) and (3.2) arises because we have sought a two-sided interval. Sometimes a one-sided interval should be used—for example, when considering compliance with upper or lower limits in water quality standards. That simply means that one of the limits is banished to $-\infty$ (when considering an upper limit) or to $+\infty$ (when considering a lower limit). Then the area in the banished tail is zero, and so we obtain the following formulae:

$$\text{Upper confidence limit:} \quad \bar{X} + t_{1-\alpha, f} \left(\frac{S}{\sqrt{n}} \right), \quad (3.3)$$

$$\text{Lower confidence limit:} \quad \bar{X} - t_{1-\alpha, f} \left(\frac{S}{\sqrt{n}} \right). \quad (3.4)$$

The translation of two-sided intervals to one-sided intervals is straightforward, replacing $\frac{\alpha}{2}$ by α, and so need not be stated for any of the following material (see Problem 3.5).

Example When $n = 60$ data are obtained from sampling cadmium concentrations in a stream the resulting sample mean and standard deviation are $\bar{X} = 23$ and $S = 16$ (both measured in parts per billion, ppb). Then, using the t-distribution table (Table A.2.2), $t_{0.975, 59} = 2.001$ and so the resulting 95% two-sided confidence interval is $23 \pm 2.001 \frac{16}{\sqrt{60}} = 23 \pm 4.133$ ppb. The 95% confidence interval half-width is 4.133 ppb. Note that this width would narrow a little (by 2.04%) were we to use the normal distribution instead of the t-distribution, giving a half-width of $1.960 \frac{16}{\sqrt{60}} = 4.049$ ppb. Now let's imagine that only one in four of these data had in fact been collected, so $n = 15$. Of course, slightly different means and standard deviations would have resulted.[10] Let's say they are $\bar{X} = 19$ and $S = 14$ ppb. Then the 95% confidence interval is $19 \pm 2.145 \frac{14}{\sqrt{15}} = 19 \pm 7.754$ ppb, so the half-width is 7.754 ppb. In this case using the normal distribution instead of the t-distribution results in the half-width $1.960 \frac{14}{\sqrt{15}} = 7.085$ ppb—an 8.6% reduction. Of course, this is an undesirable result: The wrong method has been used. If the population standard deviation is unknown, we must use the t-distribution, not the unit normal distribution, because it accounts for uncertainty in having to estimate σ by S.

Confidence interval width depends on sample size In the preceding example the width of the 95% confidence interval for 60 samples increased when recomputed for the subset of 15 samples, despite the fact that, by chance, the standard deviation of that subset decreased (it fell from 16 to 14 ppb). That is because the standard error (S_e) increases as n is reduced. This merely reflects the fact that the confidence interval width is a measure of the **precision of estimate** of the mean, *not* of the variability in the data (as we have already noted in Section 2.13.1). Simply put, the more data we have, the more precise our estimate can be.

Difference between two normal means When there are two *independent* samples from populations x and y, we may calculate a confidence interval for the difference in means as follows.

$$\bar{X} - \bar{Y} \pm t_{1-\frac{\alpha}{2}, f} \left(S_{e, \bar{x}-\bar{y}} \right), \tag{3.5}$$

where the total degrees of freedom is $f = f_x + f_y$, where $f_x = n_x - 1$ and $f_y = n_y - 1$ are the degrees of freedom for each sample. The term $S_{e, \bar{x}-\bar{y}}$ is the standard error of the difference between the means. To compute it we first calculate the pooled variance of the mean, by weighting each sample variance by its degrees of freedom; that is,

---

[10] Recall that $\bar{X}$ is an unbiased estimate of the population value μ and that S is practically unbiased also, as an estimator of σ. This means that under repeated sampling we can expect the values of $\bar{X}$ and S to bounce around their true values, *regardless of the sample size*.

$$S_p^2 = \frac{f_x S_x^2 + f_y S_y^2}{f}. \tag{3.6}$$

Then the standard error of the difference is given by the square root of the sum of the two individual variances of the sample means[11]—that is, by

$$S_{e,\bar{x}-\bar{y}} = \sqrt{\frac{S_p^2}{n_x} + \frac{S_p^2}{n_y}} = S_p\sqrt{\frac{1}{n_x} + \frac{1}{n_y}}. \tag{3.7}$$

Corrections may be made if it is held that the two populations have different variances (Zar 1996, p. 131 [349]). For equal sample sizes, each of size n, these equations simplify considerably. The pooled variance becomes

$$S_p^2 = \frac{1}{2}(S_x^2 + S_y^2), \tag{3.8}$$

and the standard error is

$$S_{e,\bar{x}-\bar{y}} = S_p\sqrt{2/n}. \tag{3.9}$$

Accordingly, the confidence two-sided interval equation is simply

$$\bar{X} - \bar{Y} \pm t_{1-\frac{\alpha}{2},f}\left(S_p\sqrt{\frac{2}{n}}\right), \tag{3.10}$$

where $f = 2(n-1)$.

Comparing the one-sample and two-sample cases In the one-sample case [Equation (3.1)] we can consider the confidence interval to be for a difference between the mean and zero. In the two-sample case for equal sample sizes [Equation (3.10)], let's imagine that the true mean of the second population (y) is also zero—unbeknownst to the sampler. Then, comparing these two equations, we see that the second confidence interval is greater than the former by a factor of $\sqrt{2}$. This widening simply reflects the fact that in the two-sample case we do not know the true value of the second mean. The uncertainty that this inclusion introduces causes the confidence interval to be wider.

Paired samples Another situation can arise when the samples from the two populations are *paired*. That is, the samples are not independent and have some inherent association; there is a one-to-one relationship between each item of the first sample and a corresponding item in the second sample. Simultaneous (or near-simultaneous) sampling at control sites and at impact sites can be viewed as paired sampling. In that case it is the mean of the differences between samples that is of interest, rather than the difference between the means (as presented above). So we have $n_x = n_y = n$ paired samples and the confidence interval is

---

[11] Mathematical statistics results show that the variance of the difference between independent random variables equals the sum of the variances of those variables.

$$\overline{X-Y} \pm t_{1-\frac{\alpha}{2},f}\left(\frac{S_d}{\sqrt{n}}\right), \tag{3.11}$$

where $\overline{X-Y}$ is the mean of the differences, $f = n-1$ and S_d is the standard deviation of the differences. The term in parentheses is the standard error of the differences:

$$S_{e,\overline{x-y}} = \frac{S_d}{\sqrt{n}}. \tag{3.12}$$

Note that with paired samples the mean of the differences is in fact the same as the difference of the means (i.e., $\overline{X-Y} = \bar{X} - \bar{Y}$), but the standard errors for the two cases are *not* the same (see Problems 3.6 – 3.8).

Advantages of pairing Several authors have noted that in environmental impact studies, simultaneous sampling at control and impact sites is highly desirable. For example, Green (1979, Chapter 3 [127]) discusses optimal sampling design for assessing the impact of an effluent discharge on a river, where samples are taken both before and after the impact occurs. The upstream site is a control. If one analyzes the differences in these data, treating them as pairs, many of the difficulties of "pseudoreplication" raised by Hurlbert (1984 [159]) are overcome (Stewart-Oaten *et al.* 1986 [302]).[12] A number of other authors have noted the benefits of simultaneous sampling at control and impact sites (Ponce 1982 [259], Millard & Lettenmaier 1986 [225], Loftis *et al.* 1987 [189], Loftis *et al.* 2001 [191]). Note that control and impact sites may not be easily found, especially in estuaries or in highly impacted aquifers with many contributing sources.

Example Simultaneous sampling of dissolved reactive phosphorus concentration (in parts per billion) downstream (X) and upstream (Y) of a drainage input on 10 occasions gives the following pairs of (X, Y) data: $(16, 12)$; $(23, 14)$; $(22, 19)$; $(17, 9)$; $(32, 22)$; $(19, 12)$; $(21, 16)$; $(12, 7)$; $(22, 14)$; $(26, 22)$. Treating X and Y as independent samples, we obtain $\bar{X} = 21.000$, $S_x = 5.558$ and $\bar{Y} = 14.700$, $S_y = 5.100$, so that the difference in means $\bar{X} - \bar{Y} = 6.300$ and the pooled standard deviation is $S_p = 5.334$. The critical t-value for a 95% confidence interval on the difference in means is $t_{0.925,18} = 2.101$ and so the interval's limits are at 6.300 ± 3.544. If we treat these data as paired, we obtain $\overline{X-Y} = 6.300$, and $S_d = 2.406$. The critical t-value for a 95% confidence interval on the mean difference is $t_{0.975,9} = 2.262$ and so the interval's limits are at 6.300 ± 1.721. In this case, pairing has reduced the width of the confidence intervals. But note that this is not always the case (see Problems 3.8 and 3.9).

"Error bars" Software packages commonly include an option for placing "error bars" on histograms. If used, it is vital that they be defined. In particular, they are

---

[12] Hurlbert observed that if one obtains independent samples, then an analysis can only reveal differences, not impacts.

typically graphed as standard deviations or as standard errors. Occasionally they are printed as two standard errors—closely approximating a 95% confidence interval. So definition of these "bars" is vital.

Recall (from Section 2.13.1) that standard deviations should be used to indicate the variability of the data, whereas standard errors have to do with the precision of estimate of a particular statistic (e.g., a mean). Therefore, Figure 3.2(a) cannot be held to show lack of data variability—Figure 3.2(b) shows that there is.

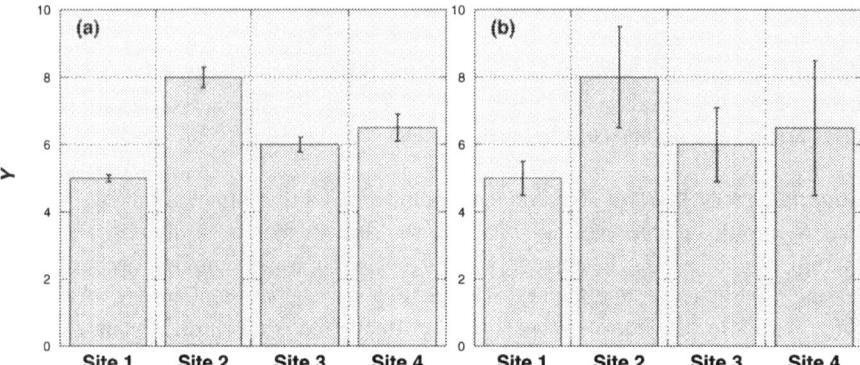

Fig. 3.2 "Error bars" for $n = 25$ observations per site: (a) standard errors, (b) standard deviations. The standard errors are one fifth of the standard deviations.

This is an issue of concern in the environmental sciences. Magnusson (2000 [195]) reports the variable and at times confusing usage of error bars—with some authors never identifying what form of "error bar" has been used. He advocates that they not be used, or at least that they be used only for data reporting (i.e., as standard deviations). Note too that there are difficulties in using "error bars" expressed as two standard deviations as surrogate statistical tests, as we shall see in Section 5.6.[13]

> **If "error bars" are reported, always define what they are. They should generally be used to depict the variability of the data, in which case they should be defined as standard deviations. If they are defined as standard errors, always report their sample size.**

3.1.2 For prediction

Confidence intervals summarize present information about a parameter, such as the mean. Prediction intervals estimate ranges in which future individual data values may lie, given the present information.

> **Prediction intervals are always wider than confidence intervals.**

---

[13]Non-overlapping of such bars in two-sample problems is equivalent to a 83% level test, *not* a 95% level test.

This is best explained in terms of confidence and prediction intervals on a linear regression. Here we seek to define the sloping line that best fits our data $(X_i, Y_i), i = 1, \ldots, n$. This line (our model) is written as $y = \alpha + \beta x$. We represent this in our data by the equations $Y_i = a + bX_i + \epsilon_i : i = 1, \ldots, n$, where a and b are point estimates of the model's values α and β, and ϵ_i represents the departures of the data from the best fit line.[14] The values of a and b are calculated by minimizing the sum of squared departures in the vertical (y) direction. This results in the "normal equations"[15] for a and b:

$$b = \frac{n\left(\sum X_i Y_i\right) - \left(\sum X_i\right)\left(\sum Y_i\right)}{n\left(\sum X_i^2\right) - \left(\sum X_i\right)^2}, \tag{3.13}$$

where (and in the following text) the summation symbols (Σ) denote summation from $i = 1$ to n, and

$$a = \bar{Y} - b\bar{X}. \tag{3.14}$$

We can now calculate yet another standard error, the **standard error of estimate**, which is in fact the standard deviation of the residuals about the regression line

$$S_{y|x} = \sqrt{\frac{\sum \left(Y_i - \hat{Y}_i\right)^2}{n - 2}}, \tag{3.15}$$

where $\hat{Y}_i$ is the ith estimate of y using the regression line; that is, $\hat{Y}_i = a + bX_i$. This term is used to calculate two-sided confidence limits for a regression line at any point X:

$$Y_{\text{upper,lower}} = a + bX \pm t_{1-\frac{\alpha}{2}, f} S_{y|x} \sqrt{\frac{1}{n} + \frac{(X - \bar{X})^2}{\sum (X_i - \bar{X})^2}}. \tag{3.16}$$

where the degrees of freedom is $f = n - 2$. This equation also suffice for the geometric mean regression, where $\hat{Y}$ is computed from that regression's equation (see footnote 37 on page 47).

Prediction limits are given by an amendment to the confidence limit equation. That is, if we want to place confidence limits on the mean of m additional measurements at a value X, the prediction limits are given by

$$Y_{\text{upper,lower}} = a + bX \pm t_{1-\frac{\alpha}{2}, f} S_{y|x} \sqrt{\frac{1}{m} + \frac{1}{n} + \frac{(X - \bar{X})^2}{\sum (X_i - \bar{X})^2}}. \tag{3.17}$$

Most often, we wish to know the prediction interval at one measurement only (so $m = 1$), in which case the first term under the square root is 1. Figure 3.3 shows

---

[14] Statisticians call these departures "errors," and so ϵ_i would be called "the error term."
[15] Although we have not (yet) invoked any normality assumption.

68 INTERVALS

the typical form of confidence and prediction intervals, discussed further in Problem 3.10.

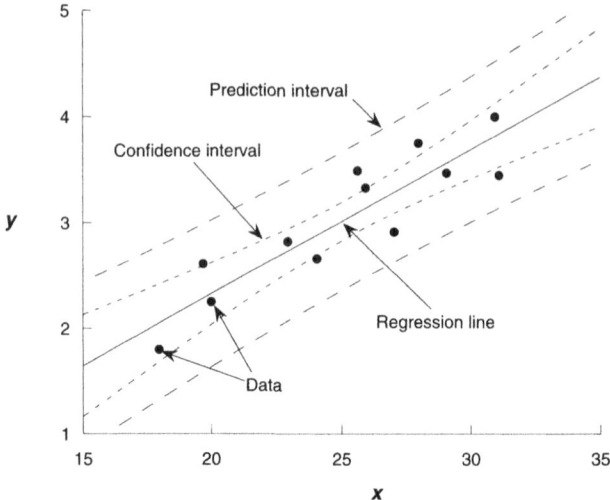

Fig. 3.3 Prediction intervals and confidence intervals for linear regression. The confidence interval indicates the likely region for the true line, while the prediction interval indicates the likely range of values of a future observation.

If you are presented with only one set of confidence intervals, it is always wise to determine whether they are confidence or prediction intervals. This can usually be simply inferred by noting that the confidence interval will encompass only some of the data, whereas the prediction interval will encompass most—if not all.

3.1.3 For percentiles

Two-sided intervals Two-sided confidence intervals for percentiles of normal distributions with unknown variance are of the form

$$\bar{X} + k_{\mathrm{LCL}} S \text{ and } \bar{X} + k_{\mathrm{UCL}} S, \qquad (3.18)$$

where k_{LCL} and k_{UCL} are lower and upper confidence limit factors read from Table A.4.1. If the data are drawn from a lognormal distribution, we may apply this equation to the logarithms of the data and then antilog the result.

Example New international guidelines for safe recreational water environments have moved away from microbiological criteria based on geometric means or medians, to the use of 95%iles (WHO 2003 [342]). As on page 61, these data are commonly drawn from distributions more like the lognormal than normal, so let's assume a lognormal parent. Say that $n = 20$ beach intestinal enterococci data have a mean (natural) logarithm of $\bar{Y} = 3.201$ and a log standard deviation $S_y = 0.65$ (again using natural logarithms). The estimated 95%ile is $X_{0.95} = e^{3.201+1.6449 \times 0.65} = 71.54$ per 100 mL. Then the lower confidence limit is $e^{\bar{Y}+k_{\text{LCL}}S_y}$. From Table A.4.1, for a 95% confidence interval, $k_{\text{LCL}} = 1.0955$ and $k_{\text{UCL}} = 2.5760$. So the lower 95% confidence limit on the lognormal 95%ile is $e^{3.201+1.0955 \times 0.65} = 50.0526$ per 100 mL, and the upper limit is $e^{3.201+2.5726 \times 0.65} = 130.7373$ per 100 mL. These limits become narrower as the number of samples increases (see Figure 3.4).

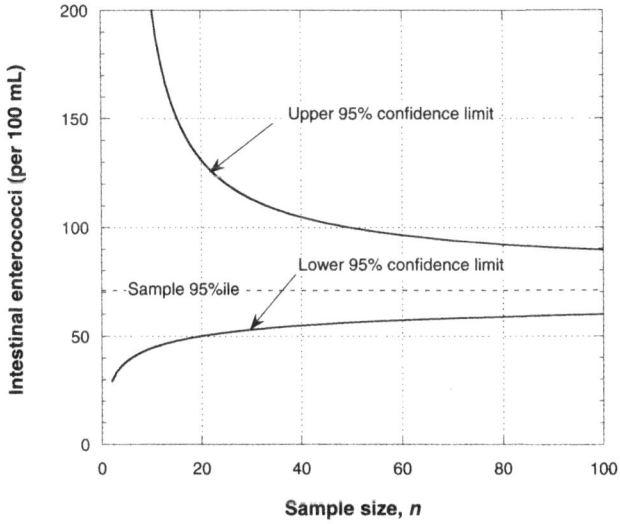

Fig. 3.4 Confidence limits for the 95%ile as a function of sample size.

Figure 3.4 shows that rather more than 20 data are needed before reaching a point of rapidly diminishing return of information for the increased effort in sampling (in contrast to the case for estimating the median, as in Figure 3.1).

One-sided intervals Similarly, one-sided confidence intervals for percentiles of normal distributions with unknown variance are of the form

$$\bar{X} + kS, \tag{3.19}$$

where the k-factor is read from Table A.4.2. As a simple example, for 12 observations the lower one-sided 95% confidence interval on the 95%ile is given by $\bar{X} + 1.0625S$ (for a huge sample size it would be very close to $\mu + 1.6449\sigma$). Again, if the data are drawn from a lognormal distribution, we may apply this equation to the logarithms of the data and then antilog the result.

Example Chemicals concentrations in drinking water may follow distributions that are more normal than lognormal. Limits on those chemicals may require that certain concentrations should not be exceeded for more than 5% of the time; that is, we have a 95%ile standard (e.g., MoH 2000 [229]). We could estimate the 95%ile directly, using the 50% level of confidence figure in Table A.4.2. For example, with $n = 15$ samples of nitrate with $\bar{X} = 18.2\,\mathrm{g(NO_3)\,m^{-3}}$ and $S = 3.22\,\mathrm{g(NO_3)\,m^{-3}}$, we would estimate the 95%ile as $X_{0.95} = 18.2 + 1.6808 \times 3.22 = 23.61\,\mathrm{g(NO_3)\,m^{-3}}$.[16] But it may be objected that protecting public health calls for a precautionary approach. After all, there is a 50% chance that this estimated value is too low—we have only 50% "confidence" in the estimated 95%ile. That provokes the question: What would that estimate be if there was only a 5% chance of it being too low? So we should have a 95% level of confidence. In that case, from Table A.4.2 we have $k = 2.5660$ and so the upper 95% confidence limit on the 95%ile is $18.2 + 2.5660 \times 3.22 = 26.46\,\mathrm{g(NO_3)\,m^{-3}}$. This higher (precautionary) result can have implications for sampling, especially given a rule that requires increased sampling effort when a 95%ile exceeds one-half of a health limit for drinking water ($50\,\mathrm{g(NO_3)\,m^{-3}}$, as in MoH 2000 [229]).

3.2 TOLERANCE INTERVALS

There are two types of tolerance intervals, β-content and β-expectation, where $100\beta\%$ is the **coverage**.[17] They are defined as follows:

β-**content** intervals are constructed so that they contain *at least* $100\beta\%$ of the population with probability $100(1-\alpha)\%$.

β-**expectation** intervals are constructed so that they contain *on average* $100\beta\%$ of the population.

In water management, we commonly deal only with the β-content type of interval. For example, monitoring frequencies in New Zealand's Drinking-Water Standards are based on the requirement to be 95% confident that critical chemical and microbiological concentrations are not exceeded for more than 5% of the time (MoH 2000 [229]).

The main difference between a confidence limit and a tolerance limit is evident for the two-sided case. A two-sided confidence interval about a parameter shrinks toward zero as the sample size gets large. On the other hand, a two-sided tolerance interval stays of finite width. For example, a 95% tolerance interval's limits shrink toward $\bar{X} \pm 1.9600\sigma$ as the sample size is increased (in the infinite limit the variance is known and the normal distribution has 95% of its mass between -1.9600 and 1.9600).

In contrast, the *one-sided β-content tolerance limit is identical to a one-sided confidence limit for a percentile* (Conover 1980, pp. 119–121 [56], Millard & Neerchal, 2001, p. 336 [226]). For example, the one-sided tolerance limit for a 95%ile gets ever closer to $\mu \pm 1.6449\sigma$ as the sample size is increased. Therefore Table A.4.2 also suffices for one-sided tolerance limits on percentiles.

---

[16] Note that the k value for 50% confidence is not 1.6449 (the 95%ile of the unit normal distribution); see Problem 3.11.
[17] The usage of β is for historical reasons—it should not be confused with the "Type II error risk" we shall meet in Chapter 4.

Table A.4.3 gives the k factors for two-sided tolerance limits on percentiles. These are also calculated from expressions of the form $\bar{X} + kS$ [i.e., Equation (3.19)]. These intervals are especially useful for identifying abrupt changes in the distribution of observations, and so lend themselves to more data-intensive situations, such as effluent monitoring.

Example Historical records of 50 total phosphorus (TP) concentrations in a wastewater treatment plant's effluent display a normal-like distribution with a mean $\bar{X} = 7.7$ g(TP) m^{-3} and standard deviation $S = 1.3$ g(TP) m^{-3}. Let's characterize this distribution by an interval that gives us 95% confidence that it contains 90% of the distribution. From Table A.4.3, $k = 1.999$ and so the interval is $7.7 \pm 1.999 \times 1.3 = (5.10, 10.30)$ g(TP) m^{-3}. In another 20 samples, 6 exceed 10.30 g(TP) m^{-3}. But we expect that no more than 2 should do so if the distribution has remained unchanged (i.e., 10% of the 20 new observations). This is signaling that the distribution has indeed changed.

3.3 CREDIBLE INTERVALS

One may say of a 95% credible interval that there is a 95% probability that the parameter of interest lies between the interval's limits. By "probability" we mean a Bayesian personal probability, in which some prior information has been incorporated by means of a prior distribution.

In many cases a credible interval has the exact same formula as does its frequentist cousin for a particular choice of prior. As discussed on page 58 this occurs when a "vague" or "diffuse" prior distribution is chosen, in which all values are equally likely (Box & Tiao 1973 [33]). For example, the formulae already presented for confidence intervals on means can be interpreted as credible intervals using vague priors (another example of using such a prior is presented in Section 5.6.1).

The two-sided α-level intervals presented in this chapter are equi-tailed. However, as already noted (page 61), this is not necessary. For symmetric distributions, such as Student's t-distribution, an equi-tailed distribution is also the shortest interval cutting off a total tail area of α, and that seems to be a desirable property. Bayesian analyses however often result in asymmetric posterior distributions, and the equi-tailed interval is not the shortest interval. Theory shows that for a unimodal distribution the shortest interval cutting off a given tail area has equal ordinates—as shown, for example, in Figure 3.5.[18]

In this figure the limits of the **HDR** (highest density region) share the same ordinate (0.0170). The HDR is somewhat narrower than the equi-tailed region, and this seems to be a desirable property for a credible interval to have. We shall meet such intervals when considering uncertainty in "most probable number" enumerations of microorganisms (in Chapter 10).

---

[18]Formally, if the interval $[a, b]$ satisfies (i) $\int_a^b f(x)\,\mathrm{d}x = 1 - \alpha$, (ii) $f(a) = f(b)$, (iii) $a \leq x^* \leq b$, where x^* is the mode of $f(x)$, then $[a, b]$ is the shortest among all intervals that satisfy (i) (Casella & Berger, 2002, p. 441 [45]).

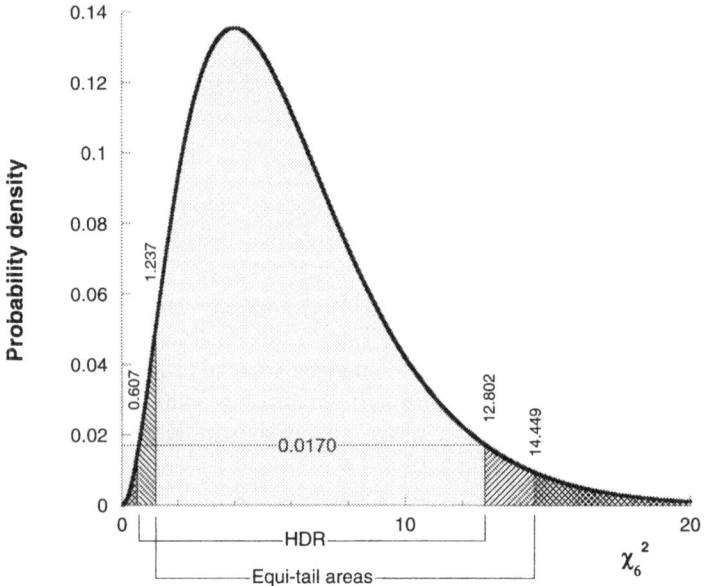

Fig. 3.5 Highest density and equi-tails regions enclosing an area of 0.05, for a chi-square distribution with 6 degrees-of-freedom. [The HDR abscissae (0.607 and 12.802) were obtained from Table B.2 of Lee (1997). The equi-tailed abscissae (1.237 and 14.449) were read from Table B.1 in Zar (1996 [349]), and they are given in many other texts.]

3.4 INTERVALS LITERATURE

Vardeman (1992 [329]) asked: "What about the other intervals?" This was based on the observation that statistical analysis should use more than just confidence intervals, yet in many disciplines these other intervals are seldom taught. For example, any substantial presentation of tolerance intervals is usually confined to engineering and quality control literatures (see especially Odeh & Owen 1980 [246]). Even then some intervals are not discussed (e.g., two-sided confidence intervals on percentiles). The correspondence between one-sided tolerance intervals and confidence intervals on percentiles is also seldom addressed, yet water quality standards are often based on them (e.g., microbiological standards for the quality of recreational water, USEPA 1986 [324]). Bayesian tolerance intervals are as yet confined to the specialist literature (Guttman 1970 [129]). There is also ongoing debate about the relative merits of confidence intervals and hypothesis tests, for example, in the public health literature (Thompson 1987 [314], Walker 1986 [333], Poole 2001 [261]).

Problems

3.1 In the section on interpreting confidence intervals in a Bayesian manner (page 58), it is observed that the prior distribution invoked postulates that "large" positive values of lake chlorophyll *a* are more likely than are small values. Explain why this is so, and discuss its implications for the common claim that a uniform prior distribution is "uninformative."

3.2 Derive Equation (3.2) from Equation (3.1).

3.3 Consider the set of 10 *E. coli* data for beach water, expressed as numbers per 100 mL of river water: 7, 15, 22, 432, 122, 52, 25, 14, 43, 33. What are the appropriate 95% confidence limits for the central of the distribution these data have been drawn from?

3.4 Why is a nonparametric confidence interval for a central tendency statistic based on the median, not the mean?

3.5 Derive one-sided confidence interval equations for the difference between two independent normal means, where each sample is of size n.

3.6 Show that the standard error for the difference in means between two independent samples each of size n is $S_{e,\bar{x}-\bar{y}} = \sqrt{\frac{1}{n(n-1)} \left[\sum X_i^2 + \sum Y_i^2 - n\left(\bar{X}^2 + \bar{Y}^2\right) \right]}$.

3.7 Show that the standard error of the mean difference of n paired samples is $S_{e,\overline{x-y}} = \sqrt{S_{e,\bar{x}-\bar{y}}^2 - \frac{2}{n(n-1)} \left[\sum X_i Y_i - n\bar{X}\bar{Y} \right]}$.

3.8 Is the standard error for n paired samples always less than the standard error for two independent samples each of size n?

3.9 What happens to the width of the confidence intervals for paired and unpaired samples in the example given on page 65 if the eighth X value is changed from 12 to 28 parts per billion of DRP?

3.10 Confidence intervals and prediction intervals for a linear regression have a waist, where the distance between the between the upper and lower limit is at a minimum. Where does this occur?

3.11 When calculating one-sided confidence (or tolerance) intervals from the formula $\bar{X} + kS$, the value of k (Table A.4.2) for 50% confidence for any percentile (except the median) differs from the equivalent percentile of the unit normal distribution. For example, when estimating the 95%ile with $n = 20$ data, $k = 1.6712$, only becoming the unit normal percentile (1.6449) for infinite sample size. Can you explain this?

3.12 If you were implementing a calculation procedure for one-sided confidence (or tolerance) limits on normal percentiles some provision would have to be made for the k values for sample sizes not reported on (Table A.4.2). How could this difficulty be resolved?

4
Hypothesis testing

One cannot go far in water resources science and management before coming across claims that the results of some investigation are "statistically significant," or even just "significant." Such claims arise after a hypothesis test of some sort has been conducted. To many, when pressed for clarification, this is a mystery—just what does this result mean?

The mechanics of performing tests are well enough understood, but their philosophical basis usually is not, nor perhaps their interpretation. A major reason for this state of affairs is that the subject itself is often presented in texts and taught in applied environmental statistics courses in a "cut-and-dried" manner, as if there are no major unresolved issues, or alternative approaches. For example, the substantial debate in the statistical, ecological, and psychological literatures about the role and appropriateness of "point-null" hypothesis tests is not mentioned at all (a summary of that literature is given in Section 4.11). Nor is another debate, between proponents of the **frequentist** (**classical**) approach and promoters of likelihood or **Bayesian** approaches—most water resources scientists and managers are exposed only to the frequentist school of thought. So questions go unasked and, therefore, unanswered. Other questions that do get asked, and maybe not answered very well, include: When can I use a one-sided test? How do I best incorporate the **precautionary approach**? When should nonparametric methods be used? What is the role of randomization tests? What is likelihood? When is it appropriate to use an **equivalence test**?

This chapter seeks to elucidate these questions. It does so by first presenting the general frequentist approach to hypothesis testing. Then we proceed to one-sided tests. These provide an excellent vehicle for discussing issues of **burden of proof** that arise when seeking to implement the precautionary approach. We also introduce the notions of **detection probability** and **detection regions**. Following that, we discuss "point-null" hypothesis tests and then the rather newer field of equivalence testing. We then turn attention to statistical power, multiple comparisons, nonparametric methods, randomization tests, likelihood, and Bayesian methods. Readers interested in the calculation procedures will find them in Chapter 6, and some (including "equivalence test" procedures) are available at http://www.niwa.co.nz/services/statistical/.

4.1 THE CLASSICAL APPROACH

We want to be able to "detect" important differences. Hypothesis tests do that by seeking to come to a conclusion about differences between *population parameters*, usually the populations' means. For example, if a hypothesis of "no difference" between those parameters is rejected, then a statistically significant result is declared, and it is said that something has been "detected."

In performing a test we assume that samples that we gather have been drawn from a statistical population, being an entity "out there" that we never see. For example, consider using monthly sampling of total phosphorus in river water in order to estimate the annual phosphorus load from the river's watershed and to compare it with some baseline value, or with another river's load. If we take one liter samples from a river with an annual average flow of $10 \, \text{m}^3 \, \text{s}^{-1}$, then the 12 liters of water we have measured is but one part in 4×10^{-11} of the total volume of water that passed the monitoring site over the year (that volume is over 3×10^{11} liters). The phosphorus concentration in that volume is the population from which we are drawing samples. And we had better be careful to avoid errors in sampling and laboratory analysis, given that we have sampled such a tiny part of that population.

The phosphorus concentration in that water of course varies throughout the year. We can characterize such variations by a statistical distribution whose shape is determined by its parameters. For example, the commonly used normal distribution's shape is characterized by just two parameters—its mean and variance—or the three-parameter lognormal distribution, as shown on Figure 2.8.

4.2 THE MAIN STEPS IN THE CLASSICAL APPROACH

In the classical (frequentist) mode of statistical inference we make probability statements about data, not about hypotheses (one does that using the Bayesian mode). In spite of this, we actually want to make some statement about the tenability of competing hypotheses. The procedure that has been devised to do this enables an investigator to use data in a "decision rule" to choose between competing hypotheses. In devising this rule the hypothesis under test (we will denote this by "**H**") is assumed to be true. Then, "under **H**," if the probability of getting data at least as extreme as has been obtained is in some sense "small," **H** can be rejected. "Small" means less than the test's specified "significance level." To use the rule, one uses a "test statistic," which is calculated from the data.

The procedure therefore has these five main steps:

1. State **H**, the hypothesis to be tested, in terms of differences between a population parameter and a standard value (in a one-sample test) or between two or more population parameters (in a multi-sample test).

2. State α, the significance level.

3. Perform sampling from the population(s).

4. Apply the test's decision rule to determine whether to reject **H**.

5. If **H** is rejected, then accept **K**, its complementary alternative. If **H** is not rejected, then we may accept it, but only if it is one-sided or interval—not if it is "point-null."

These steps will be discussed in turn. But first it is necessary to define our major terms, as in Table 4.1.

Table 4.1 Matrix of possible results from a hypothesis test; definitions of confidence level, significance level, power, and error risks

	Test result	
True state (unknowable)	Do not reject **H** (if point-null), or Accept **H** (if one-sided or interval)	Reject **H**
H is true, **K** is false	Correct decision Probability = $1 - \alpha$ *Confidence level*	Incorrect decision Probability = α *Type I error risk, Significance level*
H is false, **K** is true	Incorrect decision Probability = β *Type II error risk*	Correct decision Probability = $1 - \beta$ *Power of the test*

Many writers denote *all* tested hypotheses as "H_0" (calling their alternative hypotheses "H_1" or "H_A") and refer to H_0 as the "null" hypothesis. In the common case where the tested hypothesis posits an equality between parameters (e.g., $\mu_x - \mu_y = \delta$), some writers refer to H_0 as the "point-null" hypothesis—because it is true at only a single point on the axis of differences (that point is δ, commonly taken as zero). That terminology is adopted in this text, so that there is no ambiguity in the use of the word "null."[1] But we avoid use of "H_0" and "null" altogether in cases where the tested hypothesis is not "null"—that is, where **H** postulates a one-sided inequality (e.g., $x < y$) or postulates a (two-sided) statement of inclusion or exclusion of a random variable from an interval (e.g., $|x - y| > \delta$). This is because dictionaries

---

[1] This does make our language a little clumsy.

define "null" to mean nonexistent, or empty, and that is consistent only with an empty hypothesis—that is, postulating that the difference between quantities of interest is exactly zero. The two-sided point-null hypothesis fits that description, whereas one-sided tests and two-sided interval tests do not. We also note that use of "**H**" and "**K**" for the tested hypothesis and its alternative is consistent with the standard work on this subject (Lehmann 1986 [183]—who also avoids using "null").

In this table, **K** is the complementary alternative to **H**, in the sense that one or the other is always true. For example, if **H** postulates that the population mean (μ) is no greater than zero, then we would write "**H**: $\mu \leq 0$" and so we would also write "**K**: $\mu > 0$".[2] In these statements the colon (:) denotes "is such that." These hypotheses are stated as inequalities, and so they are "composite." On the other hand, were **H** to state an equality (e.g., **H**: $\mu = 0$), then we call it "simple."[3]

Table 4.1 shows that there are two kinds of error that we may make when drawing a conclusion from a hypothesis test. The Type I error is made if we reject a hypothesis that is in fact true, and the risk of this happening is the significance level. Conversely, the Type II error is committed when we fail to reject a false hypothesis. We may commit one, but not both, of these errors when we perform a test. We may of course make a correct decision, in which case no decision error has been made. But we never know if in fact **H** is true—if we did we would have certainty and no need of statistics at all—so whatever decision we make, we never know for sure if we have made it in error or if it is correct.

4.2.1 What is meant by a decision?

The cells in the body of Table 4.1 describe the possible outcomes of a test of **H** as a "decision." We must now consider the meaning of this—what manner of decision is this? There are three possibilities:

- What action should I take?
- Do I believe this hypothesis?
- What are the relative merits of the hypotheses?

All three are relevant to water quality management.

There is a branch of statistical inference (the "decision theoretic" approach) that identifies "decision" exclusively with "action"—that is, with the first possibility. This has been particularly advocated by one of the developers of the famous "Neyman–Pearson lemma"—on which much hypothesis testing theory is based—who said:

> The problem of testing a statistical hypothesis occurs when circumstances force us to make a choice between two courses of action: either take step A or take step B...,

---

[2] Some statistical literature does not take this stance—for example, testing two hypotheses positing equalities; such hypotheses do not explicitly contemplate the many inequalities that routinely occur (and there are infinitely many of them for continuous variables).

[3] Strictly, a simple hypothesis must also "completely specify" the distribution—it would if the distribution is normal and the variance is known.

and

> ...to accept a hypothesis H ... does not mean that we necessarily believe that the hypothesis H is true.

(Neyman 1950, pp. 258, 259; cited by Royall 1997 [275]). Neyman considered situations where the desirability of two courses of action depends on the distribution of some variable x. **H** then denotes the set of distributions where action "A" is to be preferred, and **K** denotes the set of distributions where action "B" is to be preferred. The partitioning between these sets takes account of the expected consequences of error. That is, one can act *as though* a hypothesis were true, without necessarily believing that it is. A good example (from Royall 1997 [275]) is a physician whose patient presents symptoms that indicate a very low (but not zero) probability of a serious contagious disease, it being much more plausible that the symptoms indicated an illness that would be of no consequence to the community. In spite of this, the physician may feel compelled to prescribe treatment or specialist referral for the improbable illness, the decision being dictated by the consequences of the most plausible inference being wrong. There are obvious parallels in water quality management—for example, where a contagious disease microorganism *may* be present in water, and people are showing possible symptoms of an associated disease (e.g., SARS).

Neyman called this "inductive behavior." This mode *is* relevant to water quality management—for example, in deciding what to do when it seems that a water treatment plant appears to have been in breach of its permit. But it is not the only mode that is relevant; the other two are relevant as well—action is not always required. For example, development of theories for the role of disturbances (e.g., floods) in rivers may be motivated by an individual's beliefs or hunches (the second mode). In other words, research outcomes we seek are conclusions, not decisions (Tukey 1960 [320], Stewart-Oaten 1996 [301]). It may be even more appropriate to consider the relative merits of competing hypotheses (the third mode). And note that we have used the term "merits," rather than "evidence." This is deliberate, as will be explained later (Section 4.10).

4.3 COMMENTARY ON THE FIVE STEPS

4.3.1 Step 1: State the hypothesis to be tested

The nature of the problem at hand should dictate the hypothesis to be tested. These fall into three categories.

One-sided hypotheses A one-sided hypothesis asserts that a population parameter is either above or below a certain value (e.g., a standard value or another population's mean). Such a hypothesis is appropriate when there is particular interest in, and consequence from, a difference being in a particular direction (as explained below). Conversely, there is no particular interest in, or consequence from, a difference in the opposite direction.

Not all statisticians agree with this, and it is important to know why. For example, a recent statistical-rules-of-thumb text (van Belle 2002, p. 14 [327]) discourages investigators from using one-sided tests. Yet more, of a similar persuasion, say that one-sided hypotheses are appropriate only when a difference in one direction is impossible. On the other hand, another recent text, on "statistical evidence," has a section asking "What is wrong with one-sided tests?" (Royall 1997, p. 116, [275]) and advocates their use in many situations where differences may occur in either direction. The issue has all to do with *consequences* of a test result. It arises particularly in conducting clinical trials (e.g., Fleiss 1987 [97]). Ethical considerations can dictate that investigation should be able to discover either an improvement or deterioration as a consequence of the administered drug. In such cases, one-sided tests are quite inappropriate. However, in many water resource management issues, one-sided tests are appropriate. In such cases it does not matter whether the difference in the inconsequential direction is impossible or not.

For example, in assessing compliance with a percentile environmental standard expressed as a percentile of time (not of samples), we are only interested to know if there has in fact been a breach of standard in deciding whether to take any action. This is a one-sided problem, because the true percentile may be above or below the standard's value, and we want to make an inference about that. Either "breach" or "compliance" may actually be true. This is discussed in Chapter 8. Similarly, in assessing the degree of agreement between two methods of assaying a microbiological organism's concentration, we may want to be sure that the appropriate metric is above a certain cutoff value, not that it equals that quantity (or any other quantity for that matter). We discuss such measures in Chapter 11.

One-sided tests allow for the incorporation of either a **precautionary approach** or a **permissive approach** to the burden of proof.

Point-null hypotheses These hypotheses state that the population parameter *equals* a certain value (e.g., a standard value or another population's mean). The interest in these tests is because either improvement or deterioration is of concern—either of them have potentially important consequences; hence these hypotheses are two-sided.

Point-null hypothesis tests are routinely used, but may not always be appropriate—their outcome may not produce much useful information. To see why, let's first consider continuous variables, such as the mean of a normal distribution or the geometric mean of a lognormal distribution. Now consider the chance that this population mean could *exactly* equal a particular value (which means equal to the zillionth decimal place). Remember, this is a population whose parameters that we can only estimate, so the numerical value of the population mean is not limited by the accuracy of any measuring apparatus (as it would be were we considering means of sample data). Therefore, even though the mean must be a number, there are an infinite number of numbers that it may actually be. So it is exceedingly unlikely that it will in fact equal any value we may specify, to the point where we might say that there is zero possibility of this equality being true. Of course if we held this possibility to be zero for all of those numbers, then the mean could not take on any value, which is not true. So it is appropriate to regard the possibility that a point-null hypothesis is true as "vanishingly small." Formal explanation of such a nonzero but vanishingly small quantity lies beyond ordinary number systems, and one must invoke a nonstandard analysis to attempt to resolve it.[4]

An immediate consequence of this is that rejecting a single point-null hypothesis tells us very little, because we already knew that it wasn't tenable.[5] It doesn't even necessarily tell us that whatever difference was present was in some sense "small" (though power analysis can help on this matter, as discussed in Section 4.5). This is the point made by many, including Cohen (1994 [52]), who uses the phrase "nil" hypothesis—because the possibility of equality between continuous variables is so remote that testing it conveys nil information. As a consequence, we cannot logically "accept" the point-null hypothesis, only "not reject" it (this is why Table 4.1 has "Do not reject **H** if point-null" in its header). It is also why some texts make statements such as:

> ... failing to reject a null hypothesis is not 'proof' that the hypothesis is true. It denotes only that there is not sufficient evidence to conclude that it is false.

(Zar 1984, p. 45 [348]). An example of the difficulties imposed when a point-null hypothesis is accepted is given in Section 4.3.5. Some have lamented that the very common failure to recognize this aspect of point-null testing gives rise to erroneous claims of "no effect", "no trend", or "no difference" (e.g., Parkhurst 2001 [253]).

This observation may at first appear to be startling (and note that it does not apply to tests of one-sided or interval hypotheses). So it is important to realize that it does not constitute a case for abandonment of point-null hypothesis tests. There are many situations in which these tests are appropriate, as discussed in Section 4.6— particularly when comparing the relative merits of hypotheses. The situation is not quite the same for discrete variables, where the point-null hypothesis can in fact be true, because in a given range there are only a finite number of numbers that a discrete parameter actually may be. So there is a (usually small) chance that the hypothesis might be true. Nevertheless, we will also denote hypotheses that posit equality for discrete populations' parameters as "point-null", and we will denote them by H_0.

Point-null hypothesis tests do not easily allow for the incorporation of either a precautionary or a permissive approach to the burden of proof, as discussed in Section 5.2.4. In particular, at small sample sizes they tend to be permissive but ultra-precautionary at large sample size. That is why it may be held that tests of point-null hypotheses are inappropriate when the intent is to prove "safety" or "no effect" (Bross 1985 [34], Millard 1987 [224], Dixon & Garrett 1994 [74]).

---

[4] Issues in nonstandard analysis are summarized by Davis & Hersh (1981, pp. 237–254 [65]).

[5] Increasing attention is being paid to a "three-valued logic" in interpreting the result of a point-null hypothesis test (Harris 1997a, 2001 [137] & [139]). In this approach, rejecting the hypothesis admits the conclusion that the difference between tested quantities lies in the direction evident in the data; to not reject the hypothesis merely means that one is uncertain of the direction of that difference. This approach has been noted to be "far from the standard usage in formal logic" (Krantz 1999 [175]). Yet eminent statisticians have recently endorsed it (Jones & Tukey 2000 [170]). A number of issues remain to be resolved in this approach (such as calculating statistical power).

Interval hypotheses—equivalence tests These derive from the same concern that point-null hypothesis tests are aimed at—that is, cases where either an improvement or a deterioration is of interest and either of them have potentially important consequences; hence interval hypotheses are also two-sided. But **H** specifies an interval, not a point. The basic idea is that one specifies this region, the equivalence interval, in which differences are held to be small enough to be considered "equivalent". For example, if taxonomic richness of a stream's benthic communities downstream of an outfall is within 20% of its upstream values, then the two communities could be considered as having equivalent richness. Tests of these interval hypotheses are therefore generally known as "equivalence tests".

The general nature and properties of interval tests have been known for some time (Lehmann 1959 [182], Ferguson 1967 [90]). However, it was not until advances in the evaluation of clinical drugs trial data in the 1980s that they were developed into practical form and came into routine use by drugs testing agencies (Schuirmann 1987 [283], Chow & Liu 1992 [46], Wellek 2003 [340]). While they are now well established in that field, they are much less well known in environmental science and management, where point-null hypothesis tests are overutilized. Others hold a similar view:

> ...we think equivalence testing approaches are underutilized. We often see examples where statisticians and non-statisticians are testing the wrong hypotheses, apparently stuck in a mode of thinking based on null hypotheses of no-difference. (Hauck & Anderson 1996 [142]).

Equivalence tests allow for the incorporation of either a precautionary or a permissive approach to the burden of proof, as discussed in Section 5.3.

4.3.2 Step 2: State the acceptable error risk

The hypothesis test's decision rule works by seeking to constrain the Type I error risk to a maximum value of α. So the procedure requires specification of α (but not the other error risk, β). The practical effect of this is that the significance level should be stated before a test is carried out. This is because a larger value of α can make it possible to reject **H**, whereas a smaller value will not. So on some occasions an investigator could fail to obtain a "statistically significant" with the usual significance level (5%), but gain one merely by increasing it to 10%.

For example, as will be shown in the next chapter, the decision rule for a one-sided test of the inferiority hypothesis **H**: $\mu \leq \mu_0$ with known variance is that **H** should be rejected if $T > Z_{1-\alpha}$. Here, T is the test statistic calculated from the data and $Z_{1-\alpha}$ is the $100(1-\alpha)$%ile of the unit normal distribution (i.e., the value of that distribution's abscissa where the right-tail area = α). Say that the data show that T = 1.52. For a 5% level test (i.e., $\alpha = 0.05$), $Z_{0.95} = 1.6449$, as may be read from Table A.1.2. In that case, **H** would not be rejected. But if we were to select a 10% significance level ($\alpha = 0.10$), then $Z_{0.90} = 1.2816$ and so **H** would be rejected. Note that the data have not changed; the barrier has been lowered.

Hence, a statement of statistical significance must be accompanied by the significance level of the test. It should also be accompanied by a statement of the hypothesis tested and the sample size (because, as will be seen, the power of a test tends to increase with sample size so that smaller differences can be detected with larger sample sizes).

4.3.3 Step 3: Sampling

The key feature of sampling a population is that the data should be representative of that population. Generally this calls for random sampling, to avoid any bias. For example, regular sampling may be affected by a cyclical pattern, such as the diurnal cycle of dissolved oxygen or pH in a river with substantial aquatic plant growths. But this does not mean that one must necessarily sample randomly from the time of day, including late at night! That would only be appropriate if one wishes to make inferences about what happens to dissolved oxygen over 24-hour periods. Often our interest is narrower than that (and difficulties of night-time sampling can also be a factor). For example, in routine trend monitoring it can be appropriate to monitor at about midday on each sampling occasion. This will reduce the variability in the data (because the diurnal cycle has been removed) and make long-term trends easier to detect.

In other words, first define the population you wish to make inferences about, and then devise an appropriate sampling strategy.

Gilbert (1987 [118]) gives a detailed discussion on systematic versus regular sampling, and their various combinations (e.g., systematic random sampling, where the sampled domain is split into defined blocks over time or space, but sampling within each block is random).

4.3.4 Step 4: Apply the decision rule

Hypothesis tests always require that some assumptions be met, or at least that they are not strongly violated. As a typical example, we may need to have independent samples from normal populations, and if more than one population is sampled, they may need to have the same variance. Statistical theory then shows that, under such assumptions, and when **H** is true, certain test statistics derived from our data will follow particular distributions. These are known as "sampling distributions." Typical examples are the unit normal distribution, Student's t-distribution, the chi-square distribution, and the F-distribution.

The decision rule is then derived from the sampling distribution. This is best seen by way of the one-sided 5% level example given in Section 4.3.2, in explaining Step 2, where $T = 1.52$. If we assume that we are sampling from a normal distribution with known variance (σ^2), we compare the value of the test statistic T with $Z_{1-\alpha}$ where $Z \sim n(0, 1)$ (meaning that Z is distributed as the unit normal distribution, which has zero mean and unit variance). In the example, $Z_{1-\alpha} = 1.6449$; if T exceeds this value, then **H** is rejected. But T does not exceed 1.6449, and so **H** is not rejected (this is the case shown on Figure 4.1).

This decision rule has two entirely equivalent forms. The first, as already presented, compares T with the critical value ($Z_{1-\alpha}$), and because T is less than $Z_{1-\alpha}$ (i.e.,

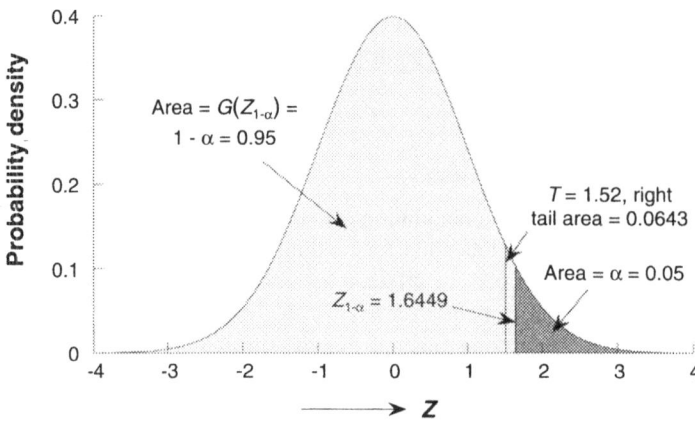

Fig. 4.1 Abscissa (Z), ordinate (pdf), and critical Z-value for a one-sided test ($Z_{1-\alpha}$), and critical region (α) for the unit normal distribution in a 5% level test.

$1.52 < 1.6449$), **H** is not rejected. (Furthermore, being a one-sided test, **H** can actually be accepted.) The other form of the decision rule is to compute the right-tail area of the unit normal distribution that is cut off by T and compare it with α (the "critical region"). In Figure 4.1 we see that the right tail area cut off by T is 0.0643, greater than α, and so once again we conclude that **H** should be accepted. The right-tail area is known as the p-value (as used in this text), sometimes written P-value, and occasionally called the "associated probability" or the "observed significance level." It is defined as

$$p = \Pr(\text{reject } \mathbf{H} \mid \mathbf{H} \text{ is true}), \tag{4.1}$$

and so it is the probability of committing the Type I error (shown on Table 4.1). This second form of the decision rule shows the basis of the decision rule more clearly. That is, the tested hypothesis will be rejected if there is only a small probability (p) of getting data at least as extreme as has been obtained tested *if* that hypothesis is true. The hypothesis test's decision rule is therefore based on rejecting **H** if p is sufficiently small—that is, less than α.

> **The most important message here is that the p-value is calculated *assuming that H is true*. So it is a probability statement about data, given that a hypothesis is true; it is not a statement about the truth of a hypothesis, given our data. Furthermore, it includes consideration of data that *have not been obtained*, being all possible data more extreme than that obtained.**

Rationale for the p-value being a tail area The preceding paragraph details the general mechanics of implementing the decision rule, but one should be clear about its rationale. Because it is an area under a probability density function, the

p-value is a probability. In particular it is the probability of obtaining results at least as extreme as have been obtained. That is, it considers samples that have not been obtained (being all those that would give more extreme values of T, and hence smaller p-values). The rationale for this definition of p is seldom explained in texts (but is sometimes criticized by adherents of Bayesian methods—Jeffreys 1961 [164], Lee 1997 [179]); statements about p are generally confined to its operational aspects only. But in essence the rationale is as follows:

> ...if we were to regard the data under analysis as just decisive against H_0 then we would have to regard more extreme samples also as evidence against H_0. Therefore, P is the probability of declaring there to be evidence against H_0 when it is in fact true and the data under analysis are regarded as just decisive (Cox 1987 [61]).

So if our data cause T to lie in the rejection region, we have reason to reject **H**.

There is also a practical reason for defining the p-value in this way: As noted earlier, it would be vanishingly small were it to be the probability of obtaining exactly the value T. So a decision rule based on the probability of obtaining our data merely by chance would always be rejected, because that probability, which is essentially zero, would always be smaller than the significance level.

This all means that the p-value is not the probability of obtaining our data merely by chance (i.e., by bad luck), as is sometimes stated. Instead, it is the probability of obtaining, by mere chance, data *at least as extreme as we got* (Poole 1988 [260]). It is also the reason that we used "merits" rather than "evidence" when considering the relative support data may give to alternative hypotheses. How can data you didn't obtain be included in "evidence"? Such a posture would not be entertained in the Courts.

p-values for point-null tests For a test of H_0 we have to consider two tail areas, as shown for a new dataset in Figure 4.2 (with a Z-value above $Z_{0.95}$ but below $Z_{0.975}$). Comparing Figures 4.1 and 4.2, we see that the magnitude of the T value in the latter (1.78) would cause the one-sided hypothesis (**H**) to be rejected (because $T > 1.6449$ and thus cuts off a right-tail area less than 0.05). However, the point-null hypothesis H_0 would not be rejected. Being a two-sided test, we have already acknowledged that a difference in either direction would be of concern, and that we did not know in what direction that difference would lie before gathering the data. Therefore we have to compare our data with *two* critical regions, one in each tail of the sampling distribution. To do that, we must consider the right-hand tail area of the sampling distribution cut off by $|T|$ and the left-hand tail area of that distribution cut off by $-|T|$. It is conventional to split α equally between those two tails (this is not logically necessary; we merely require the sum of the two critical areas to equal α). For such an equi-tailed test we simplify the decision rule by comparing the right-tail area cut off by the absolute value of our test statistic ($|T|$) with $\frac{\alpha}{2}$; equivalently we could compare twice the area cut off by $|T|$ (the p-value) with α.

The important point here is that even though the data will result in a test statistic in a particular direction (positive or negative), one still has to compare it with *both* tails of the sampling distribution. The traditional view is that one cannot turn a two-sided

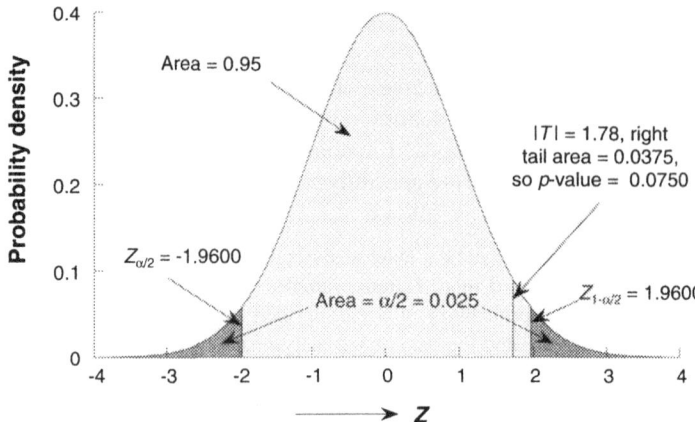

Fig. 4.2 Abscissa (Z), ordinate ("pdf"), critical Z-values for a two-sided test ($Z_{\frac{\alpha}{2}}$ and $Z_{1-\frac{\alpha}{2}}$), and critical regions ($\frac{\alpha}{2}$) for the unit normal distribution in a 5% level test.

test into a one-sided test once the data are at hand and the direction of the difference in sample parameters is known.

4.3.5 Step 5: Rejecting, and accepting (maybe!)

If the test outcome is to reject **H**, one can always accept its alternative (**K**), because it is always possible that the alternative hypothesis is true. The converse is not necessarily true. In particular, a point-null hypothesis should never be accepted as true, because the chances of it being so are vanishingly small.[6] On the other hand, we can accept a hypothesis that can be true, i.e., a one-sided or interval hypothesis.

Example: Drugs trials Here we have an example where a hypothesis may be accepted (it comes from the medical sciences, but the approach taken can be expected to become more popular in environmental sciences). Equivalence tests are often used in assessing the efficacy of new drug formulations. If a physiological response (e.g., area under a concentration-time curve for blood samples) lies within certain bounds, the new drug may be declared to be "equivalent" to a standard formulation. But a precautionary stance is taken in which the hypothesis tested postulates that the true area is *beyond* those bounds (i.e., is outside the interval). If a test fails to reject that hypothesis it is held to not be equivalent to the standard formulation. That is, the tested hypothesis is accepted. If the hypothesis is rejected, equivalence is inferred.

---

[6]But, as stated in Section 4.2.1, it may be accepted for the purposes of making a decision about what to *do*—that is, using the inductive behavior mode of inference, where one chooses action (or non-action) as though the hypothesis is true. And in those cases it is common to pose two point hypotheses, rather than at least one of them being composite—with the result that **H** and **K** do not cover all possible values of the parameter of interest.

This methodology has fruitful applications to water quality management (as described in Chapter 5), but it has yet to be used routinely.

Care in interpreting p values In addition to the need to always report the tested hypothesis and sample size along with p-values, four further related issues need to be addressed:

- *What p-values are not*—the probability that a point-null hypothesis is in fact true.

- *The need for accompanying information*—especially when comparing the results of a number of tests.

- *p-values can be influenced by the samplers' intentions*—a most surprising attribute.

- *Beware of "peeking"*—where the intention is to stop once a "statistically significant" result is obtained.

All of these issues imply that p-values must be treated with caution, especially when performing tests of the point-null hypothesis—as advised by notable authors such as Gibbons & Pratt (1975 [116]), Berger (1986 [15]) and Schervish (1996 [281]). This is reinforced by recent opinion from prominent statisticians who have opined that many interpretations of p-values are "completely misleading" (Sellke *et al.* 2001 [285]). We now discuss these issues in more detail.

What p-values are not The p-value is not the probability that the tested hypothesis is true. In Section 4.3.4 we defined a p-value to be the probability, under **H**, of getting data at least as extreme as has been obtained. It bears repeating that "under **H**" means that in calculating the p-value the tested hypothesis is *assumed to be true*. Therefore the p-value is not the probability that **H** actually is true. That is, p-value $= \Pr(D^+|\mathbf{H}) \neq \Pr(\mathbf{H}|D)$, where "$D$" denotes data, "$D^+$" denotes all data at least as extreme as D, and "|" means "given the following information." Such an inequality may in fact be turned into an equality for one-sided tests (Edwards *et al.* 1963 [79], DeGroot 1973 [67], Pratt *et al.* 1995 [263]), and even approximately so for equivalence tests, as we shall see when discussing Bayesian approaches. But it cannot be turned into an equality for $\mathbf{H_0}$: in this case the p-value cannot be the probability that $\mathbf{H_0}$ is true.

Comparing p-values There is a tendency for p-values to become smaller as the sample size increases, because the value of the test statistic T is a function of n, the sample size. This functional dependence is typically of the form $T \propto \sqrt{n}$. For example, in a single-sample test of the point-null hypothesis $\mathbf{H_0} : \mu = \mu_0$, the test statistic is defined as $T = |\bar{X} - \mu_0|/S_e$, where $\bar{X}$ is the sample mean, $S_e = S/\sqrt{n}$ is the standard error of the mean, and S is the sample standard deviation (test statistics for various tests are defined in Chapter 5). More simply, we can combine these two formulae to write $T = \sqrt{n}|\bar{X} - \mu_0|/S$. Now $\bar{X}$ and S are unbiased estimators of their population counterparts (μ and σ), or nearly so (in the case of S). So while their actual values would be different were larger samples to have been selected, their values under repeated sampling would, on average, tend to their true values. Accordingly we could expect that, on average, an increased sampling effort from a stationary population for which the inferiority hypothesis is false will tend to result in higher values of T, because of the presence of the $\sqrt{n}$ term in the equation for the test statistic. This will push T further into the right tail of its sampling distribution (Figure 4.2), so that a smaller right-tail area (i.e., p-value) is cut off.[7]

As a consequence,

> p-values have a useful role in comparing the relative measure of support for a hypothesis in varying situations, *but only if the sample sizes are comparable*. In any event, when a p-value is reported, always report what has been measured, along with the number of samples used to compute it.

Otherwise, differences in p-values might be merely a function of the sample size, regardless of the size of differences being tested. Comparison of p-values between tests using quite different sample sizes can be quite misleading. Such comparisons are rather common when results of different studies into the same phenomena are being contrasted. On the other hand, comparing p-values from results of tests of comparable sample sizes can be informative as to the relative strength of information, for example on trends in a multi-site monitoring network where there has been a similar sampling effort at each site (e.g., in a multi-site water quality trend analysis, Smith *et al.* 1996 [288]).

Your intentions may matter! Consider a situation where we wish to test whether a proposed new laboratory chemical analysis method is biased in some way (i.e., whether measurement error manifests itself as a tendency to overestimate or to underestimate the true value of the concentration of the chemical in a reference solution).[8] An unbiased analysis method would be expected to return about equal numbers of "unders and overs" when compared against a "gold standard" method. We could address this issue using a two-sided hypothesis test; it should be two-sided because the bias may be upward or downward and either would be of concern. In such situations it is often considered that useful information will be gained by testing the point-null hypothesis $\mathbf{H_0} : \theta = \frac{1}{2}$, where θ is the true proportion of unders or overs. If $\mathbf{H_0}$ were rejected, bias might be claimed to be evident.

Now let's say that we have decided to perform 20 chemical analyses, and, having done that, only five of these were "overs." The sampling distribution of such an independent series of unders and overs "under $\mathbf{H_0}$" is known (from standard theory) to be the binomial distribution. And because the p-value considers results at least as extreme as was obtained, we need to include in it the probabilities of getting five or fewer "overs" *and* 15 or more "overs" (15 or more overs would cast as much doubt

---

[7] Note that for one-sided hypotheses and equivalence hypotheses, p can *increase* for larger sample sizes. This happens if the tested hypothesis is true. See Section 5.4.
[8] This example is based on material in Berger & Berry (1988 [16]), though some details have been changed.

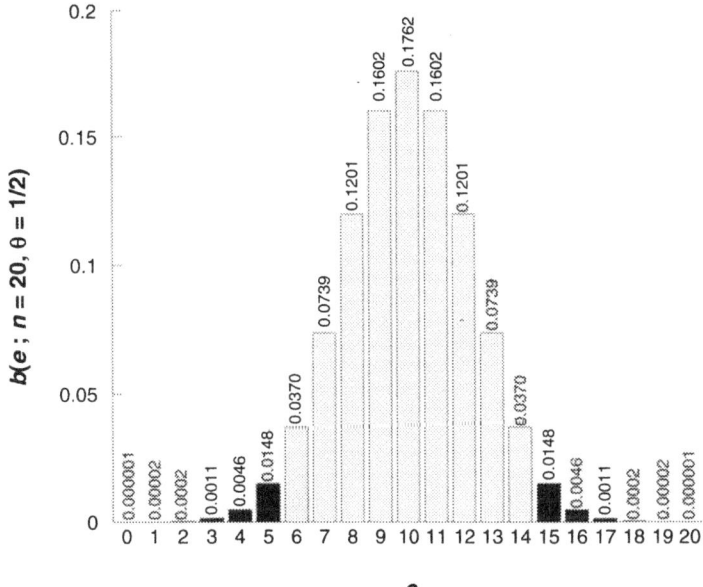

Fig. 4.3 The binomial distribution $b(e; n, \theta)$ for the number of "unders" or "overs" (e) in $n = 20$ independent laboratory analyses each with a probability of $\theta = \frac{1}{2}$ of producing an "under" or an "over." The same distribution would result for the number of heads or tails that would result were a coin tossed 20 times. The p-value, having obtaining five "overs," is 0.0414 is the sum of all the shaded bars—that is, $2B(5; 20, \frac{1}{2})$, where $B(e; n, \theta) = \sum_{i=0}^{e} b(i; n, \theta)$ is the cumulative binomial probability of getting up to e "unders" or "overs" in n random trials (see Section 6.2.5).

on $\mathbf{H_0}$ as would five or less). We can therefore use standard tables of this distribution (Table A.3.1) to compute the p-value as 0.0414—see Figure 4.3.

But let's say we had chosen to perform the chemical analyses until we had obtained at least five "overs" and five "unders," rather than necessarily stopping at 20 analyses—not an unreasonable design. This design can result in exactly the same dataset as under the "stop at 20 observations" design. That is, if we had obtained 15 unders and four overs in the first 19 observations and then on the 20th analysis obtained an over, we would have the same data as before and would stop. But a different sampling distribution would then be obtained—the negative binomial distribution.

Note that the applicable distribution is not a matter of choice. Once we assume that sampling is random these distributions are forced upon us by the study's design. As explained in Section 6.2.5, the distribution of dichotomous outcomes follows the binomial distribution if the sampling design is to cease once n trials are performed, regardless of the number of times a particular outcome occurs. On the other hand, the negative binomial distribution applies if sampling is to cease once k particular outcomes are obtained, no matter how many trials that requires (see Section 6.2.7).

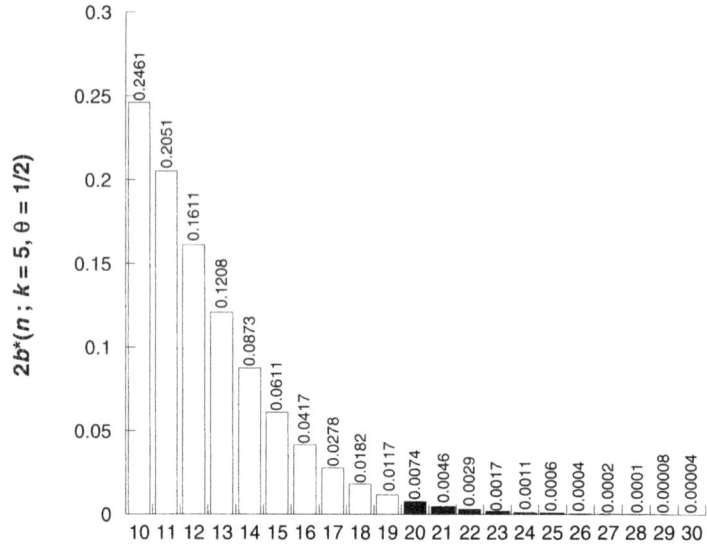

Fig. 4.4 The negative binomial distribution $2b^*(n; k, \theta)$ for the number of chemical analyses (n) required before at least $k = 5$ "unders" *and* at least $k = 5$ "overs" are obtained for independent laboratory analyses each with a probability $\theta = \frac{1}{2}$ of producing an "under" or an "over". The same distribution would result if a coin was tossed until at least five heads and at least five tails had been observed. The p-value, having obtained at least five "unders" and "overs," is 0.0192, which is the sum of all the shaded bars, i.e., $2\mathcal{B}^*(20; 5, \frac{1}{2})$, where $\mathcal{B}^*(n; k, \theta) = \sum_{i=n}^{\infty} b^*(i; k, \theta)$ is the probability of getting k "unders" *or* k "overs" in n or more trials (see Section 6.2.7, and Problem 4.3 for the calculation of p).

So for the second design we calculate the p-value for $n = 20$ as 0.0192 (see Figure 4.4).

Here we have a very unexpected result: The same data can give rise to more than one p-value! The appropriate p-value depends on the intention of the analyst in carrying out the laboratory tests (i.e., to stop at 20 analyses or to stop once at least 5 unders and overs had been obtained). How then can one or other of these values constitute "evidence" for or against bias? Why should the p-value depend on the intentions of the analyst as to when to stop the investigation? And what if those intentions cannot be determined (e.g., the analyst couldn't remember or had died!).

While tempting to say that the reason is that the p-value has this behavior is solely that it considers data more extreme than were obtained, this is not so. We are sampling from discrete distributions, so there is a nonzero probability of obtaining the data we got. These two different distributions give two different probabilities of getting those data (five overs in 20 samples), i.e., 0.0148 and 0.0074.[9] Considering data more extreme than were obtained adds to the problem.

As to practical consequences, there are two. First, using a p-value as "evidence" for or against tested hypotheses is problematical, when it may take on different values depending on the on the investigator's intentions for stopping (if indeed they are known). What can be done, as stated earlier (Section 4.3.5), is to use the p-value for similar tests and sampling designs to compare the merits of competing hypotheses.

Second, the type of example given here (binomial versus negative binomial) is somewhat unusual— usually only one candidate sampling distribution is applicable. Nevertheless, the investigator needs to appreciate this surprising behavior.

Peeking We now consider a situation where the p-value as a tail area really does matter. It arises when a "research hypothesis" is posed as the alternative hypothesis in a single point-null hypothesis test.

Paraphrasing remarks made by Cornfield (1966 [59]), let us consider this situation. An experimenter makes 20 observations in the expectation that they would permit the rejection of a point-null hypothesis at the 5% significance level. If this result occurs the intention is to publicize the confirmation of the research hypothesis in a report or in a paper. However, if after those samples are obtained it is found that $p > 0.05$, then the hypothesis would not be rejected, and this would be unexpected. Nevertheless, it is still believed that the hypothesis should be rejected, i.e., that the research hypothesis is true. That provokes this thought: "How many more observations would be required to have reasonable certainty of rejecting the hypothesis?" The surprising result is that under these circumstances no amount of additional sampling would logically permit rejection at the 5% level. While this result is of more importance in clinical trials than it is in environmental sampling, we do need to understand the issues involved, and their implications.

Why do we get this unexpected result? The reason is as follows. If the point-null hypothesis is true, there is a 0.05 chance of rejecting it after the first round of observations.[10] To this chance must be added the probability of rejecting it after the second round, given failure to reject it after the first.[11] This increases the total chance of erroneous rejection to above 0.05. Therefore no amount of additional observations can be collected which would provide evidence against the hypothesis equivalent to rejection at the 5% level. And, as we shall see, this is entirely because p includes the probability of getting results even more extreme than have been obtained.

As an example, consider the case discussed previously, where chemical analyses are to be stopped once 20 samples are to hand. But we amend it to a two-stage design by analyzing these 20 data and stopping if—and only if—a statistically significant result is obtained at the 5% level (i.e., we obtain $p < 0.05$). Otherwise a further

---

[9]That these two probabilities are in ratio 2:1 is a coincidence.
[10]It will in fact be a little less than 0.05, because we are dealing with a discrete variable.
[11]Strictly, it is not admissible to recalculate the p-value using just the larger sample size, because the investigator's intentions were to stop at the smaller sample size *if* a hoped-for result ($p < 0.05$) were to be obtained and, failing that, to carry out further sampling.

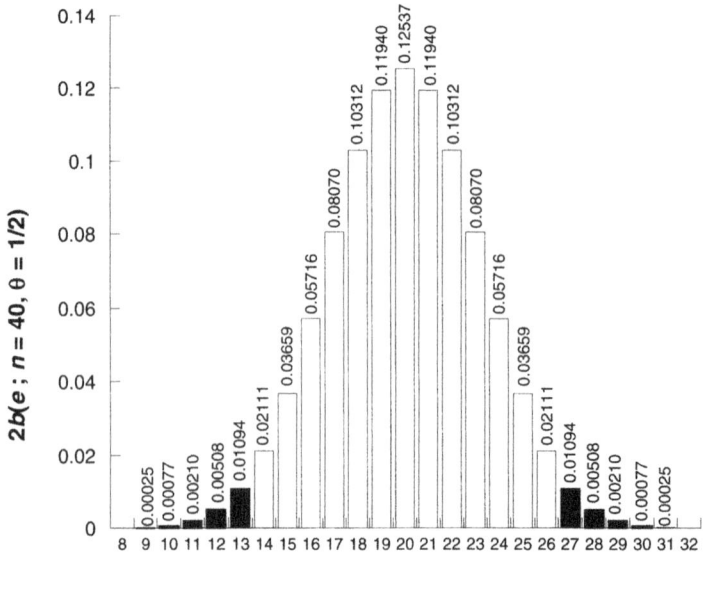

Fig. 4.5 The binomial distribution $b(k; n, \theta)$ for the number of "unders" or "overs" (k) in $n = 40$ independent laboratory analyses each with a probability of $\theta = \frac{1}{2}$ of producing an "under" or an "over." The same distribution would result for the number of heads or tails that would result were a coin tossed 40 times. The p-value for a single-stage test, having obtained 13 "overs", is 0.0385, which is the sum of all the shaded bars, i.e., $2B(13; 40, \frac{1}{2})$.

20 samples will be taken and the test is to be repeated. In other words, we will be peeking.

Let's imagine that after 20 observations between six and 14 "overs" have been obtained (so there would be between 14 and six "unders"), in which case we would land in the non-shaded part of Figure 4.3. These cases all give $p > 0.05$ (e.g., if we were to have obtained either six or 14 "overs," we would have $p = 0.1154$). Therefore, under the two-stage design, another 20 samples must be taken. More imagination: Let's say that after 40 samples we obtain either less than 14 or more than 26 "overs," and so we would land in one of the shaded parts of Figure 4.5. Now if the design had been to sample 40 samples without peeking, and we had obtained such results, we would be able to claim a statistically significant result, and hence bias.

However, as noted by Cornfield (and many others) the p-value for the two-stage design must account for the earlier peek. If the tested hypothesis is true, then there was a 5% chance of it having been rejected on the first peek (this didn't happen). To this chance we must now add the probability of rejecting the hypothesis after the second round. Therefore, to the p-value that we would have obtained at the first peek had we rejected the point-null hypothesis (i.e., 0.0414), we must add the joint probability of landing (a) in the one of the shaded parts of Figure 4.5, and (b) in the non-shaded part of Figure 4.3. This gives a p-value for the two-stage test of 0.0799. Problem 4.6 invites the reader to derive the formula for this p-value.[12]

Peeking is not necessarily an issue in statistical data analysis, but it is if the investigator plans multiple stopping points in a data collection and evaluation program. An example of that is given on page 235. Royall (1997 [275]) notes that this issue has assumed great importance in clinical trials, whereby the amount of Type I error risk that can be used at each planned "stop" of a trial is stated before the trial commences (ethical considerations often demand that a long trial should have some "stops"). He also shows that using the likelihood approach, rather than frequentist p-value procedures, avoids this problem (and, therefore, Bayesian approaches would also avoid it).

4.4 AN "EVIDENCE-BASED" APPROACH (FOR DICHOTOMOUS DATA)

Here is an approach where peeking does not matter and where Bayes' rule can be used, based on the "evidence-based medicine" (EBM) approach.

Consider a situation where the presence of a pesticide in a finished drinking-water, in any detectable amount, is considered to be a hazard. What is to be made of a positive laboratory test's result during monitoring of the water supply, given that the pesticide has seldom (if ever) been detected before? After all, laboratory procedures have "false positive rates". Maybe the pesticide wasn't really present at all? But if that false positive rate is low (say, 5%), then it may be thought that it is very likely that the pesticide really was present. A survey of experienced drinking-water quality personnel revealed that the majority were of this view when presented with this information (Hrudey & Rizak 2004 [158]).

This is a situation that can be examined using Bayes' rule [Equation 2.14] on dichotomous variables **H** (the hazard was truly present) versus **K** (the hazard was in fact absent—a false positive result was obtained). So we'll need to use Equation (2.14) for the mutually exclusive hypotheses **H** and **K**. Let's also denote the event of obtaining a positive laboratory test result by E and so the false positive rate for the laboratory test can be denoted by the conditional probability $\Pr(E|\mathbf{K})$. What we want to know is the probability of **H** being true, given that a positive result has been obtained, i.e., the posterior probability $\Pr(\mathbf{H}|E)$. Therefore, in Equation (2.14), "A" takes the role of E, "B_1" takes the role of **H** and "B_2" takes the role of **K**. Bayes' rule is thus in the form already seen on page 16:

$$\Pr(\mathbf{H}|E) = \frac{\Pr(\mathbf{H})\Pr(E|\mathbf{H})}{\Pr(\mathbf{H})\Pr(E|\mathbf{H}) + \Pr(\mathbf{K})\Pr(E|\mathbf{K})}. \quad (4.2)$$

To implement it we need two further items [we already know that the laboratory test's false positive rate is $\Pr(E|\mathbf{K}) = 0.05$]:

---

[12]That formula is somewhat complex, especially when compared to the much simpler formula used for a single-stage test. That complexity would appear to be another barrier to full interpretation of p-values under peeking.

- $\Pr(\mathbf{H})$, the prior probability for $\mathbf{H}$ [because $\mathbf{H}$ and $\mathbf{K}$ are mutually exclusive, covering all possibilities, we have $\Pr(\mathbf{K}) = 1 - \Pr(\mathbf{H})$].

- $\Pr(E|\mathbf{H})$, the test's "sensitivity". This is related to the test's false negative rate, $\Pr(\tilde{E}|\mathbf{H})$, where "$\tilde{E}$" denotes a negative laboratory result. That is, $\Pr(E|\mathbf{H}) = 1 - \Pr(\tilde{E}|\mathbf{H})$.

We'll assume that the test's false negative rate is 0.10 and so its sensitivity is $\Pr(E|\mathbf{H}) = 0.90$. For the prior probability we'll take $\Pr(\mathbf{H}) = 0.001$, reflecting the historical rarity of the presence of the pesticide. The calculated posterior probability—the probability of hazard, or pesticide prevalence probability—is thus

$$\Pr(\mathbf{H}|E) = \frac{0.001 \times 0.90}{0.001 \times 0.90 + 0.999 \times 0.05} = 0.0177. \quad (4.3)$$

The probability of hazard is less than 2%, a result that often surprises.[13] The reason that it is so low lies in the value of the chosen prior probability. Even though the test has high sensitivity, the low pesticide prevalence value means that most of the time a positive result is obtained when the pesticide is actually absent.[14] But were we to have used a higher prior probability, say $\Pr(\mathbf{H}) = 0.1$, the probability of hazard would be a lot higher (66.67% in this case). This reflects a major issue in using this Bayesian approach. Whereas in medicine there are very good grounds for assuming that the historical value of disease prevalence cannot rapidly change (i.e., $\Pr(\mathbf{H})$ is stable), the prevalence of the pesticide may be thought to change rapidly in the inflow to the water treatment plant (e.g., if a new agricultural land-use in the water supply catchment begins to use the pesticide). Or, plant processes may suddenly change and become less efficacious. So there isn't one "correct" answer to this problem; there are many answers, depending on the adopted prior pesticide prevalence value.

One way around this, for water quality determinands that are stable, is to not analyze all the collected sample. Then, if a positive result is obtained one can take another portion of that collected sample and analyze it. If that result is positive then, even for very low prior probabilities, the posterior probability becomes very much larger (see Problems 4.7 and 4.8.)

4.5 POWER ANALYSIS

In Chapter 5 we will examine "power curves" for various hypothesis tests. In essence these characterize the future performance of a test in terms of the unknown parameters—for example, population means and standard deviations. This is an example of "before-trial betting," as discussed in the Introduction to Chapter 2. They are therefore most useful in planning sampling programmes, because they can be used to

---

[13] In EBM this probability is called the "Positive Predictive Value", where $\mathbf{H}$ is concerned with the presence of disease.

[14] Of course the test may occasionally fail to detect the pesticide when in fact it was present; the second term on the denominator of Equation (4.3) accounts for that possibility.

show what population effects could be *regularly* detected (e.g., 80% of the time). In particular, one can determine the number of samples one needs to regularly detect an "effect size" that is deemed to be practically important. There are texts and software available for such calculations (e.g., Cohen 1988 [50], Cohen 1992 [51], Thomas & Krebs 1997 [313]), including the author's *Detection Calculator*.[15] These curves are often used (particularly in medical research) to calculate the desirable sample size, the number of subjects needed to be able to routinely detect an important effect. That is, they may be used for **experimental design**. But in many water quality studies, data are accumulated, and so, with a point-null hypothesis test, ever-smaller effects may be detected as yet more data are obtained. This important issue is addressed further in Section 5.3.5.

Some writers also present procedures for calculating the power of a performed test (e.g., Zar 1996 [349]). This is often advocated for situations in which the tested hypothesis has not been rejected and there is a desire to determine the probability of having made a Type II error. This is an example of "after-trial evaluation." It is performed by substituting sample statistics for population parameters in the power calculation. In fact such an analysis tells the investigator no more than is already available from the p-value and is essentially a violation of the concept of power—which attaches probability to the long-run frequency of given outcomes, not to the particular outcome from a given set of data. For such reasons applying power analysis to a performed test has recently been actively discouraged (Gerard *et al.* 1998 [111], Hoenig & Heisey 2001 [153]). Power is a consideration in designing studies—"before-trial betting"—not in analyzing results.

Finally, we will soon be discussing nonparametric methods, yet power analysis requires a parametric model. However, power calculation procedures are available for some nonparametric tests (Noether 1987 [243], Hamilton & Collings 1991 [135], Mumby 2002 [234]), but in general a Monte Carlo analysis (see Section 4.8) can be used. Alternatively, we can use the transformation that seems most likely to result in near-normal distributions and calculate approximate power using available parametric methods.

4.6 MULTIPLE COMPARISONS

Special considerations of the risks of error arise when comparing more than two samples, such as in an analysis of variance (ANOVA). In these analyses we typically have a hypothesis of equality between three or more means. Under the assumption that these samples have come from distributions with the same variances but with different means, we can develop an α-level test of differences in means, based on differences in variances—as depicted in Figure 4.6. If that analysis rejects the hypothesis we then may proceed to examine the individual pair-wise differences. But herein lies a trap. The overall analysis has restricted the Type I error rate (of falsely rejecting the hypothesis, when in fact it is true) to $100\alpha\%$. If we now perform each of these

---
[15] http://www.niwa.co.nz/services/statistical.

comparisons at the $100\alpha\%$ level, the overall risk of error is enhanced—a simple compounding of error risk occurs so that the overall error rate among the individual comparisons is inflated.

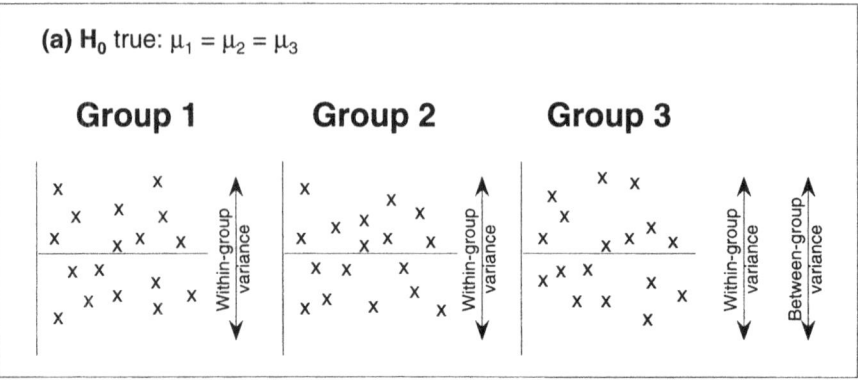

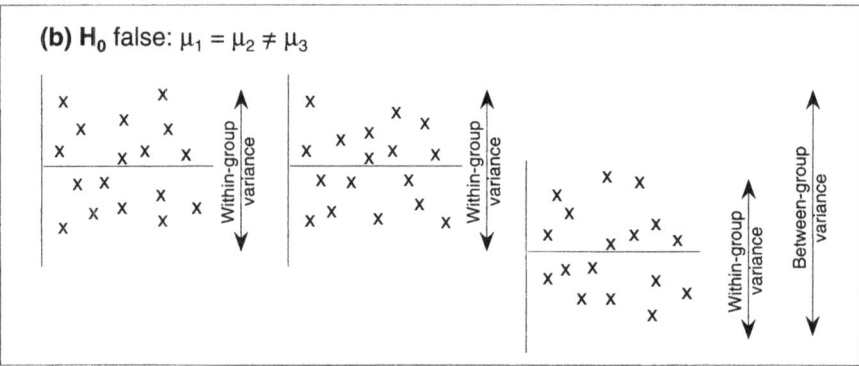

Fig. 4.6 ANOVA: inferring differences in means from differences in variances.

A similar situation arises when multiple correlations are presented. Testing a hypothesis for each correlation at a level α will inflate the overall Type I error rate.

The usual way around this issue has been to adjust each comparison-wise permissible Type I error rate so as to maintain the overall rate at $100\alpha\%$. A common approach is to use the "Bonferroni correction" in which the comparison-wise rate is set to α/m, where there are m comparisons to be made.[16] This keeps the risk of making *at least one* Type I error below a nominated risk level (usually $\alpha < 0.05$), *regardless of the number of tests made*. As such it has the unfortunate effect of discouraging the publication of large studies. That is, the more data one includes in a reported study the more onerous each comparison becomes. It can happen too often that a particular

---

[16]The Bonferroni correction is the most drastic way of doing this. There are a multitude of other correction procedures (e.g., Zar 1996, Chapter 11 [349]), but these are not our concern here.

comparison can fail to be detected in a large dataset, but be detected in a subset of it. That is, statistical power is lost as the (reported) sample size increases, and that is as unexpected as it is undesirable.

Noting this, a new and more powerful approach has recently been advocated—the false discovery rate ("FDR", Benjamini & Hochberg 1995 [14], Sabatti et al. 2003 [279], García 2003 [107]). This is defined as the expected proportion of true hypotheses rejected out of the total number of rejections—it considers how many of α-level rejections may actually be in error.[17] By focusing on the expected proportion of false rejections, rather than on the presence of at least one error, it presents a much more satisfactory solution to the problem of multiple testing—regaining statistical power whilst avoiding an abundance of false-positive results. It is a step-down method (Bonferroni is a single-step method) in which we first order the p-values obtained from all the comparisons and then use a decision rule with a cutoff p-value that depends on the dataset. The FDR procedure includes the following steps:

- Put the p-values in an ascending sequence $p_{(i)}$, so $p_{(1)}$ is the lowest, and $p_{(1)} \leq p_{(2)} \leq \cdots \leq p_{(m)}$.

- Step down from the largest (from $i = m$ to $i = m - 1$, to $i = m - 2$, etc.) until, for the first time, $p_{(i)} \leq iq^*/m$.

- Denote that cutoff value of i by k.

- Reject all tests with p-values up to this cutoff, i.e., $p_{(1)}, \ldots, p_{(k)}$.

where q^* is the false discovery rate and $p_{(k)} \leq kq^*/m$ is the cutoff value. Note that if we were to set $q^* = \alpha$ the FDR constraint requires that $p_{(i)} \leq i\alpha/m$ and this can never exceed α (because $i \leq m$). Indeed, the p-value cutoff is constrained to be between the Bonferroni comparison-wise value (α/m) and α. In this way, setting the FDR to the significance level ($q^* = \alpha$) the FDR indeed considers only the rejections.

Example: Small number of comparisons Consider a case where 15 comparisons are made resulting in these ordered $p_{(i)}$ values: 0.0001, 0.0002, 0.0064, 0.0075, 0.0101, 0.0178, 0.0298, 0.0366, 0.0499, 0.1231, 0.3764, 0.4621, 0.6984, 0.7647, and >0.99994.[18] Using the FDR procedure with $q^* = 0.05$, the first p-value to satisfy the FDR constraint is $p_{(6)}$, because $p_{(6)} = 0.0178 \leq \frac{6}{15} q^* = 0.020$. This is three times as many rejections as would have been obtained using the Bonferroni (each comparison would need to be made at a significance level $0.05/15 = 0.0033$), but just two-thirds of as many as would have been obtained using uncorrected α values for each comparison.

Example: Large number of comparisons Jowett & Richardson (2003 [171]) have published a study on 21 fish communities at about 1100 New Zealand river sites.

---

[17] Of course in a point-null hypothesis test on continuous variables, one could argue that it is impossible to commit the Type I error because the hypothesis is always false, and so cannot be falsely rejected.
[18] This example is based on a case presented by Benjamini & Hochberg (1995 [14]).

Correlation of the species' abundance with 27 environmental variables was sought (e.g., run, riffle, pool, pasture, forest,...). The resulting (Spearman) correlation matrix is therefore very large (it has 567 cells) and therefore the Bonferroni correction for tests of the hypothesis (H_0: there is in fact no correlation whatsoever) would require each comparison to be conducted at a significance level $\alpha = 0.000088$ in order to maintain the overall significance level at $\alpha = 0.05$.[19] The number of "statistically significant" results produced by the Bonferroni procedure is 99, whereas the FDR and unadjusted-α procedures return 179 and 210, respectively. These results, in which the FDR detections are about double those made by the Bonferroni procedure, suggest that FDR is a more appropriate approach than the two extremes represented by the alternative (Bonferroni) and unadjusted-α alternative. A further attraction is that the same number of FDR detections occur by splitting the dataset in two (e.g., according to odd or even ranks), performing the FDR on each, and adding the results. This means that the ability to detect a particular difference is independent of the number of other comparisons made, a property not shared by the Bonferroni approach.

Note that from Section 4.3.5 we could expect that the p-values derived in this example to be very low, because there are so many sites. Indeed they are. For example, a point-null hypothesis test on a Spearman rank correlation coefficient of 0.06 with 1100 data returns a p-value of 0.047; with just 100 data, p rises to 0.55. So if we use the same correlation data with only 100 sites, the picture changes dramatically.[20] For all 21 fish species we obtain 5, 9 and 41 rejections from the 567 tests for Bonferroni, FDR and unadjusted-α procedures, respectively. This is a considerable loss in pattern identification *if* we are using the p-value as its measure.

4.7 NONPARAMETRIC TESTS

The procedures discussed so far all rely on a statistical model in which distributions are characterized by parameters (e.g., μ and σ for the normal distribution). But if the distribution that data have been drawn from seem too irregular to be described by such distributions, or if there are simply too few data to tell,[21], or nondetects, a nonparametric approach may be more useful and appropriate. The essence of this approach is to replace data values by their relative ranks and then perform calcula-

---

[19]The authors' Spearman's rank correlation coefficient values (r_s) have been translated into p-values using the formula $p \approx 2[1 - G(Z)] = 2G(-Z)$ where $Z = |r_s|\sqrt{(m-2)/(1-r_s^2)}$. This formula was derived by approximating the standard error of r_s by the standard error for Pearson's correlation coefficient r (Zar 1996, p. 373 [349])—that is, by $S_{e,r_s} \approx \sqrt{(1-r_s^2)/(m-2)}$. For a large sample size, $\mathbf{H}$ is rejected at level α if $|r_s|/S_{e,r_s} > Z_{1-\alpha/2}$. The required formula is obtained by simple manipulation of this formula, noting that $G(Z_{1-\alpha/2}) = 1 - \alpha/2$, replacing α by p, and using Equation (2.19).

[20]The actual correlation data would not be the same of course (an accurate analysis would require a Monte Carlo study taking repeated random subsets of size 100 from the original 1100), but we lose no generality by considering this simplified case.

[21]The degree to which data appear to conform to particular distributions is often examined with "goodness-of-fit" tests. These also test a form of point-null hypothesis and thus suffer the same problems as noted earlier—large departures from the postulated distribution may go undetected with small sample size; trivial departures may be detected with large sample size. Wellek (2003, pp. 3, 177–203 [340]), in noting that such procedures really establish lack-of-fit, has proposed some equivalence procedures for these problems.

tions on those ranks (so these methods are sometimes called "distribution-free"). This changes the nature of the tested hypotheses somewhat—for example, turning a parametric hypothesis on the population mean to a hypothesis on the population median. When sampling from a known distribution, these methods are never as powerful as parametric methods, but may come rather close. But in situations where the wrong statistical model is used, the nonparametric method may outperform its parametric cousin.

Example: Fecal contamination in freshwater The data for somatic coliphage displayed on Figure 2.5 can be subjected to an analysis of variance. That analysis considers differences between the five watershed types. If that is done on the raw data the result is a p-value of 0.2645: The hypothesis of no difference *between the sample means* is not rejected. Given that there are $n = 727$ data in total, that seems surprising, because you would think that the test should be powerful enough to confirm differences all-too-obvious on the graph. But the boxplot shows the data plotted on logarithmic scales, in which case they are reasonably symmetrical about the geometric mean or median—but not about the mean. The normality assumption inherent in ANOVA has been seriously breached by the skewness of the data, so analysis of logarithms of the data seems called for. Doing that returns a p-value ≤ 0.0001, indicating differences *between the geometric means* of the Somatic phage concentrations over the five watershed types.[22] A similar p-value is obtained using a nonparametric ANOVA (the Kruskal–Wallis test). But this time of course the differences are *between the medians*. This is consistent with the closeness of the geometric means and medians, as seen on Figure 2.5 (the geometric mean and median of lognormal populations are identical—see Section 6.2.2).

While this example indicates that a data transformation approach can be very beneficial, one is not always free to do it. Instead, a nonparametric approach is effectively *demanded* for some water quality data, especially those containing "greater than" censored values (e.g., microbiological assays, biochemical oxygen demand tests). Were a parametric approach to be used in such circumstances, it is usually not obvious how to assign numbers to these high unknown values—but they can be assigned ranks.

Water quality studies have increasingly used nonparametric procedures for trend detection (for which details are given by Gilbert 1987 [118], Helsel & Hirsch 2002 [147]). In using them it is important to note that they are not free of assumptions.[23] For example, in using a Wilcoxon signed ranks test (the nonparametric equivalent of a paired t-test), the distribution of the difference is assumed to be symmetric (Conover 1980, p. 281 [56]). If that assumption is untenable one would have to use the less powerful sign test (Gilbert 1987, p. 244 [118]). For detailed descriptions of these tests see Conover (1980 [56]) or Iman & Conover (1983 [161]).

---

[22] Recall the the geometric mean is the antilog of the mean of the logarithms of our data.
[23] The thought that some may hold them to be so provoked Johnson (1995 [166]) to describe these methods as "statistical sirens."

4.8 RANDOMIZATION TESTS

These tests are focused more directly on samples than they are on populations. As such inferences may be made about the pattern of effects in those samples, leaving generalization to other subjects to be justified by other nonstatistical arguments (Edgington 1987 [76]). They achieve this by multiple resampling (i.e., reshuffling) of the samples' observations many times to discover the form of the sampling distribution of a test statistic (e.g., Manly 1997 [197]). These tests work on the basis of the idea that if a point-null hypothesis is true, then any random rearrangement of the observations should be equally likely. So they too are especially useful in situations where the underlying distribution is unknown. Once again, these techniques are not free of assumptions. For example, a hypothesis of equality between means typically invokes an assumption that variances are identical also. Randomization methods include the **bootstrap** (random resampling with replacement) and the **jackknife** (sampling a number of times, each time eliminating a different datum). **Monte Carlo** methods are also encompassed in this topic, and are finding increasing use in water quality studies—especially for risk assessment calculations (see Chapter 9). In this case, multiple random sampling is made from distributions assigned to key input variables in a model. In this way the profiles of uncertainty in each of these inputs is melded into a profile of uncertainty for the output. This has the added attraction of providing a means of advising water managers of the probabilities of certain outcomes happening, giving a substantial amount of information upon which they may make decisions on a balance of risks.

Example Consider a river with rate of flow and total phosphorus concentration upstream of a wastewater treatment plant inflow denoted by Q_u and C_u, respectively, and with corresponding inflow values Q_i and C_i. Simple mass balance, assuming complete mixing of the inflow with the river water,[24] shows that the downstream concentration of total phosphorus is

$$C_d = \frac{Q_u C_u + Q_i C_i}{Q_u + Q_i},$$

where the subscript d denotes "downstream." If this the standard for this downstream phosphorus concentration is expressed in terms of a 95%ile, we need a means of calculating it given distributions for the four input variables (Q_u, Q_i, C_u, C_i). It is sometimes possible to combine their distributions analytically—for example, if all are lognormal (Warn & Brew 1980 [339]). But in general it is simplest to perform Monte Carlo simulations by repeatedly sampling these four distributions, performing the mass balance equation each time, to build up an empirical distribution function of C_d from which the 95%ile can be taken. This approach is used widely in the UK water industry (Murdoch 2001 [235]).

---

[24]Mixing is generally incomplete at downstream sites where standards are to be applied; see Rutherford (1994 [278]) for river mixing calculation procedures.

4.9 BAYESIAN METHODS

Bayes' rules for probabilities have been introduced in Section 2.1. When the "events" in those rules [Equations (2.13) and (2.14)] are assigned to hypotheses, we are thereby enabled to talk about the probability that a hypothesis is true. This probability assignation is at the heart of the controversy surrounding applications of Bayesian methods. A strict frequentist view of probability means that a hypothesis is either true or false, with probability zero or one; one doesn't attach an intermediate probability to that. But, as discussed in the Introduction to Chapter 2, there seems no one best way to view probability. Furthermore, some statement about the status of the tested hypothesis is actually desired—even a frequentist hypothesis test concludes with some form of words concerning it. Accordingly, Bayesian methods have a role, sometimes a very useful one, in analyzing data. A spectacular example is the use of such methods made by Alan Turing during the Second World War, in breaking the German "Enigma code" (Hodges 1983 [151]).[25] The methodology is well described elsewhere (e.g., Berry 1996 [22], Lee 1997 [179], Carlin & Louis 2000 [40], Press 2003 [264], Bolstad 2004 [28]).

A simple example of a Bayesian approach has already been given (the estuary deposition problem on page 15). In that case we had hypotheses **H** and **K** denoting complementary probabilities about the deposition resulting from erosion, with event **E** denoting data. In that case we obtained posterior probabilities (0.982 and 0.222), under two very different assignments of prior probabilities. That is the sum total of the result from such an analysis. It does not invoke any considerations of p values, Type I or Type II error risks, and it does not consider data more extreme than that obtained (as does a p-value—see the boxed text on page 84). These are attractive features, but the strong influence of prior probabilities in this case is not (though the example exemplifies an extreme case of it). Fortunately this sensitivity to prior probabilities problem is seldom so severe: when there is a larger amount of data available the posterior probabilities resulting from different assignments of prior probabilities become rather close.

More complex examples arise when the data and/or the hypotheses refer to continuous variables. In that case some of the probabilities in Equations (2.13) and (2.14) become probability densities and summations become integrals. These are fully documented in texts (especially Press 2003 [264]) and are not repeated here (though we use one of these, for discrete data and continuous hypotheses, in Chapter 8). However, three important points must be made.

First, the general form of Bayes' rule can be stated as

$$\text{posterior} \propto \text{prior} \times \text{likelihood}.$$

where $\propto$ means "proportional to," "prior," and "posterior" are subjective probabilities that obtain before and after gaining data, and "likelihood" is a description of that data

---

[25] Alan Turing (1912–1954) was a founder of modern computer science.

(see also Section 4.10).[26] It takes the form of a probability mass function (for discrete data) or a probability density function (for continuous data). While it "takes the form" it does not in fact measure probability. That is because a likelihood function is regarded as a statement about the distribution's parameter(s) for fixed values of data, rather than what we have seen to date—a statement about the data for fixed value(s) of the parameter(s). As such it does not encompass an area of unity and is not a probability. Rather, it serves as a statement about the distribution of parameters, give our data.

Second, when making calculations using the above general form we have to marry two distributions: the prior and the likelihood. That can pose computational difficulties, many of which have only recently been overcome with the advent of powerful software (e.g., WINBUGS[27]). But in some cases the integration can be made remarkably simple. This occurs when we choose *conjugate distributions* for the prior and likelihood. For example, if we take a beta distribution for the prior and a binomial distribution for discrete data, then the posterior distribution is also a beta distribution, whose parameters are calculated very simply (as demonstrated in Chapter 8). Other examples of conjugate prior and likelihood distributions are normal/normal and Poisson/gamma.

Third, there are great disparities between the p-value outcome from a classical hypothesis test and the posterior probability outcome from a Bayesian analysis for point-null tests (e.g., Berger & Sellke 1987 [19]). But this is not necessarily true for one-sided tests—in some cases the p-value and the posterior probability can actually be numerically identical (Edwards *et al.* 1963 [79], DeGroot 1973 [67]). This is particularly the case for tests of continuous variables where the prior distribution is unimodal and has half its mass in each of the **H** and **K** hypothesis regions (Casella & Berger 1987 [44]). So discrepancies between classical and Bayesian outcomes for point-null tests should not be taken to imply that discrepancies will occur for other types of hypotheses.[28]

> **For some one-sided tests the p-value and the Bayesian posterior probability can be the same numerical value.**

4.10 LIKELIHOOD APPROACH

---

[26]The term "likelihood" has been used since 1925 (Fisher 1925 [91]): "What has now appeared is that the mathematical concept of probability is inadequate to express our mental confidence or diffidence in making such inferences, and that the mathematical quantity which appears to be appropriate for measuring out order of preference among different possible populations does not in fact obey the laws of probability. To distinguish it from probability I have used the term '*Likelihood*' to designate this quantity"—cited by Edwards (2000 [78]).

[27]The BUGS acronym stands for *Bayesian inference using Gibbs sampling*, where Gibbs sampling is a technique for performing the integrations. It is available at www.mrc-bsu.cam.ac.uk/bugs/.

[28]The extent of this discrepancy for interval tests seems not to have been reported—but note that Schervish (1996 [281]) demonstrates that, for variance-known cases, whereas p-values for tests of one-sided hypotheses are "coherent", those for tests of point-null and equivalence hypotheses ($\mathbf{H_0}$ and $\mathbf{H_e}$) are not.

When addressing the relative merits of competing hypotheses, the likelihood approach offers strong advantages (Edwards 1972 [77], Dupont 1983 [75], Royall 1997 [275]). It uses the likelihood function alone as a measure of support for the hypotheses, and so it avoids the p-value's inclusion of data more extreme than that obtained (Bayesian methods also avoid the use of these data). This seems consistent with the notion that the outcome of a likelihood procedure really could be called "evidence," whereas it is difficult to see how that term could be used for classical procedures' p-values—how can "evidence" include considerations of data that have not been obtained?[29] Restricting attention to only those data actually obtained makes it possible to conform to the "Likelihood principle," which essentially asserts that two samples that generate identical likelihood functions are equivalent as evidence.

Therefore, likelihood inference is a major departure from classical inference.[30] But it also departs from Bayesian inference, in that it completely avoids the use of prior distributions, and thus removes the subjective features associated with that.

Likelihood inference has played very little part in water science and management, where "classical" methodology has been so dominant. Some methods of calculating credible intervals for MPN methodology are based on likelihood calculations (discussed in Chapter 10). It seems time that it should be used more widely—especially for issues where the *relative* merits of hypotheses are in view (such as model selection, as discussed below). But it may not be useful when the merits of a *particular* hypothesis is required, such as assessing compliance of a wastewater effluent with standards.

4.11 HYPOTHESIS TESTING LITERATURE

4.11.1 The point-null hypothesis testing debate

The proponents of "significance testing" (Fisher 1935–1971, 1959 [93, 94]) and of "hypothesis testing" (Neyman & Pearson 1928, 1933 [241, 242]) had substantial public disagreements about the manner in which these p-value-based procedures should be employed (Inman 1994 [162], Goodman 1993 [122]). Others joined the fray early on (Berkson 1938, 1942 [20, 21]). Since that time, there has been a wealth of material in various literatures criticizing and occasionally defending point-null hypothesis tests. This has mostly occurred in fields seldom read by environmental professionals, such as psychology and medicine. For example, criticisms have been raised (in date order) by: Rozeboom (1960 [276]), Nunally (1960 [244]), Carver (1978 [41]), Loftus (1991 [192]), McBride *et al.* (1993 [208]), Carver (1993 [42]), Nester (1996 [239]), Bower (1997 [30]), Johnson (1999 [167]), Goodman (1999a [123]), Germano (1999 [113]), Germano (2001 [114]). Defenses of point-null tests

---

[29] A Bayesian statistician, Jeffreys (1961 [164]) has characterized this as: "What the use of P implies, therefore, is that a hypothesis that may be true may be rejected because it has not predicted observable results that have not occurred."

[30] Although both schools of inference do use the likelihood function in some way—many classical tests have been derived by the "maximum likelihood method".

have been given by Fleiss (1986 [96]), Frick (1995 [102]), Chow (1996 [47]), Hagen (1997 [134]), Muliak *et al.* (1997 [233]) and by Harris (1997b [138]). Of course there is also a vast amount of literature in which point-null hypothesis tests are routinely used and not criticized.

Variations of testing procedures have been proposed to account for some of the difficulties. For example, adjusting the significance level to maximize power (Cascio & Zedeck 1983 [43]), Mapstone (1995 [201]) and standardizing tail-area probabilities (Good 1982 [121]).

It is fair to say that the level of debate continues.

4.11.2 The p-value culture

Some statisticians have criticized the over-reliance on p-values as evidence for or against useful hypotheses, especially when point-null testing is performed. Nelder (1999 [238]) has opined that: "The most important task before us in developing statistical science is to demolish the P-value culture, which has taken root to a frightening extent in many areas of both pure and applied science, and technology...". In medical science, Goodman (1999a [123]), in promoting a Bayesian approach (in Goodman 1999b [124]), has noted that p-values and hypothesis tests are widely perceived as a mathematically coherent approach to inference, regardless of the "p-value fallacy ... the mistaken idea that a single number can capture both the long-run outcomes of an experiment and the evidential meaning of a single result." Concerns expressed in these papers, and in many others (Gibbons & Pratt 1975 [116], Goodman 1983 [122], Berger 1986 [15], Berger & Sellke 1987 [19], Schervish 1996 [281]) are echoed in this book. Most importantly, failure to gain $p < 0.05$ in a point-null test, signifying "*statistical*" insignificance, does not necessarily mean that population differences or trends are *environmentally* insignificant.

A related issue is the commonly expressed dissatisfaction with the dichotomous approach to interpreting hypothesis test results: "reject" or "fail to reject." A common rejoinder is: "Why not just publish the p-value and let the reader decide?" Such advocacy contains an implied acceptance of the notion that p-values serve as useful measures of evidence for informative hypotheses. But they do not constitute such evidence, at least not for tests of point-null hypotheses. They may be useful for ranking the relative merits of point-null hypotheses—if sample sizes are comparable—but "evidence" for particular point-hypotheses they are not.

There is also dissatisfaction with the use of point-null hypothesis tests and associated p-values for model selection (Burnham & Anderson 2002 [37], Johnson & Omland 2004 [168]). This is based on the fundamental observation that progress in science occurs when *competing ideas* about processes in nature are tested against observations (Hilborn & Mangel 1997 [149]). As a consequence, we make progress by testing the *relative merits* of competing models. This moves the modeler away from an arbitrary probability threshold for a single model (e.g., p-value < 0.05, where the power of the test depends on the sample size and so the test is more likely to "validate" a model if its power is low), toward measures that rank the merits of the competing models. These measures (likelihood ratios and "information criteria") are based on likelihood theory, and some use Bayesian methods. They can account for both model parsimony (i.e., number of coefficients or parameters) and the sample size. Another alternative to using point-null hypothesis tests for a single model is to use equivalence tests: Robinson & Froese (2004 [270]) have applied tests of the inequivalence hypothesis to validation of an ecological model. This is a subject in its own right, not pursued further in this text, other than to make the following observations.

> **Use information criteria for model selection;**
> **consider using equivalence tests for validation of single models.**

Problems

4.1 A study has been of the effect of animal faecal residues in recreational water on the health of those swimming in that water (Calderon *et al.* 1991 [39]). The study sought to see if there was an association between the concentration of fecal indicators and pathogens and swimmers' health. It did so by measuring faecal material concentrations in the water over a period, during which time some swimmers kept diaries of health status. The study concluded that there was "no association"—the p-value for the appropriate hypothesis test ($p = 0.059$) was above the critical limit ($\alpha = 0.05$). Discuss the merits of that conclusion.

4.2 Show that the p-value for Figure 4.3 is $B(5; 20, \frac{1}{2}) + \mathcal{B}(15; 20, \frac{1}{2}) = 2B(5; 20, \frac{1}{2})$. As demonstrated in chapter 2, $B(k; n, \theta)$ is the cumulative binomial probability of getting up to k successes in n random trials each with a probability θ of success; similarly, $\mathcal{B}(k; n, \frac{1}{2})$ is the probability of getting k *or more* successes.) Hint: Use results obtained for Problem 6.7.

4.3 Show that the p-value for a single-stage sampling regime (with 20 observations) referred to in Figure 4.4 is $2B^*(20; 5, \frac{1}{2})$.

4.4 Discuss the special negative binomial case $[B^*(2k - 1; k, \frac{1}{2}) = \frac{1}{2}]$ demonstrated in Problem 6.14 in terms of the probabilities shown on Figure 4.4.

4.5 Similarly, show that the p-value for the single-stage sampling regime (with 40 observations) referred to in Figure 4.5 is $2B(13; 40, \theta)$.

4.6 [Difficult problem] Show that the p-value for the two-stage sampling regime referred to in Figures 4.3 and 4.5 (with 20 observations and possibly a further 20 observations if the first set fails to attain statistical significance) is $p = 2B(5; 20, \frac{1}{2}) + b(6; 20, \frac{1}{2})B(7; 20, \frac{1}{2}) +$
$\sum_{i=7}^{13} \left\{ b\left(i; 20, \frac{1}{2}\right) \left[B\left(13 - i; 20, \frac{1}{2}\right) + \mathcal{B}\left(27 - i; 20, \frac{1}{2}\right) \right] \right\} +$
$b(14; 20, \frac{1}{2})\mathcal{B}(13; 20, \frac{1}{2}) = 0.0799$.

4.7 What is the relationship between laboratory false positive and false negative rates, and the Type I and Type II error rates (α and β) for hypothesis tests, as shown on Table 4.1?

4.8 Using the drinking-water example in Section 4.4, show that the posterior hazard probability after a sequence of n positive laboratory test results is $\frac{P_0(1-\beta)^n}{P_0(1-\beta)^n+(1-P_0)\alpha^n}$, where α is the test's false positive rate, β is its false negative rate, and P_0 is the prior hazard probability adopted before these n results became available. What is the posterior hazard probability given a sequence of three positive results, a false negative rate of 10%, a false positive rate of 5%, and a prior hazard probability 0f 0.001?

5
Detection

Hypothesis tests can be used to detect a difference that would be deemed to be important. Such deeming cannot be determined by statistical means, nor can it be done by statisticians; rather it is a difference that environmental professionals (sometimes after community consultation) would regard as *practically* important. What statisticians can do is to show how statistical methods can be used to detect such differences. To do so, let's introduce two quantities that may be unfamiliar, the **detection probability** and **detection regions**.

Depending on the circumstances, detection probabilities may be shown by either the **power curve** or the **OC curve** of the appropriate test. These curves describe what a future test may be able to detect, in terms of population parameters. That is, they show how a future test will behave were the population statistics to take on certain ranges of values. In contrast, **detection regions** display what the test can detect with data already obtained, using only sample statistics (being estimates of those parameters). In this way we see both the "before-trial betting" (detection probability) and "after-trial evaluation" (detection regions) of tests, as advocated by Edwards (2000 [78]).

In this chapter we will be using the population mean as the main parameter of interest, considering mostly one-sample and two-sample cases; that is,

(a) the difference between a normal mean (μ) and a standard value (μ_0), or

(b) the difference between the means of two normal distributions.

In the two-sample case (case b), the samples are assumed to be independent and the populations' distributions are assumed to have a common variance. In either case, the variances may be known or unknown (which is the more likely case). Results are approximately correct (i.e., methods are "robust") for cases where distributions are somewhat non-normal—this is of great importance, given that we never know for sure just how close to normal our populations are.

We first consider one-sided tests followed by two-sided tests—specifically tests of the point-null hypothesis and of interval hypotheses (known as equivalence tests). Finally, we end with more general approaches to detection currently finding favor, in which tests are de-emphasized. Many of the procedures described here for calculating it are accessible as a web-based calculator (*The Detection Calculator*).[1] But first we must introduce three new terms.

Power Curves, Effect Size, OC Curves   Recall from Table 4.1 that statistical power is the probability that a test will *reject* a false hypothesis. The power curve simply plots this probability (of rejecting **H**) over all possible states of nature (Figure 5.1 is an example). So it is defined as α (the Type I error rate) when **H** is true and as $1 - \beta$ (the power of the test) when **K** is true.

The "state of nature" is best described by the effect size, denoted by Δ. In the numerator of Δ, we have $\mu - \mu_0$ (for a one-sample test) or $\mu_x - \mu_y$ (for a two-sample test). In either case, the denominator contains the population standard deviation (σ, the square root of the population variance). The effect size is the most appropriate variable against which to measure statistical power, because it expresses the potentially important difference in means in terms of the populations' variability.

The operating characteristic curve (OC curve) is the complement of the power curve. It shows the probability of *accepting* **H**, versus effect size. Therefore it shows the confidence level when **H** is true and the Type II error rate when **K** is true. OC curves are often preferred over power curves in industrial and engineering statistics literature, presumably because they emphasize acceptance rather than rejection (e.g., Miller & Freund 1985 [228]).

Both power curves and OC curves are presented as detection probabilities in the following sections, for both single-sample (of size n) and two-sample (of sizes n_x and n_y) cases. For two-sample cases we will assume, without loss of generality, that each sample is of the same size ($n_x = n_y = n$). In these cases n is the number of observations *in each sample*. However, the two-sample tests may be implemented with unbalanced sample sizes ($n_x \neq n_y$)—we only invoke balanced sample sizes to simplify the presentation of detection probabilities.

5.1   ONE-SIDED HYPOTHESIS TESTS

5.1.1   Defining detection probability

In general, one should test the hypothesis that would be false were an important effect size to be present. For example, an ecologist may view a change in means exceeding 20% of the true standard deviation to be practically important, in which case the "important effect size" is $\Delta = 0.2$. In another case one might only be

---
[1] http://www.niwa.co.nz/services/statistical/

concerned if a shift in means equal to the true standard deviation occurred, in which case $\Delta = 1$.[2] "Detection" occurs when that hypothesis is rejected, because one can have confidence that such an effect has indeed been detected. This convention—of testing a hypothesis that would be false if an important effect size is present—is used here (and in *The Detection Calculator*). First, note that this convention means that detection probability curves and power curves are synonymous for one-sided tests. That is,

$$\text{detection probability} = D = \text{Pr(reject } \mathbf{H}). \tag{5.1}$$

Consider an unknown-variance one-sample case where we wish our test to have at least 90% probability of detecting a positive unit effect size [i.e., $\Delta = \frac{\mu-\mu_0}{\sigma} = 1$], assuming that such an effect, if present, would also be of practical importance. We also want to keep the significance level at the traditional 5% level; that is, there is to be only a 5% chance of "detection" when in fact the true mean was actually at some unimportant value, μ_0. So we want $\alpha = 0.05$ at $\Delta = 0$ and detection probability ≥ 0.90 at $\Delta = 1$.[3] This important difference is positive, so we need to test the inferiority hypothesis $\mathbf{H}$: $\mu \leq \mu_0$, because to reject $\mathbf{H}$ would be to accept the superiority hypothesis ($\mathbf{K}$: $\mu > \mu_0$). From Figure 5.1 (or using *The Detection Calculator*) we find that we need to have a sample size of $n = 11$ to achieve these requirements. (Actually, with $n = 11$ the detection probability is 92.45% (> 90%), but it is insufficient (89.75%) with only $n = 10$ data, as the figure shows.)

Note that there is no logical reason against testing $\mathbf{H}'$, the complement of $\mathbf{H}$ (i.e., $\mathbf{H}'$: $\mu \geq \mu_0$), to get the same result. But life gets rather more complicated were we to do that, as shown in Figure 5.2. Understanding why that is so is instructive for tests in general.

The thin curve in Figure 5.2 shows the probability of rejecting $\mathbf{H}$, for a 5% level test, as in Figure 5.1. As expected, the curve demonstrates a 5% probability of rejecting $\mathbf{H}$ when in fact it was (just) false (i.e., when $\Delta = 0$) and a probability of rejecting $\mathbf{H}$ of least 90% when $\Delta = 100\%$. That is, there is a high probability of detecting an important effect, were it to exist. In order to achieve the same result for a test of $\mathbf{H}'$, we would have to perform a 95% level test (as in the thick curve). Why 95%? Because when $\Delta = 0$ we want only a 5% probability of rejecting $\mathbf{H}$, which is a 95% probability of rejecting the complementary hypothesis $\mathbf{H}'$. Now the thick line's ordinate is the probability of *rejecting* $\mathbf{H}'$, but to compare results to the thin line we want the ordinate to be the probability of *accepting* $\mathbf{H}'$, because Pr(accept $\mathbf{H}'$) = Pr(reject $\mathbf{H}$). Accordingly, we also plot 1 − Pr(reject $\mathbf{H}'$) = Pr(accept $\mathbf{H}'$) in Figure 5.2. This is the OC curve for a test of $\mathbf{H}'$. So the test of $\mathbf{H}$ is depicted by the power curve and is also depicted by the OC curve for the test of $\mathbf{H}'$ (at the 95% level). To

---

[2]This case can apply when considering detection of time trends, in which case we want to detect if the magnitude of the trends exceeds one standard deviation of the **de-trended** and **de-seasonalized** water quality random variable.
[3]Equivalently, when $\Delta = 0^+$ (just greater than zero), $\beta = 1 - \alpha = 0.95$; when $\Delta = 1$ we want $\beta = 0.1$.

110 DETECTION

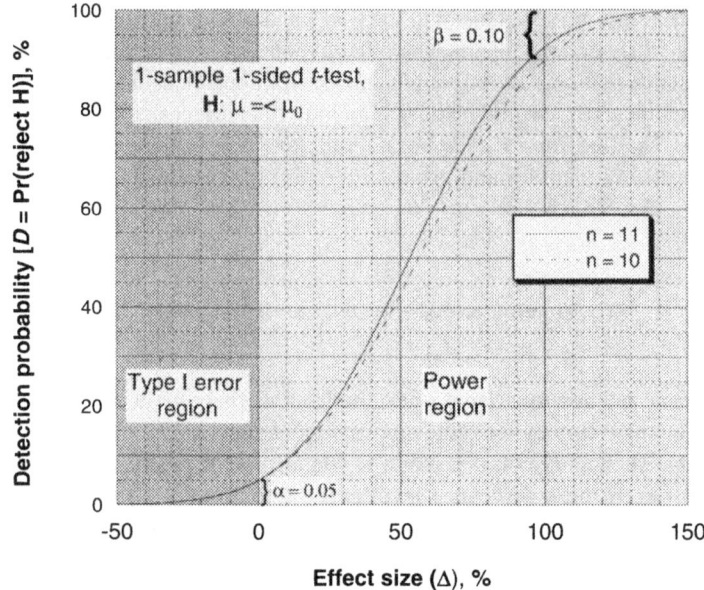

Fig. 5.1 Detection curves for two sample sizes for a one-sided variance-unknown 5% level test of the inferiority hypothesis **H**: $\mu \leq \mu_0$. This hypothesis is true in the "Type I error region" but false in the "Power region."

keep matters simple we avoid using OC curves for one-sided tests (and for point-null hypothesis tests)—but we cannot avoid doing so when examining equivalence tests.[4]

5.1.2 What burden of proof?

In the above example, let's assume that $\Delta = 1$ corresponds to an environmental standard (we are assuming that σ is constant). Then what we see in Figure 5.1 is the precautionary approach—there is only a small chance of failing to detect a breach of standard (when $\Delta > 100\%$). The Type II error risk, of "letting the guilty go free", or "slipping through the net," has been minimized. But this means that the Type I error risk, of "convicting the innocent" or "crying wolf," is magnified. For example, if Δ were in fact only 50%, there is still approximately a 45% chance of concluding that the standard has been breached.

If the Type I error risk is to be minimized then μ_0 should be made equal to the standard. It is then a simple matter to show that Type I errors are minimized. This shifts the burden of proof away from the environmental user, onto the environment itself.

So what burden of proof should be used? There is no statistical theory that answers that question. It is a matter of costs and benefits of these alternative postures; this is

---
[4] As discussed in Section 5.3.

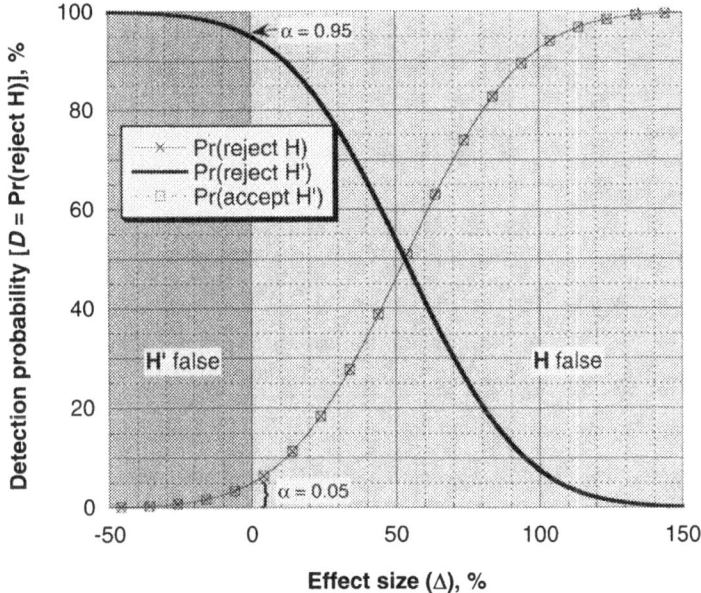

Fig. 5.2 Probabilities of accepting and rejecting complementary hypotheses **H** and **H**′ (one-sided, single sample, variance unknown) for testing the inferiority hypothesis at the 5% level, with $n = 11$ observations. The OC curve is the dashed line (hidden under the thin solid line, but marked by boxes).

a question of policy. We discuss this further when considering percentile standards (Chapter 8).

5.1.3 Decision rules and detection probabilities

The main calculation elements of the tests for the four cases of interest are given in Tables 5.2–5.4, for inferiority and superiority hypotheses, variance known and unknown, using the definitions given in Table 5.1.

Table 5.2 states the form of the hypotheses that may be tested, and their alternatives. Table 5.3 gives the test's decision rules and detection probabilities.[5] In this table: $Z_{1-\alpha}$ is the $100(1-\alpha)$%ile of the unit normal distribution; $G(x)$ is the cumulative unit normal probability, of getting a value no greater than x (as described in chapter 2.5.1); γ is the "noncentrality parameter"; t' is the noncentral t variable; and $t_{1-\alpha,f}$ is the critical abscissa of Student's t-distribution cutting off a right-tail area of α with f

---

[5] Note that in the known-variance case the test is UMP (Uniformly Most Powerful—Mood & Graybill 1963, Section 12.4 [230]), and in the unknown-variance case the test is UMPU (Uniformly Most Powerful Unbiased)—Stuart *et al.* 1999, p. 225 [304]). These are highly desirable properties, generally implying that these tests are the best that can be obtained for these hypotheses.

112 DETECTION

Table 5.1 Definition of population variables

Term	Meaning
One sample	
μ	Population mean
μ_0	"Standard value" of μ (usually zero)
Two samples	
δ	$\mu_x - \mu_y$, the difference between population means
δ_0	"Standard value" of δ (usually zero)

degrees of freedom.[6] Table 5.4 gives associated information necessary to implement the decision rules and compute the detection probabilities (the same formulae also apply to point-null hypothesis tests). In this table we have these additional terms: Δ is the effect size; σ is the true standard deviation; S_e is the standard error of a single mean; n is the sample size; $\bar{X}$ is the sample mean; $S_{e,\bar{x}-\bar{y}}$ is the standard error of the difference in means; d is the difference between sample means ($\bar{X} - \mu_0$ in the one-sample case, $\bar{X} - \bar{Y}$ in the two-sample case); T is the test statistic; f is the degrees of freedom. Further explanation, and the development of the tables, is given in Sections 6.4.1 and 6.4.2.

Table 5.2 Inferiority and superiority hypotheses for one-sided tests

	Inferiority	Superiority
One sample	**H**: $\mu \leq \mu_0$; **K**: $\mu > \mu_0$	**H**: $\mu \geq \mu_0$; **K**: $\mu < \mu_0$
Two samples	**H**: $\delta \leq \delta_0$; **K**: $\delta > \delta_0$	**H**: $\delta \geq \delta_0$; **K**: $\delta < \delta_0$

Example—Houhou inferiority test, variance unknown From sampling on seven occasions of taxonomic richness upstream (x) and downstream (y) of a mining discharge to the Houhou Stream (on the West Coast of New Zealand's South Island) we have the following upstream/downstream data: 13, 11; 15, 10; 16, 11; 15, 12; 19, 11; 14, 12; 20, 14 (from Quinn *et al.* 1992 [266]; seven replicates were taken at each site on each occasion). We calculate $\bar{X} = 16.000$, $S_x = 2.582$; $\bar{Y} = 11.571$,

---

[6]When the hypothesis is only just true (e.g., when $\mu = \mu_0$), the test statistic follows Student's t-distribution. But, when calculating the detection probability, we need to know the distribution of that statistic when $\mu \neq \mu_0$. This is the *noncentral* t-distribution (denoted as t'). The t-distribution is a special case of the t'-distribution, namely the t-distribution with zero noncentrality parameter, as discussed further in Section 6.3.1.

Table 5.3 Decision rules ("Reject **H**") and detection probabilities [D, Equation (5.1)] for one-sided inferiority and superiority tests[a,b]

Hypothesis	Variance known	Variance unknown
Inferiority	Reject **H** if $T > Z_{1-\alpha}$ $D = G(-Z_{1-\alpha} + \gamma)$	Reject **H** if $T > t_{1-\alpha,f}$ $D = \Pr(t' < -t_{1-\alpha,f}| - \gamma)$
Superiority	Reject **H** if $T < -Z_{1-\alpha}$ $D = G(-Z_{1-\alpha} - \gamma)$	Reject **H** if $T < -t_{1-\alpha,f}$ $D = \Pr(t' < -t_{1-\alpha,f}|\gamma)$

[a] T and γ are defined on Table 5.4; $G(x)$ is the cumulative unit normal distribution; t' is the noncentral t-variate.
[b] Note that the test can be performed with unbalanced sample sizes ($n_x \neq n_y$), using Equation (3.7) for the standard error (in the definition of T in Table 5.4), and setting $f = n_x + n_y - 2$.

$S_y = 1.272$. It is expected that mining would have caused a decrease in taxonomic richness—a deleterious impact. Furthermore, there would be no concern or consequence if there were to be an increase (i.e., a higher taxonomic richness at the downstream site). Therefore, from Section 4.3.1, we should perform a one-sided test. We will do this at the 5% significance level. Also, from Section 5.1.1, we should test the inferiority hypothesis (**H**: $\mu_x \leq \mu_y$), because it would be false were an important deleterious effect to be present. That is, if **H** were to be false then the taxonomic richness at the upstream site (x) would be greater than at the downstream site (y). Treating the data as two independent samples we obtain the pooled standard deviation as $S_p = \sqrt{(2.582^2 + 1.272^2)/2} = 2.035$ [using Equation (3.8)] and the standard error of the difference in means as $S_{e,\bar{x}-\bar{y}} = 2.035\sqrt{2/7} = 1.088$ [using Equation (3.9)]. The test value of the difference in means (δ_0) should be taken as zero—we are interested only in a decline or fall in richness, not its rise above a nonzero critical value (or fall below it). So the test statistic (from Table 5.4) is $T = d/S_{e,\bar{x}-\bar{y}} = 4.429/1.088 = 4.071$. Applying the decision rule in Table 5.3 we find that we should reject **H** if $T > t_{1-\alpha,f}$. Now, also from Table 5.4, $f = 2(n-1) = 12$. Consulting the t-distribution table (Appendix A.2.2) we find that $t_{0.95,12} = 1.782$. So the decision rule requires us to reject the inferiority hypothesis. That is, the upstream site's taxonomic richness is most probably greater than that at the downstream site. (Problem 5.2 repeats this analysis for treating the data as paired samples.)

5.1.4 Detection regions

To keep matters simple, and without loss of generality, we will now consider a one sample case with known variance, with a zero standard value ($\mu_0 = 0$). Tables 5.2 to 5.4 show that detection will occur when testing the inferiority hypothesis only if $T > Z_{1-\alpha}$; that is, when $S_e < \bar{X}/Z_{1-\alpha}$. Accordingly the line $S_e = \bar{X}/Z_{1-\alpha}$ separates the detection and non-detection regions, as shown for a 5% level test in

Table 5.4 Effect size, noncentrality parameter, standard error, test statistic, and degrees of freedom for one-sided inferiority and superiority tests and for point-null hypothesis tests

Effect size, noncentrality parameter	Standard error, test statistic, degrees of freedom	
	Variance known	Variance unknown
One sample		
$\Delta = \dfrac{\mu - \mu_0}{\sigma}$	$S_e = \dfrac{\sigma}{\sqrt{n}}$	$S_e = \dfrac{S}{\sqrt{n}}$
$\gamma = \Delta\sqrt{n}$	$T = \dfrac{\bar{X} - \mu_0}{S_e}$	$T = \dfrac{\bar{X} - \mu_0}{S_e}$
		$f = n - 1$
Two independent samples[a]		
$\Delta = \dfrac{\delta - \delta_0}{\sigma}$	$S_{e,\bar{x}-\bar{y}} = \dfrac{\sigma}{\sqrt{n/2}}$	$S_{e,\bar{x}-\bar{y}} = \dfrac{S_p}{\sqrt{n/2}}$
$\gamma = \Delta\sqrt{n/2}$	$T = \dfrac{d - \delta_0}{S_{e,\bar{x}-\bar{y}}}$	$T = \dfrac{d - \delta_0}{S_{e,\bar{x}-\bar{y}}}$
		$f = 2(n - 1)$

[a] S_p is the pooled standard deviation, calculated from Equation (3.6). For n paired samples, the standard errors are $S_{e,\overline{x-y}} = \dfrac{\sigma}{\sqrt{n}}$ (for variance known) and $S_{e,\overline{x-y}} = \dfrac{S_d}{\sqrt{n}}$ (for variance unknown), where S_d is the standard deviation of the differences. Also, the degrees of freedom becomes $f = n - 1$.

Figure 5.3(a). In a like manner, when testing the superiority hypothesis, the line $S_e = -\bar{X}/Z_{1-\alpha}$ separates these two regions, as also shown in Figure 5.3(b). Similar results obtain for the variance-unknown and two-sample cases.

From Tables 5.2 to 5.4, we note that the standard error S_e on the vertical axes of these graphs is proportional to the sample standard deviation (pooled, in the two-sample case), and to the reciprocal square root of the sample size. Applying this to the above figures we see that for either test, smaller differences between the sample mean and the standard value will be detected as the sample variability decreases or as the sample size increases (either of which cause S_e to decrease).

5.1.5 Non-normality and unequal variances

When it seems plausible that data have been derived from normal distributions, but with rather unequal (unknown) variances, we have the "Behrens–Fisher" problem. Some parametric procedures are available for this problem. Welch's test (sometimes called the Satterthwaite approximation) can be used (e.g., Snedecor & Cochran 1980 [291], Zar 1996 [349]). This has the attraction of requiring only an adjustment to the

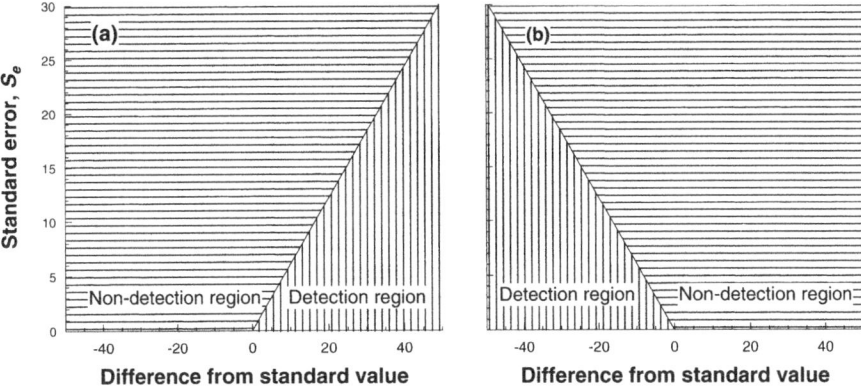

Fig. 5.3 Detection regions for one-sided variance-known tests at the 5% significance level: (a) testing the inferiority hypothesis, (b) testing the superiority hypothesis.

degrees of freedom of the critical t-value in Table 5.3, and so the general features of the decision rules and detection regions discussed above are affected only a little. This also holds for tests of point-null hypotheses.

If there is no concern about unequal variances but the data indicate substantial non-normality, a simple randomization procedure is available (Manly 2004 [199]). However, when there is concern about both non-normality and differences in variances, it may not be possible have a straightforward procedure to reliably test for differences in means. A computationally-intensive scheme may be the only parametric option in this case (Manly & Francis 2002 [200], Manly 2004 [199]).

5.2 POINT-NULL HYPOTHESIS TESTS

These tests postulate that the difference between a mean and a standard value (one sample case) or between the difference in two populations means and a standard difference value (two-sample case) are exactly zero. That is, the point-null hypothesis ($\mathbf{H_0}$) postulates an equality. Such tests are always two-sided.

5.2.1 Defining detection probability

A point-null hypothesis test postulates that $\mathbf{H_0}$ is true. As discussed in Chapter 4, we are not free to consider the other side of the coin (as we may with one-sided tests and equivalence tests)—that $\mathbf{H_0}$ may be false and $\mathbf{K}$ is true— the chance of $\mathbf{H_0}$ being true is vanishingly small, so we would never reject $\mathbf{K}$)! Accordingly for point-null hypothesis tests, the detection probability is the same as statistical power. That is,

$$\text{detection probability} = D = \Pr(\text{reject } \mathbf{H_0}). \qquad (5.2)$$

Example Consider an unknown-variance one-sample case where we wish our test to have at least 90% probability of detecting a unit effect size, *in either direction* [i.e., $\Delta = \pm|\mu - \mu_0|/\sigma = 1$], assuming that such an effect, if present, would also be of practical importance. We also want to keep the significance level at the traditional 5% level; that is, there is to be only a 5% chance of "detection" when in fact the true mean was at some unimportant value, μ_0. So we want $\alpha = 0.05$ at $\Delta = 0$ and detection probability $D \geq 0.90$ at $\Delta = \pm 1$. From Figure 5.4 (or using *The Detection Calculator*) we find we need to have a sample size of $n = 13$ to achieve these requirements. [Actually, with $n = 13$ data the detection probability is 91.07% (> 90%), but it is insufficient (88.29%) with only $n = 12$ data—as the figure shows.]

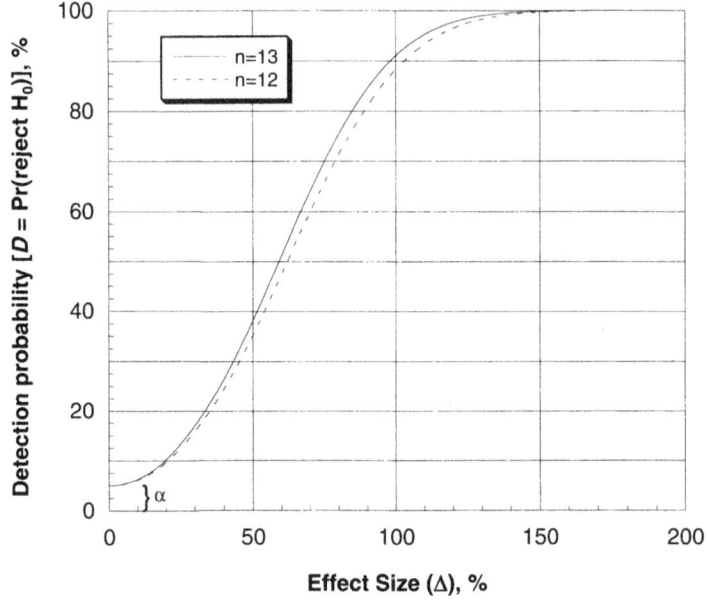

Fig. 5.4 Detection curve for a one-sample variance-unknown 5% level point-null hypothesis test.

5.2.2 Power depends on sample size

The detection curve gets steeper as the sample size (n) increases by one (Figure 5.4). For larger increases the change in steepness is more dramatic (Figure 5.5).

Figure 5.5 shows the detection probability for either sign of the effect size, to demonstrate that point-null hypothesis tests tend to detect ever-smaller effect sizes as the sample size is increased—so much so that trivial effects could be routinely detected. For this reason some writers have suggested that it is a mistake to take too many samples when intending to perform point-null hypothesis tests, because trivial differences may be routinely detected (e.g., Fleiss 1981 [95]). This reinforces the point made earlier about p-values for point-null hypothesis tests (Section 4.3.4); a

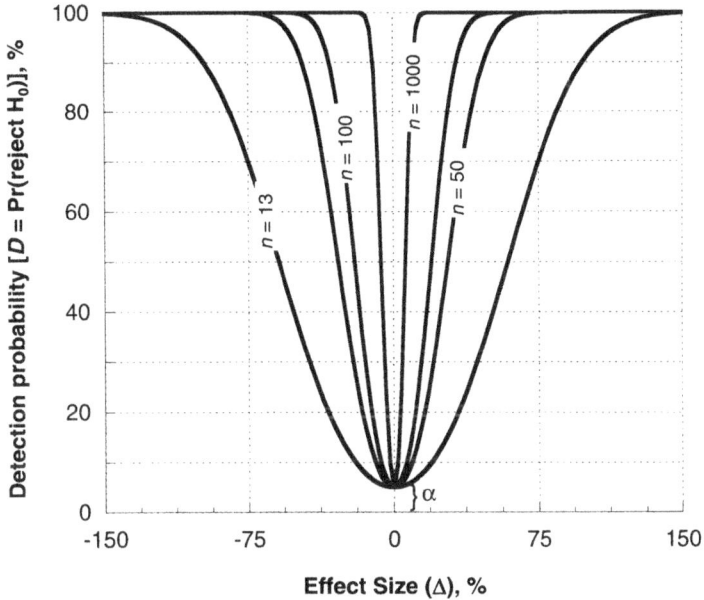

Fig. 5.5 Effect of increasing sample size on detection curves (power) of a one-sample unknown-variance 5% level point-null hypothesis test. The Type I error "region" is the line $\Delta = 0$ (where $\mathbf{H_0}$ is true) and the power region is everywhere else (where $\mathbf{H_0}$ is false).

point-hypothesis that is assumed to be true in order to calculate a p-value will eventually be found wanting if there are enough data, precisely because it is false.[7] This feature is strongly related to the width of the 95% confidence intervals for the sample sizes used on Figure 5.5 as shown below.[8]

Table 5.5 Sample size and confidence interval width

Sample size n	Width of 95% confidence interval
13	$1.2087s$
50	$0.5685s$
100	$0.3986s$
1000	$0.1241s$

---

[7]Rather than counseling against the collection of "too many" samples, a better strategy may be to test a more realistic hypothesis in terms of a meaningful difference.
[8]The confidence limits on the mean for a one-sample case (see Chapter 3) are $\bar{X} \pm t_{1-\frac{\alpha}{2}, n-1} S/\sqrt{n}$ and so the width is $2t_{1-\frac{\alpha}{2}, n-1} S/\sqrt{n}$. See also the discussion of confidence interval width being dependent on sample size (page 63).

5.2.3 Decision rules and detection probabilities

The main mathematical features of the tests for the four cases of interest are given in Tables 5.6 and 5.7.[9] Table 5.7 gives the test's decision rules and detection probabilities.[10]

Table 5.6 Hypotheses for point-null hypothesis tests

One sample	Two samples
H₀: $\mu = \mu_0$; **K**: $\mu \neq \mu_0$	**H₀**: $\delta = \delta_0$; **K**: $\delta \neq \delta_0$

Table 5.7 Decision rules ("Reject **H₀**") and detection probabilities (D) for point-null hypothesis tests[a,b]

Variance	Decision rule and detection probability
Known	Reject **H₀** if $\|T\| > Z_{1-\frac{\alpha}{2}}$
	$D = G(-Z_{1-\frac{\alpha}{2}} + \gamma) + G(-Z_{1-\frac{\alpha}{2}} - \gamma)$
Unknown	Reject **H₀** if $\|T\| > t_{1-\frac{\alpha}{2}, f}$
	$D = \Pr(t' < -t_{1-\frac{\alpha}{2}, f} \| -\gamma) + \Pr(t' < -t_{1-\frac{\alpha}{2}, f} \| \gamma)$

[a] T and γ are as defined for one-sided tests, on Table 5.4; $G(x)$ is the cumulative unit normal distribution; t' is the noncentral t-variate.
[b] Note that the test can be performed with unbalanced sample sizes ($n_x \neq n_y$), using Equation (3.7) for the standard error (in the definition of T in Table 5.4), and setting $f = n_x + n_y - 2$.

Example—Houhou point-null test, variance unknown Continuing the Houhou Stream example given in Section 5.1.3, suppose a change in taxonomic richness between upstream and downstream sites in *either* direction would be of practical environmental concern. In such situations it is common to advocate that a point-null hypothesis test should be conducted, at the 5% significance level, in which case the hypothesis to be tested is **H**: $\mu_x - \mu_y = 0$. Using the test's decision rule (Table 5.7)

---

[9] The definitions of effect size, noncentrality parameter, standard error, test statistic and degrees of freedom for one-sided tests also apply to point-null hypothesis tests, hence we use Table 5.4 for those.
[10] Note too that these tests are UMPU (uniformly most powerful unbiased, Stuart *et al.* 1999, p. 225 [304]), even when the variance is unknown. This is a highly desirable property, implying that these tests are the best that can be obtained for these hypotheses.

we find that **H** should be rejected if $|T| > t_{0.975,12}$. Indeed, **H** is rejected, because $|T| = 4.071$ and $t_{0.975,12} = 2.179$ (from the Student's t-distribution table in Table A.2.2). A statistically significant result can now be claimed. Just what that actually means is a matter we will return to in Section 5.5.

5.2.4 Detection regions

As in Section 5.1.4, we keep matters simple, without loss of generality, by considering a one sample case with known variance and with a zero standard value ($\mu_0 = 0$). Table 5.7 shows that detection will occur when testing the point-null hypothesis only if $|T| > Z_{1-\frac{\alpha}{2}}$; that is, when $S_e < -\bar{X}/Z_{1-\frac{\alpha}{2}}$ or $S_e > \bar{X}/Z_{1-\frac{\alpha}{2}}$. Accordingly the lines $S_e = \pm \bar{d}/Z_{1-\frac{\alpha}{2}}$ separate the detection and nondetection regions, as shown for a 5% level test on Figure 5.6. Similar results obtain for the variance-unknown case.

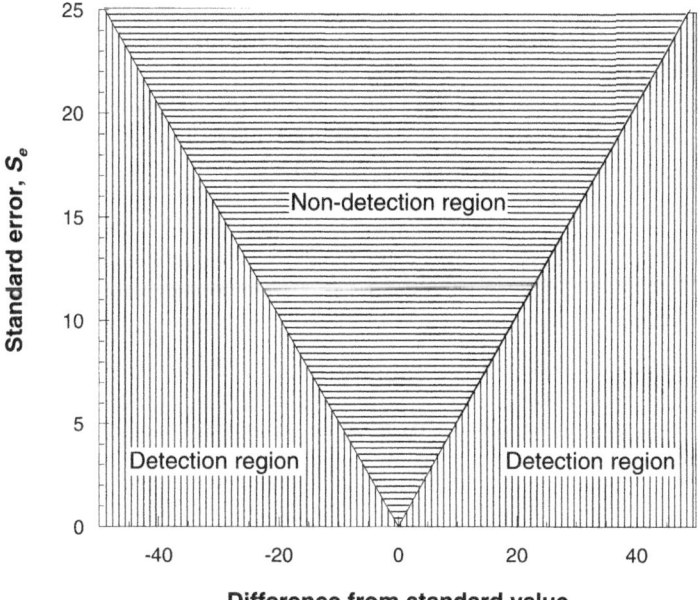

Fig. 5.6 Detection region for the variance-known test of the point-null hypothesis at the 5% significance level.

From the above tables we note that the standard error S_e on the vertical axis of this graph is proportional to the sample standard deviation and to the reciprocal square root of the sample size. Applying this to Figure 5.6, we see that smaller differences between the sample mean and the standard value will be detected as the sample variability decreases or as the sample size increases (either of which cause S_e to decrease).[11] That is, at large sample variability or small sample size important

---
[11] These observations parallel those made in respect of one-sided tests, in Section 5.1.4.

differences may not be detected, whereas at small sample variability and large sample size trivial differences will be detected. This means that, depending on the sample size the point-null hypothesis test swings from permissive to ultra-precautionary. As a consequence, merely stating that a "statistically significant" result has been obtained may not be very helpful. We discuss this further in Section 5.5.

5.3 EQUIVALENCE TESTS

We must first distinguish two approaches to test equivalence. The first is where the hypothesis to be tested postulates that the difference lies *beyond* the equivalence interval. We denote this hypothesis by H_i, with the subscript denoting "inequivalence." Tests using this form of the hypothesis are often referred to as bioequivalence tests. Their essential nature is *precautionary*, because equivalence is only inferred if H_i can be rejected beyond reasonable doubt (i.e., data are "convincing"). This is also called the **proof of safety** approach, commonly used in clinical trials for new drug formulations.

The second is where **H** postulates that the difference lies *within* the equivalence interval. We use H_e for this hypothesis, with the subscript denoting "equivalence." Tests using this form of the hypothesis are much less common in the statistical literature. Their essential nature is *permissive*, in that they assume equivalence unless data are convincing to the contrary. Inequivalence is only inferred if H_e can be rejected beyond reasonable doubt. This is also called the **proof of hazard** approach, used when the "producer's risk" is minimized.

5.3.1 Defining detection probability

We detect an important difference by rejecting H_e, or by accepting H_i. This means that the detection probability is given by the power curve for the test of H_e, but as the operating characteristic curve (OC curve) for the test of H_i.[12] Therefore we have

$$\text{detection probability} = D = \Pr(\text{reject } H_e) \text{ or } \Pr(\text{accept } H_i). \quad (5.3)$$

Consider an unknown-variance two-sample case where we test the permissive equivalence hypothesis H_e. We define the equivalence interval limits at $\Delta = \pm\frac{1}{2}$, meaning that a difference up to half the true standard deviation is held to be unimportant; that is, within that interval the populations' means are regarded as "equivalent." Let's denote the true difference between the populations' means by δ, and the midpoint of the equivalence interval as δ_0. We wish our test to have at least 90% power of detecting a unit effect size [i.e., $\Delta = \pm(\delta - \delta_0)/\sigma = 1$], assuming that such an effect, if present, would also be of practical importance (the Power region lies beyond the equivalence interval). We also want to keep the significance level at the

---

[12] Much of the equivalence test literature reports the power curve for the test of H_i, not the OC curve, but in order to compare the two types of equivalence test in terms of detection it is better to depict their performance using the OC curve for the test of H_i and the power curve for the test of H_e.

traditional 5% level; that is, there is to be only a 5% chance of "detection" when in fact the true mean was at the edge of the equivalence region. Note that this level is to apply at both edges of the equivalence interval—this is a two-sided procedure. So we want $\alpha = 0.05$ at $\Delta = \pm\frac{1}{2}$ and detection probability $D \geq 0.90$ at $\Delta = \pm 1$. From Figure 5.7 (or, better still, using *The Detection Calculator*) we find we need to have a sample size of $n = 70$ to achieve these requirements. (Actually, with $n = 70$ data the detection probability is 90.27% (>90%), but it is insufficient (89.93%) with only $n = 69$ data—see the figure's insert.)

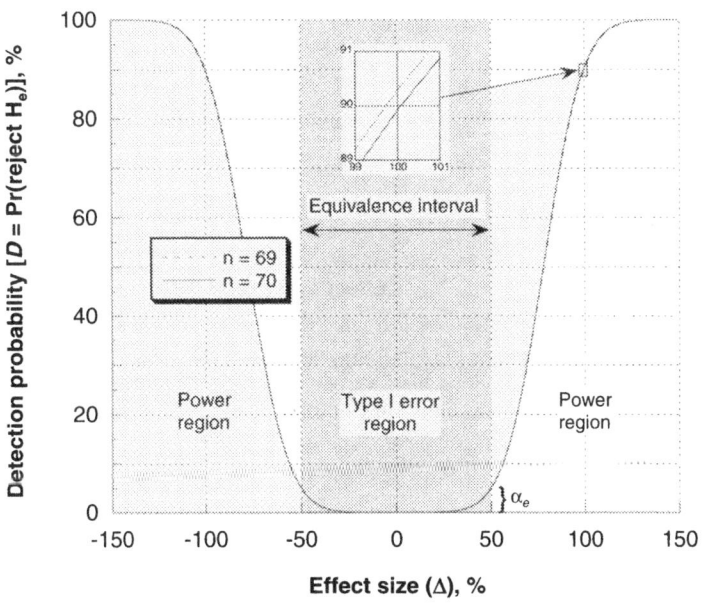

Fig. 5.7 Detection probability (variance unknown, 5% significance level) for testing the equivalence hypothesis (the proof of hazard approach) is a power curve. $\mathbf{H_e}$ is false in the "Power region," outside of the equivalence interval. The test's Type I error risk is α_e.

Now consider the unknown-variance case where we test the precautionary inequivalence hypothesis $\mathbf{H_i}$, again using $n = 70$. In this case the tested hypothesis is false somewhere *inside* the equivalence interval. Figure 5.8 shows this case. It also shows what happens with larger and smaller sample sizes, both for testing $\mathbf{H_e}$ and for testing $\mathbf{H_i}$.

Considering first the test of $\mathbf{H_e}$, we see that the effect of increasing the sample size is to bring the detection curve ever closer to the edges of the equivalence interval. It cannot shrink within the interval however, as does a test of the point-null hypothesis—which can be seen as a test of an interval with zero width.

The detection curves for the test of $\mathbf{H_i}$ show that equivalence will only be regularly inferred when the difference is some way *within* the equivalence interval, where the detection probability is low. This does not happen with only 20 samples: the detection probability within the equivalence interval is very high, even if there is no difference

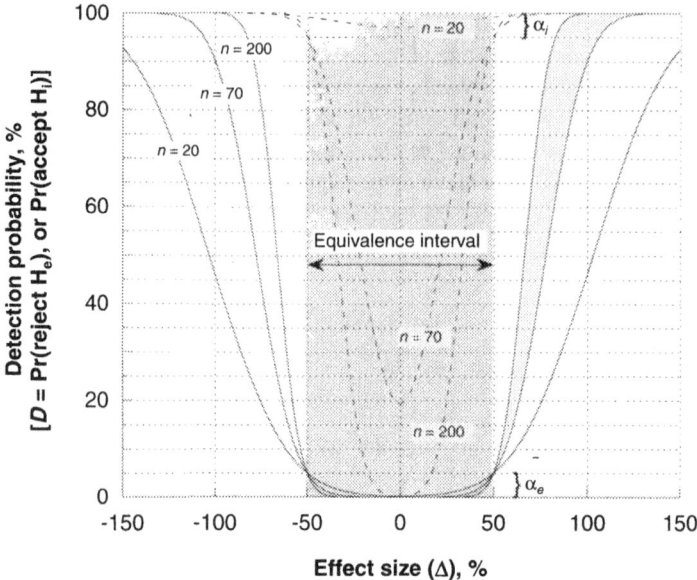

Fig. 5.8 Detection probability (variance unknown, 5% significance level) for testing the equivalence hypothesis $\mathbf{H_e}$ (the proof of hazard approach) and the inequivalence hypothesis $\mathbf{H_i}$ (the proof of safety approach). Solid lines are power curves, the probability of rejecting $\mathbf{H_e}$; the dashed lines are OC curves, the probability of accepting $\mathbf{H_i}$. The Type I error risks for each test are denoted by α_e and α_i, respectively.

at all (where $\Delta = 0$)—demonstrating that proof of safety is harder than proof of hazard (as noted by Bross 1985 [34]). Note too that the $n = 20$ line does not maintain the 5% Type I error level at the edge of the equivalence interval (and in fact the detection probability for the test of $\mathbf{H_e}$ with $n = 20$ is very slightly above 5%). That is because there are no "UMPU" interval tests of normal distributions when the variance is unknown.[13] As a result, such tests do not always maintain their "size" (i.e., guaranteeing that the values of α_e and α_i are maintained at the edge of the equivalence interval). If the variance is known then a UMP test of $\mathbf{H_i}$ and a UMPU test of $\mathbf{H_e}$ can be derived (see Section 6.4.5). Figure 5.9 repeats the cases shown in Figure 5.8, but for known variance, using these UMP and UMPU tests. The difference in detection probabilities between unknown (Figure 5.8) and known (Figure 5.9) variances are minimal, except for the $n = 20$ case for testing $\mathbf{H_i}$. These tests are all of size α.

[13]The UMP and UMPU properties (uniformly most powerful, and uniformly most powerful unbiased—a subset of UMP) are highly desirable; tests that have them are the best that can be derived. Stuart et al. (1999, page 225 [304]) show that there are no UMPU interval tests for the means of normal distributions with unknown variance (and therefore no UMP tests either). The particular tests used in this text adopt the two one-sided test "TOST" procedure for tests of $\mathbf{H_i}$ (Schuirmann 1987 [283]) and its complement for tests of $\mathbf{H_e}$ (McBride 1999 [205]). These tests have generally desirable properties over a wide range of effect sizes, although more specialized tests can outperform them in some regions.

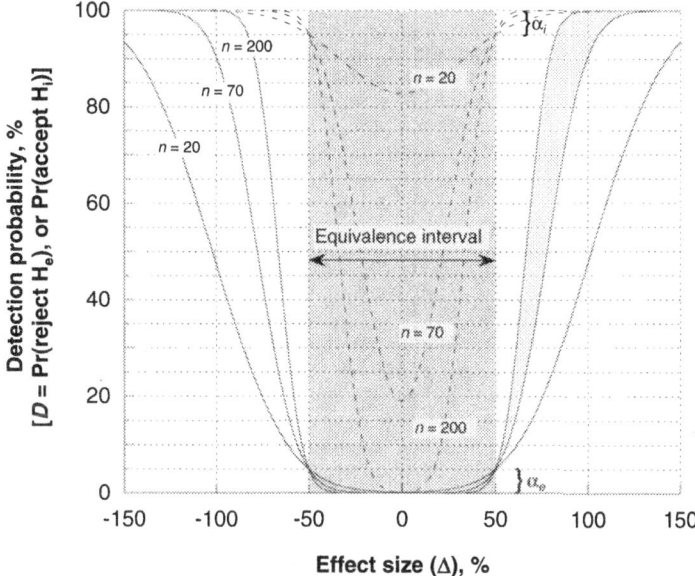

Fig. 5.9 Detection probability (variance known, 5% significance level) for testing the equivalence hypothesis H_e (the proof of hazard approach) and the inequivalence hypothesis H_i (the proof of safety approach). Solid lines refer to testing H_e, dashed lines refer to testing H_i. The Type I error risks for each test are denoted by α_e and α_i, respectively.

5.3.2 Decision rules and detection probabilities

The principal mathematical features of the tests for the four cases of interest are given in Tables 5.8 to 5.10. Table 5.8 gives the explicit form of the hypotheses that can be tested while Table 5.9 gives the tests' decision rules and detection probabilities. Table 5.10 gives associated information necessary implement the decision rule and to compute the detection probabilities. Figure 5.10 interprets these details in terms of symmetric confidence intervals, using the TOST ("two one-sided test") procedure. As explained in Section 6.4.7 these are $100(1 - 2\alpha)\%$ confidence intervals, *not* $100(1 - \alpha)\%$ confidence intervals, but the test is at level α.

Example—Houhou equivalence test, variance unknown Continuing again the Houhou Stream example given in Section 5.1.3, an ecologist offered an opinion that were taxonomic richness at the downstream site to differ from that at the upstream site by more than 20%—in either direction—an environmental impact could be inferred as a result of the mining discharge between the two sites. This means that we should now carry out equivalence tests. To do so, note that, as before, we have the standard error of the difference in means $S_{e,\bar{x}-\bar{y}} = 1.099$. The lower and upper bounds of the equivalence interval are $\delta_L = -0.2 \times \bar{X} = -0.2 \times 16.0 = -3.2$ and $\delta_U = 3.2$. The left test statistic (Table 5.10) is therefore $T_l = (d - \delta_L)/S_{e,\bar{x}-\bar{y}} = (4.429 +$

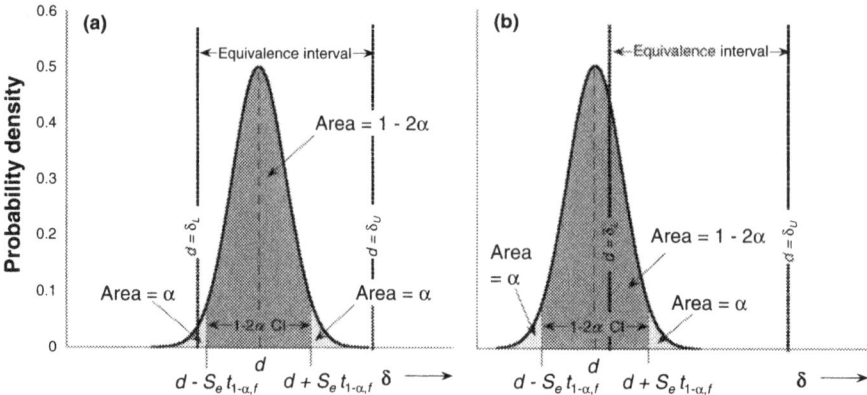

Fig. 5.10 Declarations of "equivalence" when using the two-sample two one-sided (TOST) procedure to test: (a) the inequivalence hypothesis (the $1-2\alpha$ confidence interval is completely contained by the equivalence interval), (b) the equivalence hypothesis (the $1-2\alpha$ confidence interval intrudes into the equivalence interval).

Table 5.8 Equivalence and inequivalence hypotheses

	Equivalence	**Inequivalence**
One sample	$\mathbf{H_e}$: $\mu_L \leq \mu \leq \mu_U$ $\mathbf{K_e}$: $\mu < \mu_L$ or $\mu > \mu_U$	$\mathbf{H_i}$: $\mu < \mu_L$ or $\mu > \mu_U$ $\mathbf{K_i}$: $\mu_L \leq \mu \leq \mu_U$
Two samples	$\mathbf{H_e}$: $\delta_L \leq \delta \leq \delta_U$ $\mathbf{K_e}$: $\delta < \delta_L$ or $\delta > \delta_U$	$\mathbf{H_i}$: $\delta < \delta_L$ or $\delta > \delta_U$ $\mathbf{K_i}$: $\delta_L \leq \delta \leq \delta_U$

Note that a test on ratios of means can be translated to a test on differences of means by taking logarithms of those ratios (Berger & Hsu 1996 [18]). Wellek (2003 [340]) argues that equivalence tests are best posed in terms of ratios of means.

$3.2)/1.088 = 7.012$. Similarly the right test statistic takes the value $T_r = (4.429 - 3.2)/1.088 = 1.130$. (Note that $T_l > T_r$—this always occurs, see the footnote to Table 5.10.) Now, from Table 5.9 we need to make comparisons with $t_{0.95,12} = 1.782$. We would reject the equivalence hypothesis if *either* $T_l < -1.782$ or $T_r > 1.782$ (using the Student's t-distribution table in Appendix A.2.2). Neither of these conditions holds true, so we would not reject $\mathbf{H_e}$. On the other hand, if we were testing the inequivalence hypothesis $\mathbf{H_i}$ we would reject it only if $T_l > 1.782$ *and* if $T_r < -1.782$. Only the first of these two required conditions actually holds, so $\mathbf{H_i}$ would not be rejected either. We will discuss these apparently conflicting findings in Section 5.5.

Table 5.9 Decision rules (Reject H_e) and (Reject H_i) and detection probabilities (D = Reject H_e and D = Accept H_i) for equivalence hypothesis tests[a,b]

Variance	Decision rule and detection probability

Testing the equivalence hypothesis, H_e

Known   Reject H_e if $|T| > c_e$, where $G(c_e + \gamma_U) - G(-c_e + \gamma_U) = 1 - \alpha$
$$D = G(-c_e + \gamma) + G(-c_e - \gamma)$$

Unknown[c]   Reject H_e if $T_l < -t_{1-\alpha,f}$ or $T_r > t_{1-\alpha,f}$
$$D = \Pr(t' \leq -t_{1-\alpha,f}|\gamma + \gamma_U) + \Pr(t' \leq -t_{1-\alpha,f}|\gamma_U - \gamma)$$

Testing the inequivalence hypothesis, H_i

Known   Reject H_i if $|T| \leq c_i$, where $G(c_i + \gamma_U) - G(-c_i + \gamma_U) = \alpha$
$$D = G(-c_i + \gamma) + G(-c_i - \gamma)$$

Unknown[d,e]   Reject H_i if $T_l \geq t_{1-\alpha,f}$ and $T_r \leq -t_{1-\alpha,f}$
$$D = 1 - Q_f(-t_{1-\alpha,f}, \gamma - \gamma_U; 0, R) + Q_f(t_{1-\alpha,f}, \gamma + \gamma_U; 0, R),$$
where $R = \frac{\Delta_U}{t_{1-\alpha,f}}\sqrt{\frac{nf}{2}}$

[a] T, T_l, T_r, γ, and γ_U are defined on Table 5.10; $G(x)$ is the cumulative unit normal distribution; t' is the noncentral t-variate; c_e and c_i are critical test statistic values, obtained by solving the variance-known equations given in the table: c_e is always greater than c_i.
[b] Note that the test can be performed with unbalanced sample sizes ($n_x \neq n_y$), using Equation (3.7) for the standard error (in the definition of T in Table 5.10), and setting $f = n_x + n_y - 2$.
[c] Equivalently, rejection occurs if $T < -t_{1-\alpha,f} - \gamma_U$ or $T > t_{1-\alpha,f} + \gamma_U$.
[d] Q_f is the incomplete noncentral t-distribution function of Owen (1965 [250]), for which an algorithm is given in Section 6.4.9.
[e] Equivalently, rejection occurs if $T \geq t_{1-\alpha,f} - \gamma_U$ and $T \leq -t_{1-\alpha,f} + \gamma_U$.

5.3.3 Equivalence tests versus point-null tests

We have noted the swinging burden of proof for point-null tests, compared to their equivalence counterparts. Figure 5.11 demonstrates this graphically. At "small" sample size ($n = 20$) and an equivalence interval of $\Delta = (-\frac{1}{2}, \frac{1}{2})$, the point-null test is neither as permissive as the test of the equivalence hypothesis H_e nor as precautionary as the test of the inequivalence hypothesis H_i. However, at "large" sample size the point-null test is ultra-precautionary, having its detection probability everywhere higher than that of the precautionary test of the inequivalence hypothesis. At "huge" sample size ($n = 2000$) the point-null detection curve becomes closer to its limit at infinite sample size (see Problem 5.5).

Table 5.10 Hypotheses, effect size, noncentrality parameter, standard error, test statistic, and degrees of freedom for equivalence tests

Interval midpoint, effect size, noncentrality parameter[c]	Standard error, test statistic,[a,b] degrees of freedom	
	Variance known	Variance unknown

One sample

$$\mu_0 = \frac{\mu_L + \mu_U}{2} \qquad S_e = \frac{\sigma}{\sqrt{n}} \qquad S_e = \frac{S}{\sqrt{n}}$$

$$\Delta_{(L,U)} = \frac{\mu_{(L,U)} - \mu_0}{\sigma} \qquad T = \frac{\bar{X} - \mu_0}{S_e} \qquad T_{l,r} = \frac{\bar{X} - \mu_{L,U}}{S_e}$$

$$\gamma_{(L,U)} = \Delta_{(L,U)} \sqrt{n} \qquad\qquad\qquad f = n - 1$$

Two independent samples[d]

$$\delta_0 = \frac{\delta_L + \delta_U}{2} \qquad S_{e,\bar{x}-\bar{y}} = \frac{\sigma}{\sqrt{n/2}} \qquad S_{e,\bar{x}-\bar{y}} = \frac{S_p}{\sqrt{n/2}}$$

$$\Delta_{(L,U)} = \frac{\delta_{(L,U)} - \delta_0}{\sigma} \qquad T = \frac{d - \delta_0}{S_{e,\bar{x}-\bar{y}}} \qquad T_{l,r} = \frac{d - \delta_{L,U}}{S_{e,\bar{x}-\bar{y}}}$$

$$\gamma_{(L,U)} = \Delta_{(L,U)} \sqrt{n/2} \qquad\qquad\qquad f = 2(n-1)$$

[a] The test statistics have subscripts l and r, corresponding to the left and right boundaries of the equivalence interval, rather than lower (L) and upper (U) subscripts. The definition of these statistics means that T_l is always greater than T_r, so using L and U subscripts could be misleading.
[b] Note that T_l and T_r can be written in terms of T and γ_U only: $T_l = T + \gamma_U$ and $T_r = T - \gamma_U$.
[c] The "(L,U)" notation means that these subscripts are optional (e.g., $\Delta = \frac{\delta - \delta_0}{\sigma}$ and $\Delta_U = \frac{\delta_U - \delta_0}{\sigma}$). Note too that $\Delta_U = -\Delta_L$ and so $\gamma_U = -\gamma_L$. This term (γ_U) is a *partial* noncentrality parameter—in Table 5.9 we see that in the variance-unknown case there are two full noncentrality parameters: $\gamma - \gamma_U$ and $\gamma + \gamma_U$.
[d] S_p is the pooled standard deviation, calculated from Equation (3.6). See Problem 5.7 for the appropriate formula for paired samples.

5.3.4 Detection regions

Figures 5.12(a) and 5.12(b) show examples of the shape of these regions for equivalence tests.[14] Both show that the "window of uncertainty" increases with the magnitude of the standard error, where this window is the distance from the edge of the equivalence interval to the boundary of the relevant region. For example, for small standard errors the difference between sample means need not be far beyond the

---

[14] The formulae for the boundaries of these regions are developed in Section 6.4.

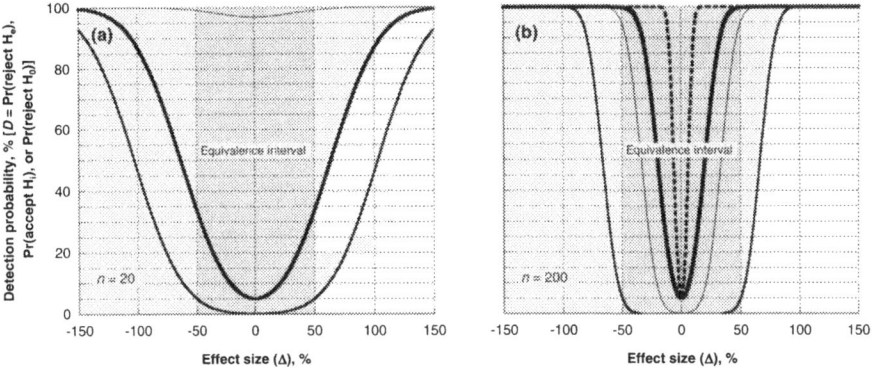

Fig. 5.11 Detection curves for the tests of H_i (thin line), H_e (medium line), and H_0 (thick line): (a) "small" sample size ($n = 20$); (b) "large" sample size ($n = 200$)—the dashed line is for the point-null test with a "huge" sample size ($n = 2000$). These are all 5%-level tests for two-sample cases with unknown variance.

equivalence interval for the equivalence hypothesis to be rejected, and inequivalence would be inferred. Likewise, the difference between sample means need not be far inside the equivalence interval for the inequivalence hypothesis to be rejected leading to an inference of equivalence. These distances (beyond and within the equivalence interval) increase as the standard error increases, as we should expect. That is, as the standard error increases the test of H_e becomes more permissive while the test of H_i becomes more precautionary. Indeed, if the variance is unknown, the inequivalence hypothesis *cannot be rejected* once the standard error reaches a particular value[15]— even if the measured difference in means is effectively zero. Once there are too few samples with too much variability, one cannot conclude that there is nothing to detect when adopting the precautionary approach. On the other hand, if the variance is known, the precautionary test is able to maintain its "size"; that is, the test can keep the Type I error risk *exactly* at α at the edge of the equivalence interval. This is manifest in Figure 5.12(b) by the "chimney on the teepee" shape, whereby the test of H_i can lead to a conclusion of equivalence, when the measured differences are sufficiently small.[16]

Perhaps the most notable feature of Figures 5.12(a)&(b) is shown by their duos. These show two areas where tests of either H_i or H_e will give the same result—detecting an important positive or negative difference, or failure to detect such a difference. But there is a region in between for which the test results will disagree. Furthermore, the width of that region increases with the standard error. This pattern is simply the manifestation of adopting a precautionary versus a permissive stance on the burden of proof. Furthermore, equivalence tests *cannot* exhibit a swinging

---

[15] This value is simply $\frac{\delta_U}{t_{1-\alpha \cdot f}}$ (see Problem 6.23).

[16] But see Problem 6.24 for the behavior of this test at larger values of S_e.

128 DETECTION

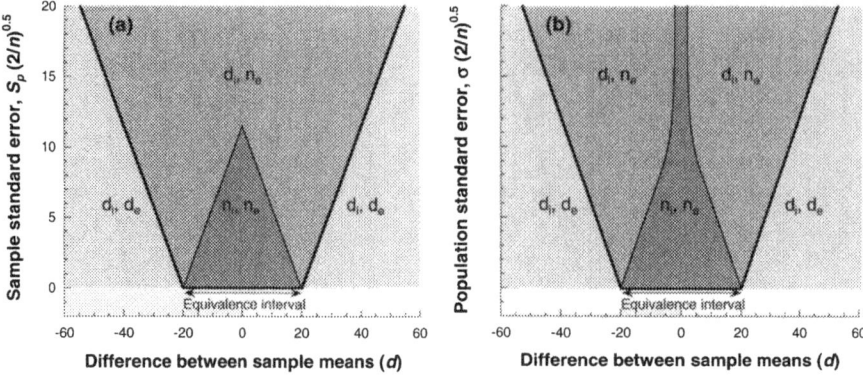

Fig. 5.12 Two-sample ($n = 10$) detection regions (at the 5% significance level) for testing the inequivalence hypothesis (thin boundary lines) and for testing the equivalence hypothesis (thick boundary lines): (a) variance unknown, (b) variance known. The equivalence interval limits are at $\delta_L = -20$ units and $\delta_U = +20$ units. The duos denote the inference to be drawn regarding detection from tests of $\mathbf{H_i}$ and $\mathbf{H_e}$, respectively. For example, (d_i, n_e) denotes detection from a test of $\mathbf{H_i}$ (i.e., $\mathbf{H_i}$ is accepted), but nondetection from a test of $\mathbf{H_e}$ (i.e., $\mathbf{H_e}$ is accepted). "S_p" is the pooled standard deviation of the differences between the two samples.

burden of proof; they are either permissive (when testing the equivalence hypothesis) or precautionary (when testing the inequivalence hypothesis). As a result, many of the difficulties that arise when using tests of a point-null hypothesis (e.g., Gray 1990 [126] Buhl-Mortensen 1999 [35]) are simply resolved by choosing the appropriate hypothesis to test. Testing the inequivalence hypothesis answers the call made by Dayton (1998 [66]) and Gerrodette *et al.* (2002 [115]), that in some cases the burden of proof should be "reversed."[17] Exhortations to consider statistical power when designing sampling programs (e.g., Fairweather 1991 [89]) are equally appropriate when adopting equivalence tests—just use the appropriate detection curves.

5.3.5 Equivalence versus the Power Approach

It may be thought that a suitably modified point-null hypothesis test can mimic the behavior of of equivalence tests. The idea is: "All we need to do is use a point-null test, ensuring that it has sufficient power to detect important differences—that is, anything beyond the equivalence interval." This is the Power Approach.[18] Figure 5.13 shows its detection region, for variance-unknown, superposed on Figure 5.12(a).

There are a number of notable features in this figure. First, as shown in Section 6.4.8, the width of the nondetection region for the Power Approach cannot extend

---

[17]These authors made this call with respect to fisheries management, but water quality impact issues can be contemplated for which a similar call can be made.

[18]This approach has been in routine use by drugs testing agencies (Schuirmann 1987 [283], Chow & Liu 1992 [46]). The formulae for the boundaries of these regions are developed in Chapter 6.4.8.

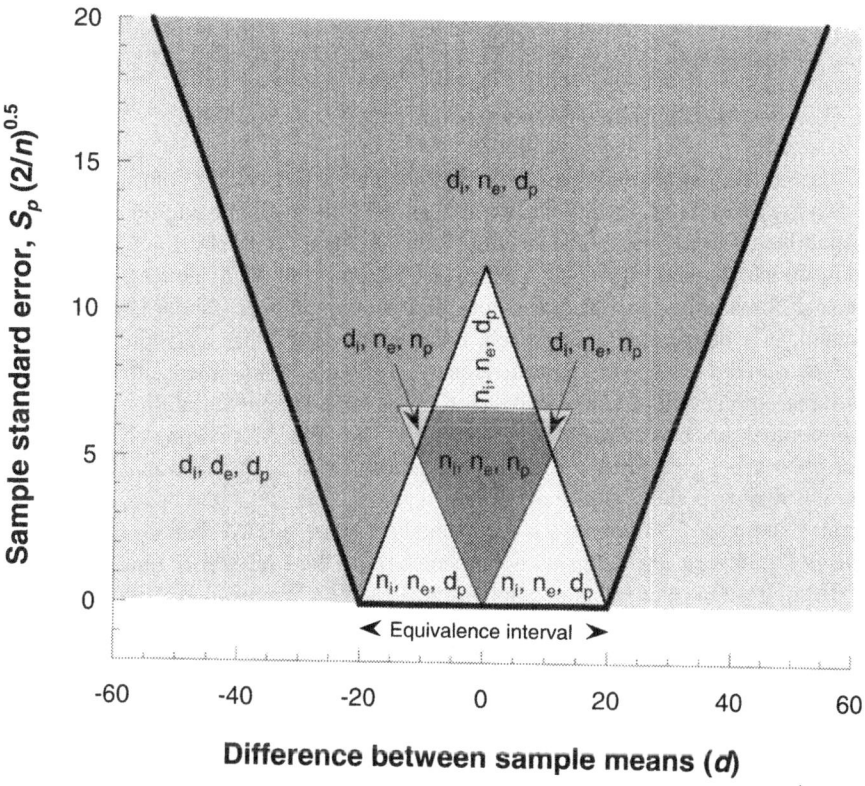

Fig. 5.13 Two-sample detection regions (variance unknown, at the 5% significance level) for testing the inequivalence hypothesis (medium boundary lines), for testing the equivalence hypothesis (thick boundary lines) and for the Power Approach (thinnest boundary lines). A minimum power of 80% is required to detect a difference beyond the equivalence interval, the limits of which are placed at $\delta_L = -20$ units and $\delta_U = +20$ units. The trios denote the inference to be drawn from tests of H_i, H_e and from the Power Approach, respectively. For example, n_i, n_e, d_p denotes nondetection from a test of H_i (i.e., "equivalence" is inferred), nondetection from a test of H_e (i.e., "equivalence" is inferred again), and detection from the Power Approach (by rejecting H_0)."S_p" is the pooled standard deviation of the differences between the two samples.

beyond the equivalence interval if the desired power is greater than 50% (and requiring at least 50% power would seem to be desirable). This is a sensible property in a precautionary approach. Second, a much less satisfactory situation is the inverted triangle shape of the nondetection region for the Power Approach. This has a number of unfortunate consequences.

1. A trivial difference will be detected if the standard error is "small" (in the "n_i, n_e, d_p" regions).

2. A larger difference will be detected as the standard error moves toward the top of the triangle.

3. Above the triangle the Power Approach detects *any* difference.

Feature 1 is explicable when one considers the behavior of a point-null hypothesis test that we have already observed; these tests are ultra-precautionary for small variability or large sample size. Feature 2 means that a difference $d = 5$ units would indicate inequivalence for a standard error of 2 units, but equivalence at a standard error of three units. The decreased precision that the higher standard error implies should, in a precautionary approach, make it *less* likely that we should infer that the measured d value implies equivalence. Feature 3 simply means that beyond a certain combination of sample size (too small) and sample variability (too large), the required power is unattainable—so the test will "detect" anything! Note too that this approach is capable of declaring equivalence when the precautionary test of the inequivalent hypothesis does not (in the "d_i, n_e, n_p" regions—the "wings on the A-Frame" in Figure 5.13), making it occasionally more permissive than that test, though still precautionary (the wings do not extend beyond the equivalence interval).

In contrast, the equivalence region for the test of the inequivalence hypothesis has a logical shape for a precautionary approach. As variability increases, or as sample size decreases, measured differences need to be ever smaller in order for equivalence to be declared. The converse is true for the permissive approach embodied in the test of the equivalence hypothesis.

> This leads to the conclusion that equivalence tests are often more appropriate than are point-null hypothesis tests when testing single hypotheses.

Recognizing that, can a point-null hypothesis test be regarded merely as an approximation to a test of a small interval? Indeed it may: Berger & Delampady (1987 [17]) show that a test of an interval (**H**: $|\delta| < \epsilon$), where ϵ is in some sense "small," may be approximated by a point-null test (**H₀**: $\delta = 0$)—but only if $\epsilon < \sigma/(4\sqrt{n})$. This condition will of course be violated for large enough n, because the interval will become too "small"—Sellke *et al.* (2001 [285]). But even if we could satisfy this condition, we are still left with a test with an unfortunately shaped rejection region.

5.3.6 A potted history of equivalence testing

Basic mathematical features of interval tests have been known for many years (Ferguson 1967 [90], Stuart *et al.* 1999 [304]—and in previous editions). But it was not until the late 1970s that much attention was paid to their practical use and implementation (e.g., Metzler 1974 [222], Spriet & Beiler 1979 [295]).[19] The 1980s saw the start of a burgeoning discussion on these tests, mostly two-sample problems for drugs trials (e.g., Westlake 1981 [341], Schuirmann 1981 [282], Mandallaz & Mau 1981 [196], Anderson & Hauck 1983 [5], Selwyn & Hall 1984 [286], Rocke 1984

---

[19]An earlier paper by Bondy (1969 [29]) has been largely overlooked (Wellek 2003, p. 50 [340]).

[271], Patel & Gupta 1984 [255], Hauck & Anderson 1984 [141], Anderson & Hauck 1985 [6], Martin Andrés 1990 [202], Hauschke *et al.* 1990 [143], Diletti *et al.* 1991 [72], Steinijans *et al.* 1992 [298], Steinijans & Hauschke 1993 [299]). Schuirmann (1987 [283]) is a seminal paper proposing a simple versatile approach ("TOST"— two one-sided tests) for the unknown-variance case, for which power calculations were reported by Phillips (1990 [258]). A full analysis of this approach, along with alternatives, is given by Berger & Hsu (1996 [18]). Extensions to multiple samples have been reported by Hauschke & Hothorn (1998 [144]). Applications to environmental science have made a belated appearance (McDonald & Erickson 1994 [216], Stunkard 1994 [305], Erickson & McDonald (1995 [87]), Garrett 1997 [109], Dixon 1998 [73], McBride 1999 [205], Cole *et al.* 2001 [55], Cole & McBride 2004 [54], Manly 2004 [199]). There are now two excellent texts on the general subject (Chow & Liu 1992 [46], Wellek 2003 [340]), where the focus is on testing the inequivalence hypothesis, rather than the equivalence hypothesis. Robinson & Froese (2004 [270]) have applied equivalence testing to validation of ecological models.

5.4 p-VALUE BEHAVIOR UNDER DIFFERENT HYPOTHESES

The fact that the p-value tends to decrease with sample size in a point-null test is pointed out in many texts. It is not so well known that p can *increase* for tests of other hypotheses as the sample size is increased. This can only happen when the tested hypothesis is in fact true. This accords with our intuition: If a hypothesis is true, the more data we have, the stronger the support for it should be. And the reason this p-value behavior doesn't occur for tests of the point-null hypothesis is that its hypothesis is virtually never true.

For example, consider a one-sided, one-sample, variance-known inferiority test of the hypothesis **H**: $\mu \leq \mu_0$, and imagine that we take a series of samples from the same stationary population, increasing the sample size each time we do so. From Table 5.4, the test statistic can be written as $T = \left(\frac{\bar{X}-\mu_0}{\sigma}\right)\sqrt{n}$. If **H** is true, then, as the sample size increases, we may expect to usually get $\bar{X} < \mu_0$, in which case T will become ever more negative (because $T \propto \sqrt{n}$).[20] This pushes T further to the left along the negative part of the x-axis on Figure 4.1. Now, from the p-value definition [Equation (4.1)] and the test's decision rule (Table 5.3), $p = \Pr(\text{reject } \mathbf{H} \mid \mathbf{H} \text{ is true}) = \Pr(T > Z_{1-\alpha} \mid \mu \leq \mu_0)$, so a decreasing (negative) value of T will cause p to increase toward 1, because in this case p is the area to the right of T. Similar arguments apply to the case of testing the superiority hypothesis (**H**: $\mu \geq \mu_0$), as well as to cases where the variance is unknown or where there are two samples.

This p-value behavior is also evident for tests of the equivalence or inequivalence hypotheses. For example, consider a one-sample, variance-unknown test of the inequivalence hypothesis $\mathbf{H_i}$: $\mu < \mu_L$ or $\mu > \mu_U$ (see Table 5.8). From Table 5.10, the two test statistics we need are $T_l = \left(\frac{\bar{X}-\mu_L}{S}\right)\sqrt{n}$ and $T_r = \left(\frac{\bar{X}-\mu_U}{S}\right)\sqrt{n}$. Also, from

---

[20] $\bar{X}$ is an unbiased estimator of μ, so its average over many samples will be close μ. And because μ_0 and σ are constants, T depends particularly on $\sqrt{n}$ only.

the test's decision rule (Table 5.9), we reject $\mathbf{H_i}$ only if $T_l \geq t_{1-\alpha,f}$ and $T_r \leq -t_{1-\alpha,f}$. So the p value is $p = \Pr(T_l \geq t_{1-\alpha,f}$ and $T_r \leq -t_{1-\alpha,f} \mid \mu < \mu_L$ or $\mu > \mu_U)$. Now let's say that $\mathbf{H_i}$ is true, and in particular that $\mu > \mu_U$. Then as the sample size increases, we may expect to usually get $\bar{X} > \mu_U$, in which case T_l will become ever larger, and so the first inequality in the p-value equation will be satisfied. However, T_r will also become increasingly larger and so the second inequality in the p-value equation will not be satisfied. Once again, this will cause p to increase as the sample size increases. Similar arguments apply to the other variance-unknown cases (testing the equivalence hypothesis, two samples).

5.5 WHAT HAS BEEN DETECTED?

In Section 5.1.3 we applied one-sided tests to the Houhou stream data and concluded that the taxonomic richness downstream of a mining discharge was most probably lower than that upstream. In Section 5.2.3 we applied a point-null hypothesis test and found a statistically significant difference between the upstream and downstream sites. In Section 5.3.2 we found that we could infer either that the upstream and downstream sites had "equivalent" taxonomic richness or that they did not. This leads to the question, Is this difference in findings a resolvable conflict? The answer is yes and no!

The one-sided and point-null hypothesis tests' results are in accord, especially if we interpret the result of the latter to mean that there are strong grounds for believing that taxonomic richness fell from upstream to downstream.[21] The conflicting results from the equivalence tests reflect the different burdens of proof adopted (permissive versus precautionary) and also the fact that the two sites would need to have differences in taxonomic richness sufficiently large for an environmentally important difference to be detected by the test of the equivalence hypothesis. For such reasons, one should not generally expect the results of these testing regimes to be in accord. One can even find situations in which a one-sided hypothesis is rejected (e.g., with $p = 0.038$ in a 5%-level test), but a point-null hypothesis on the same data at the same level is not (because its p-value would be 0.076).

> **Therefore, the most important issue is as follows:**
> **Test the appropriate hypothesis with an appropriate burden of proof.**

5.6 TESTS AND CONFIDENCE INTERVALS

For the one-sample case there is a one-to-one correspondence between calculating a $100(1-\alpha)\%$ confidence interval and performing an α-level hypothesis test. Rejection of a point-null hypothesis on the mean is equivalent to finding that the two-sided α-level confidence interval for the mean does not include the test value (usually zero).

---

[21] See footnote 5 in Chapter 4.

The reader is invited to demonstrate this in Problems 5.3 and 5.4. This correspondence is not true for two-sample cases (Cole & Blair 1999 [53], van Belle 2002 [327]), although it is often thought to be so. That is, rejection of a two-sample point-null hypothesis at the α-level is *not* equivalent to requiring that the two $100(1-\alpha)\%$ confidence intervals on each of the means do not overlap. For example, a 5%-level point-null hypothesis test on the difference between two means corresponds to a requirement that approximately 83% two-sided confidence intervals on the means do not overlap. Essentially this is because the standard error of a difference between two samples is $\sqrt{2}$ times the standard error of a single sample.

For example, consider a case where two independent samples of size n are taken from normal populations with known common variance (σ^2) but unknown means (μ_x and μ_y). Consider first the $100(1-\alpha_1)\%$ confidence interval for the difference between means ($\mu_x - \mu_y$). From Equation (3.10) the interval is $\bar{X} - \bar{Y} \pm Z_{1-\frac{\alpha_1}{2}}(\sigma\sqrt{2/n})$.[22] Because rejection of the point-null hypothesis occurs as soon as this interval does not contain zero we require that

$$|\bar{X} - \bar{Y}| > Z_{1-\alpha_1/2}\sigma\sqrt{\frac{2}{n}}. \quad (5.4)$$

Now consider $100(1-\alpha_2)\%$ confidence intervals for the individual means. From Equation (3.1) these are $\bar{X} \pm Z_{1-\frac{\alpha_2}{2}}(\sigma/\sqrt{n})$ and $\bar{Y} \pm Z_{1-\frac{\alpha_2}{2}}(\sigma/\sqrt{n})$. If the above rejection rule is to correspond to nonoverlapping of these intervals, and if $\bar{X} > \bar{Y}$, we require that the bottom of the interval on μ_x should be greater than the top of the interval on μ_y. That is, $\bar{X} - Z_{1-\frac{\alpha_2}{2}}(\sigma/\sqrt{n}) > \bar{Y} + Z_{1-\frac{\alpha_2}{2}}(\sigma/\sqrt{n})$. This may be written as $\bar{X} - \bar{Y} > Z_{1-\frac{\alpha_2}{2}}(2\sigma/\sqrt{n})$. Alternatively, if the difference in means is such that $\bar{Y} > \bar{X}$ a similar argument shows that we would require that $\bar{X} - \bar{Y} < Z_{1-\frac{\alpha_2}{2}}(2\sigma/\sqrt{n})$. These two equations can be combined to require that

$$|\bar{X} - \bar{Y}| > Z_{1-\alpha_2/2}\sigma\frac{2}{\sqrt{n}}. \quad (5.5)$$

Note that if $\alpha_1 = \alpha_2$ the two rejection rules, stated in terms of confidence intervals, are not that same—because the right-hand-side of Equation (5.5) exceeds that for Equation (5.4) by a factor of $\sqrt{2}$.

Now we know that the first rule is correct, so by equating the right-hand-sides of these two equations we can find a value of α_2 that would also make the second rule correct. That is $Z_{1-\frac{\alpha_2}{2}} = Z_{1-\frac{\alpha_1}{2}}/\sqrt{2}$. If we select $\alpha_1 = 0.05$ then, using the unit normal table (Table A.1.1), we must have $Z_{1-\frac{\alpha_2}{2}} = \frac{1.96}{\sqrt{2}} = 1.386$ and so $\alpha_2 = 0.166$. The confidence interval we require is $100(1-\alpha_2)\% \approx 83\%$.

This issue has assumed some importance in recent years with many writers encouraging analysts to move away from hypothesis test procedures to confidence interval

---

[22] This equation is the result of replacing the t-statistic by the unit normal distribution's abscissa (Z) and replacing sample variances by σ^2 in Equation (3.10); these replacements have made because the variance has been assumed to be known.

procedures (e.g., Gardner & Altman 1989 [108]), yet there is obvious room for confusion in their interpretation.

> **Overlapping confidence intervals do not necessarily imply a "nonsignificant" result. For multi-sample problems, assess "statistical significance" from tests, not from inspection of confidence intervals.**

5.6.1 A Bayesian alternative

Consider again the Houhou Stream example. Rodda & Davis (1980 [272]) have shown that a Bayesian posterior probability for sampling from normal distributions with common variance can be calculated from

$$\Pr(\delta_L \leq \delta \leq \delta_U) = F_f(T_l) - F_f(T_r), \qquad (5.6)$$

where F_f is the cumulative distribution function of Student's t-distribution with f degrees of freedom. This equation is developed by assuming an impartial prior distribution for the true difference between the means; that is, any value of δ is as likely as any other.[23] Inserting the sample data values into Equation (5.6) we obtain $\Pr(\delta_L \leq \delta \leq \delta_U) = 0.140$.

This is a rather simply-stated result when compared to those obtained from classical tests. There is no appeal to Type I error rates, confidence levels, Type II error rates or power. It is merely a bald statement as to the "probability" that the true differences in means lies within the equivalence interval. This simplicity comes at a cost—stating the prior probabilities of the mean difference and of their variance. If they are inappropriate, then the posterior probability may be also. Nevertheless, it has the attractive feature of providing water managers with additional information upon which they may make decisions.

5.7 ASSOCIATION OR CAUSATION?

From the start (see Section 1.2) we have observed that merely establishing a correlation between two variables does not constitute evidence that there is a causal relationship between them. Conversely, failure to establish a correlation does not necessarily imply a lack of causation. Likewise, obtaining or failing to obtain a statistically significant result does not necessarily say anything about causation. So what criteria should be used to establish causation?

---

[23]This is an *improper* distribution, because, being nonzero over the infinite domain, it does not encompass a unit area. Nevertheless, it can be appropriate to use improper prior distributions and, via Bayes' rule, gain a proper posterior distribution; this is the case here. One must also note that stating an impartial distribution on one parameter does not necessarily mean that the other parameter(s) of the prior distribution have, as a consequence, also been assigned impartial prior distributions. In this case there is one other parameter, the population standard deviation (σ), and it automatically has been assigned a uniform distribution on the logarithm of σ (e.g., see Lee 1997 [179])—hardly an impartial assignment!

Clarity was brought to this issue by the 1964 report to the US Surgeon General on smoking and health (USDoH 1964 [323]). This was a meta-analysis, interpreting the results of a number of studies of various sizes and with different methodologies. Five criteria were suggested: consistency of association, strength of association, specificity of association, temporal relationship and coherence. These were expanded to nine criteria a year later by Sir Austin Bradford Hill, in an inaugural address to the newly formed section on Occupational Medicine of the British Royal Society of Medicine (Hill 1965 [150]). These have become known as "The Bradford Hill criteria". While they apply particularly to epidemiological studies,[24] they have much to offer environmental science and management—in a somewhat altered form. For example, USEPA (2000 [326]) has proposed a set of sixteen modified Bradford–Hill criteria for assessing the potential impacts of multiple stressors on aquatic life. Not all of these seem appropriate for water quality studies. The set shown in Table 5.11 can be considered.

Note that only the first two criteria are *necessary* for the demonstration of cause and effect. Regarding the second, it is not necessary that the effect be absent upstream of the cause; there may be other inputs further upstream giving rise to the effect. The third, gradient of exposure, is especially important when interpreting the results of statistical models: if there is no exposure gradient (i.e., there are no "controls") an important cause may not be detected. For example, consider a town with a poor drinking water, in which all inhabitants are on the same water supply. To study potential health effects we could construct a logistic regression model (see Chapter 2.12), including drinking-water quality as one of the independent variables. The results of such a model would not find the water quality variable to be "significantly different" from zero—essentially because the regression of the health effect on the water quality term gives a flat response. This does not mean that it is not an important cause of health effects; the absence of a drinking-water quality gradient has resulted in a model with a "significant" constant term. When there is a gradient, health effects can be found. For example, Fraser & Cooke (1991 [99]) studied rates of giardiasis in a city with two water supplies of differing quality, finding statistically significant health effects between the populations on each supply.

The other items in Table 5.11 are discussed at length in other places (especially USEPA 2000 [326]). Suffice it to say that they represent a guide, rather than hard-and-fast rules in which the role of statistical methods is to account for sampling error and to assess the likely magnitude of effects.

[24] For example, "strength of association" may be indexed by odds ratios in epidemiological models.

Table 5.11 Modified Bradford–Hill criteria for evidence of association[a]

	Evidence criterion	Rationale
	Strongest	
1	Temporality	Cause must precede effect.
2	Co-occurrence	Effect must be downstream of cause.
3	Gradient of exposure	Causation cannot be inferred if there is no gradient. Effect should increase with exposure.
4	Exposure pathway	Can the cause reach the location of effect?
	Repeatability	
5	Consistency of association	Is the effect always related to the cause?
6	Experiment	Do manipulative experiments suggest causality?
	Other knowledge	
7	Plausibility	Does it make sense?
8	Analogy	Similar to other well-established cases?
9	Specificity of cause	Do we know of similar relationships?
10	Predictability	Do predicted effects from this cause actually occur?
	Lines of evidence	
11	Consistency of evidence	Consistent with available evidence?
12	Coherence of evidence	Can apparent inconsistencies be explained?

[a] Based on USEPA (2000 [326]).

Problems

5.1 In Problem 4.1 on swimmers' health a point-null hypothesis was not rejected in a 5% level test, because the data gave rise to $p = 0.059$. Discuss the possible outcomes of this study were the study to have had double the number of swimmers participating.

5.2 Repeat the Houhou Stream example given in Section 5.1.3 treating the data as paired samples. Discuss the meaning of the results.

5.3 Consider a single sample case, where we wish to make a test on the mean (μ) of a normal population, with unknown variance. We ask, is it zero? From Section 5.2, the point-null hypothesis $\mathbf{H_0}$ postulates that $\mu = \mu_0 = 0$. From Tables 5.4 and 5.7 we reject this hypothesis if $|T| > t_{1-\frac{\alpha}{2},f}$ where $T = \frac{\bar{X}-\mu_0}{S_e}$. Show that this decision procedure is the same as finding that the α-level two-sided confidence interval for μ does not include zero.

5.4 Show that the one-to-one correspondence considered in Problem 5.3 also holds for one-sided confidence interval and one-sided tests on single samples.

5.5 What is the limiting form of Figure 5.5 as $n \to \infty$? Hint: $\mathbf{H_0}$ is true only at $\Delta = 0$; $\mathbf{K}$ is true everywhere else.

5.6 What is the limiting form of Figure 5.8 as $n \to \infty$?

5.7 What modifications should be made to Table 5.10 for the paired-sample case?

5.8 In Problem 5.2 the test statistic is larger when the data are treated as paired, compared to treating them as two independent samples. This will result in a smaller p-value and it therefore seems that pairing has made the test more "powerful." It is often held that pairing *always* makes a test more "powerful," because of the removal of the $\sqrt{2}$ term from the standard error, as noted in Problem 5.7. But in Problem 3.8 we saw that the standard error for paired samples may sometimes be *greater* than is obtained by treating the data as two independent samples. In that case, if the difference is great enough, a paired-sample test could be less "powerful" than its two-independent sample cousin. Can you explain this?

5.9 Explain why the duo (n_i, d_e) doesn't appear in Figures 5.12(a) and 5.12(b).

5.10 Explain why the trios (n_i, d_e, d_p); (n_i, d_e, n_p) and (d_i, d_e, n_p) don't appear in Figure 5.13.

6

Mathematics and calculation methods

6.1 INTRODUCTION

This chapter contains mathematical details and calculation algorithms for the preceding chapters, divided into three sections:

- important distributions,
- confidence intervals and tolerance intervals,
- detection formulae.

Problems are set in each section, for which answers are given in Chapter 12.

The algorithms (calculation procedures) are useful when writing computer procedures, because they avoid the rounding errors that can occur when using standard approaches (and of which naïve users may be unaware). Typically, these errors arise because many probabilities are the quotient of very large and small numbers, such as 100! and 100^{-100} in binomial calculations. A simple (and it really is simple) remedy is to calculate such quotients by exponentiating their logarithms, turning multiplication and division of large numbers into addition and subtraction of much smaller numbers. In doing so, we need the logarithm of the gamma (Γ) function, and this is usually available (e.g., via Excel's GAMMALN function).

For example, excessive rounding errors can arise when calculating the binomial probabilities in Figure 2.11(d) using ordinary methods for the pmf formula $b(e; n, \mu) = \frac{n!}{e!(n-e)!} n^{-n} \mu^e (n-e)^{n-e}$, with $n = 100$ and $\mu = 5$, where $n! = n(n-1)(n-2)\ldots(2)(1)$ is the integer factorial function. A reliable algorithm is obtained by noting that $n! = \Gamma(n+1)$, and so the algorithm's equation is $b(e; n, \mu) =$

exp$\{\ln[\Gamma(n+1)] - \ln[\Gamma(e+1)] - \ln[\Gamma(n-e+1)] - n\ln(n) + e\ln(\mu) + (n-e)\ln(n-\mu)\}$. Each of the numbers being exponentiated is "small" (with respect to a computer's rounding error), and so the exponentiation is straightforward.

Another example is the computation of probabilities of the beta distribution. That is, noting that $B(\alpha, \beta) = \frac{\Gamma(\alpha)\Gamma(\beta)}{\Gamma(\alpha+\beta)}$, we have $B(\alpha, \beta) = \exp\{\ln[\Gamma(\alpha)] + \ln[\Gamma(\beta)] - \ln[\Gamma(\alpha+\beta)]\}$.

6.2 FORMULAE AND CALCULATION METHODS FOR IMPORTANT DISTRIBUTIONS

6.2.1 Normal distribution

The pdf (f) for a continuous random variable x is given by

$$f(x; \mu, \sigma) = \frac{1}{\sqrt{2\pi}\sigma} e^{-\frac{1}{2}z^2}, \tag{6.1}$$

where $z = \frac{x-\mu}{\sigma}$ is the normal deviate (Equation 2.15). Summary statistics are:

Mean = median = mode μ,

Variance σ^2,

Coefficient of variation $\eta = \frac{\sigma}{\mu}$,

Coefficient of skewness $g_1 = 0$.

The unit normal pdf is obtained by setting $\mu = 0$ and $\sigma = 1$. Its CDF is $F(x; 0, 1) = G(x) = \int_{-\infty}^{x} f(\xi; 0, 1) \, d\xi$, where ξ is a dummy variable of integration.[1] $G(x)$ does not have a closed form solution and a numerical approximation must be used. A simple-but-accurate procedure for its calculation is based on the Chebyshev approximation to the complementary error function reported by Press *et al.* (1992, p. 214 [265]), combined with Equation (26.2.29) of Abramowitz & Stegun (1972, p. 934 [1]) relating the unit normal CDF to that function. It proceeds by defining $w = \frac{|x|}{\sqrt{2}}$ and $t = 1/(1 + \frac{1}{2}w)$. The complementary error function ("erfcc") is given by first calculating the intermediate variable θ, where

$$\begin{aligned}\theta &= t\exp\bigl(-w^2 - 1.26551223 + 1.00002368t + 0.37409196t^2 \\ &+ 0.09678418t^3 - 0.18628806t^4 + 0.27886807t^5 - 1.13520398t^6 \\ &+ 1.48851587t^7 - 0.82215223t^8 + 0.17087277t^9\bigr).\end{aligned} \tag{6.2}$$

Then $G(x) = 1 - \frac{1}{2}$erfcc, where erfcc $= \theta$ if $x \geq 0$, else erfcc $= 2 - \theta$. This formulation has been used to produce Table A.1.1 in the Appendix at the end of this text.

---
[1] Because the unit normal distribution is used so often, it is commonly given a special symbol, either $\Phi(x)$ or $G(x)$.

6.2.2 Lognormal distribution

The pdf (f) for a continuous random variable x is given by

$$f(x;\mu_y,\sigma_y,\tau) = \frac{1}{\sqrt{2\pi}\sigma_y(x-\tau)} e^{-\frac{1}{2}\left[\frac{\ln(x-\tau)-\mu_y}{\sigma_y}\right]^2}, \tag{6.3}$$

where μ_y and σ_y^2 are the true mean and variance of the transformed random variable $y = \ln(x)$, and $x > \tau$. This pdf has three parameters, including the "shift parameter" τ. As seen on Figure 2.8 this parameter merely shifts the distribution to the right ($\tau > 0$) or to the left ($\tau < 0$), without changing its shape. The two-parameter lognormal distribution is obtained by setting $\tau = 0$. Its CDF is $F(x;\mu_y,\sigma_y,\tau) = G\left(\frac{\ln(x-\tau)-\mu_y}{\sigma_y}\right)$ (see Problem 6.1). Summary statistics for this distribution (Gilbert 1987 [118]) are:

Mean $\mu = \exp(\mu_y + \frac{1}{2}\sigma_y^2) + \tau$,

Median = Geometric mean $\mu_g = \exp(\mu_y) + \tau$,

Mode $\tilde{f} = \exp(\mu_y - \sigma_y^2) + \tau$,

Variance $\sigma^2 = \exp(2\mu_y + \sigma_y^2)[\exp(\sigma_y^2) - 1]$,

Geometric standard deviation $\sigma_g = \exp(\sigma_y)$,

Coefficient of variation $\eta = \sqrt{\exp(\sigma_y^2)-1} \Big/ \left[1 + \frac{\tau}{\exp(\mu_y + \frac{1}{2}\sigma_y^2)}\right]$,

Coefficient of skewness $g_1 = \left[\exp(\sigma_y^2) + 2\right]\sqrt{\exp(\sigma_y^2) - 1} > 0$.

Note that this distribution is always right-skewed, because g_1 is always positive. Some applications to water quality rules for shellfish-harvesting waters are given in Problems 6.2 and 6.3.

6.2.3 Gamma distribution

The pdf of a continuous random variable having a gamma distribution is

$$f(x;\alpha,\beta) = \frac{1}{\beta^\alpha \Gamma(\alpha)} x^{\alpha-1} e^{-x/\beta}, \tag{6.4}$$

for $x > 0$, and where $\alpha > 0$ and $\beta > 0$. The standard gamma function is defined as $\Gamma(x) = \int_0^\infty \xi^{x-1} e^{-t} d\xi$. Summary statistics for this distribution are:

Mean $\mu = \alpha\beta$,

Mode $\tilde{f} = \begin{cases} \beta(\alpha - 1) & \text{for } \alpha \geq 1, \\ 0 & \text{for } \alpha < 1, \end{cases}$

Variance $\sigma^2 = \alpha\beta^2$,

Coefficient of variation $\eta = \frac{1}{\sqrt{\alpha}}$,

Coefficient of skewness $g_1 = \frac{2}{\sqrt{\alpha}} > 0$.

This pdf can be calculated using $f(x; \alpha, \beta) = \exp\{(\alpha - 1)\ln(x) - \alpha\ln(\beta) - \frac{x}{\beta} - \ln[\Gamma(\alpha)]\}$. Press et al. (1992 [265]) present an accurate algorithm for evaluating the natural logarithm of the Γ function. The CDF of this distribution is $F(x; \alpha, \beta) = P(\alpha, x/\beta)$, where $P(a, x)$ is the incomplete gamma function ratio $P(a, x) = \frac{1}{\Gamma(a)} \int_0^x e^{-\xi} \xi^{a-1} d\xi$ (see Problem 6.4). Press et al. (1992 [265]) also give accurate algorithms for this function and, as noted in Section 6.1, it is available in Excel's GAMMALN function. Two special cases arise: when $\alpha = 1$ we get the **exponential distribution** and when $\alpha = \frac{f}{2}$ and $\beta = 2$ we get the **chi-square distribution**, with f degrees of freedom.

6.2.4 Beta distribution

The pdf of a continuous random variable having a two-parameter beta distribution [Be(α, β)] is

$$f(x; \alpha, \beta) = \frac{x^{\alpha-1}(1-x)^{\beta-1}}{B(\alpha, \beta)}, \quad (6.5)$$

for $0 < x < 1, \alpha > 0, \beta > 0$ and where $B(\alpha, \beta) = \frac{\Gamma(\alpha)\Gamma(\beta)}{\Gamma(\alpha+\beta)}$ is the beta function. Summary statistics for this distribution are:

Mean $\mu = \frac{\alpha}{\alpha+\beta}$,

Mode $\tilde{f} = \begin{cases} \frac{\alpha-1}{\alpha+\beta-2} & \text{for } \alpha > 1 \text{ and } \beta > 1, \\ 0 \text{ and } 1 & \text{for } \alpha < 1 \text{ and } \beta < 1, \\ 0 & \text{for } \alpha < 1 \text{ and } \beta \geq 1, \text{ or } \alpha = 1 \text{ and } \beta > 1, \\ 1 & \text{for } \alpha \geq 1 \text{ and } \beta < 1, \text{ or } \alpha > 1 \text{ and } \beta = 1, \\ \text{does not uniquely exist} & \text{for } \alpha = 1 \text{ and } \beta = 1, \end{cases}$

Variance $\sigma^2 = \frac{\alpha\beta}{(\alpha+\beta)^2(\alpha+\beta+1)}$,

Coefficient of variation $\sqrt{\frac{\alpha}{\beta}(\alpha+\beta+1)}$,

Coefficient of skewness $g_1 = \frac{2(\beta-\alpha)}{\alpha+\beta+2}\sqrt{\frac{\alpha+\beta+1}{\alpha\beta}}$.

The evaluation of this pdf is addressed in Problem 6.5. Its CDF is the incomplete beta function ratio; that is, $F(x; \alpha, \beta) = I_x(\alpha, \beta)$ where $I_x(\alpha, \beta) = \frac{1}{B(\alpha,\beta)} \int_0^x \xi^{\alpha-1}(1-\xi)^{\beta-1} d\xi : \alpha, \beta > 0$ (see Problem 6.6). Press et al. (1992 [265]) give accurate algorithms for this function.

6.2.5 Binomial distribution

This describes the distribution of a discrete variable, e. The pmf is

$$b(e; n, \theta) = \binom{n}{e} \theta^e (1-\theta)^{n-e}, \tag{6.6}$$

for $e = 0, 1, 2, \ldots, n$, and is the probability of getting e "successes" (e.g. pass, exceed,...) in n random trials, each with a probability θ of success. This distribution applies to a design in which sampling ceases once n results are obtained, regardless of how many "successes" are obtained. $\binom{n}{e} = \frac{n!}{e!(n-e)!}$ is the binomial coefficient. Summary statistics are:

Mean $\mu = n\theta$,

Mode $\tilde{f} = \begin{cases} \theta(n+1) \text{ and } \theta(n+1) - 1 & \text{for integer } \theta(n+1), \\ \lfloor \theta(n+1) \rfloor & \text{else,} \end{cases}$

Variance $\sigma^2 = n\theta(1-\theta) = \mu(1-\theta) < \mu$,

Coefficient of variation $\eta = \sqrt{\frac{1-\theta}{n\theta}}$,

Coefficient of skewness $g_1 = \frac{1-2\theta}{\sqrt{n\theta(1-\theta)}} = \frac{1}{\sqrt{\mu}} \frac{1-2\theta}{\sqrt{1-\theta}} < \frac{1}{\sqrt{\mu}}$.

where $\lfloor \ \rfloor$ denotes "integer part of." The pmf can be written in terms of the mean, as $b(e; n, \mu) = \binom{n}{e} n^{-n} \mu^e (n-\mu)^{n-e}$. This distribution can be left-skewed or right-skewed and is always **underdispersed** (its variance is always less than its mean, as the variance equation above shows). For a given mean it has less skew than the Poisson distribution (for which $g_1 = \frac{1}{\sqrt{\mu}}$; see Section 6.2.6). Indeed if $\theta < \frac{1}{2}$, it is left-skewed. To calculate the pmf two approximation methods are sometimes useful.

Normal approximation For large n, such that both $n\theta > 5$ and $n(1-\theta) > 5$, $b(e; n, \theta) \approx f(Z; \mu, \sigma)$, where $\mu = n\theta$, $\sigma = \sqrt{n\theta(1-\theta)}$, and $Z = \frac{e-\mu}{\sigma}$.

Poisson approximation For large n and small θ, $b(e; n, \theta) \approx p(e; \mu)$.

[see Section 6.2.6 for a description of $p(e; \mu)$]. These approximations (demonstrated in Problem 6.8) have often been used because binomial published tables seldom have entries for $n > 20$ and are typically for p in steps of 0.05 (e.g., Conover 1980 [56]). With the advent of readily available computing power, such approximations are not so often used. Nowadays one can use a standard feature in a spreadsheet to return binomial values, or use a stand-alone algorithm. The following recursive algorithm is particularly simple and accurate—it is exact (to within a computer's precision error).[2] It includes calculation of the CDF,

---
[2] Problem 6.9 invites the reader to derive it. An alternative method is presented in Problem 6.10.

$$B(e;n,\theta) = \sum_{i=0}^{e} b(i;n,\theta). \tag{6.7}$$

It is the probability of getting *up to* e successes in n random trials. The algorithm is as follows

1. Set $i = 0$, $\bar{\theta} = 1 - \theta$, $\lambda = n\ln(\bar{\theta})$, $b = \exp(\lambda)$, $B = b$. If $e = 0$, then stop, else set $v = p/q$ and go to Step 2.

2. Set these variables, in the order given: $\xi = v(n-i)/(i+1)$, $\lambda = \ln(\xi) + \lambda$, $b = \exp(\lambda)$, $B = B + b$, $i = i + 1$.

3. Repeat Step 2 until $i = e$.

Some useful recursive formulae (Miller & Freund 1985 [228]) are

$$\begin{align}
b(e;n,\theta) &= b(n-e;n,1-\theta) = B(e;n,\theta) - B(e-1;n,\theta), & (6.8)\\
b(e;n,\theta) &= B(n-e;n,1-\theta) - B(n-e-1,n,1-\theta), & (6.9)\\
B(e;n,\theta) &= 1 - B(n-e-1;n,1-\theta). & (6.10)
\end{align}$$

These can be used to calculate binomial probabilities from values in standard tables.

Finally, we sometimes wish to know the probability of getting e *or more* exceedances in n trials, denoted by

$$\mathcal{B}(e;n,\theta) = \sum_{i=e}^{n} b(i;n,\theta), \tag{6.11}$$

which is also equal to $I_\theta(e, n - e + 1)$ (Abramowitz & Stegun 1972, p. 945 [1]), where $I_\theta(e, n-e+1)$ is the incomplete beta function ratio (Press *et al.* (1992 [265]). So we have these useful relationships:

$$\begin{align}
\mathcal{B}(e;n,\theta) &= 1 - B(e-1;n,\theta) = 1 - B(e;n,\theta) + b(e;n,\theta), & (6.12)\\
B(e;n,\theta) &= 1 - I_\theta(e+1, n-e) = I_{1-\theta}(n-e, e+1). & (6.13)
\end{align}$$

See also Problem 6.7.

6.2.6 Poisson distribution

The pmf for a discrete variable e is

$$p(e;\mu) = \frac{\mu^e e^{-\mu}}{e!}, \tag{6.14}$$

for $e = 0, 1, 2, \ldots$. Summary statistics are:

FORMULAE AND CALCULATION METHODS FOR IMPORTANT DISTRIBUTIONS 145

Mean μ,

Mode $\tilde{f} = \begin{cases} \mu \text{ and } \mu - 1 & \text{for integer } \mu, \\ \lfloor \mu \rfloor & \text{else,} \end{cases}$

Variance $\sigma^2 = \mu$,

Coefficient of variation $\eta = \frac{1}{\sqrt{\mu}}$,

Coefficient of skewness $g_1 = \frac{1}{\sqrt{\mu}}$.

Because its variance and mean are identical, it is neither underdispersed nor overdispersed. Problem 6.11 invites the reader to develop an exact recursive algorithm for this pmf. It may also be evaluated directly using $p(e; \mu) = \exp[e \ln(\mu) - \mu - \ln \Gamma(e+1)]$,[3] where Γ is the gamma function defined as $\Gamma(x) = \int_0^\infty \xi^{x-1} e^{-\xi} \, d\xi$, which, for integer values of x, has the property that $\Gamma(x+1) = x!$ Often μ is replaced by the product of two terms, say r and t, where r denotes a first-order rate coefficient and t denotes area, dose or time. For example, in the spreading of microorganisms on enrichment media in an agar plate, r would be the number of organisms per unit area. When t denotes dose, we are considering the modeling of dose-response data, in **quantitative risk assessment (QRA)**, see Chapter 9. If t is time, we refer not to a Poisson distribution, but to a **Poisson process**. The CDF is simply $P(e; \mu) = \sum_{i=0}^{e} p(i; \mu)$, which is known as the **Poisson series**. If $e \to \infty$, the series sums to unity.

6.2.7 Negative binomial distribution

This distribution is usually derived by considering the number (n) of random trials required for the occurrence of the kth "success" (e.g., exceedance of a standard's limit), where each trial has a probability of occurrence = θ. This distribution applies to a design in which sampling ceases once k successes are obtained, regardless of how many trials are performed. Then the pmf is

$$b^*(n; k, \theta) = \binom{n-1}{k-1} \theta^k (1-\theta)^{n-k}, \qquad (6.15)$$

for $n = k, k+1, k+2, \ldots$.[4] In contrast to the binomial distribution, where k is the variable and n and θ are the parameters that must be supplied, the negative binomial distribution has n as the variable with θ and k being the parameters. It is related to the binomial distribution by $b^*(n; k, \theta) = \frac{k}{n} b(k; n, \theta)$ (see Problem 6.12). In the

---

[3] Nonrecursive probability calculations often call for the evaluation of the quotient of two very large numbers (such as μ^e and $e!$), each of which can in extreme cases exceed the ability of a common computer's number storage capacity. This difficulty is simply avoided by exponentiating the logarithm of the quotient, turning multiplication and division of large numbers into addition and subtraction of much smaller numbers.

[4] The name "negative binomial" arises from the fact that the values of $b^*(n; k, \theta)$ for $n = k, k+1, k+2, \ldots$ are successive terms of the binomial expansion of $(\frac{1}{\theta} - \frac{1-\theta}{\theta})^{-k}$. The generalization of binomial series to negative powers, and to fractional powers, was first announced by Sir Isaac Newton, 1642–1727.

special case that $k = 1$ we obtain the **geometric distribution**, such as would arise in sampling until an event first occurs; that is,

$$g(n;\theta) = \theta(1-\theta)^{n-1}. \tag{6.16}$$

Summary statistics of these distributions are:

Mean $\mu = \frac{k(1-\theta)}{\theta}$,

Mode $\tilde{f} = \begin{cases} z, z+1 & \text{for integer } z, \text{ where } z = \max\left(\mu - \frac{1}{\theta}, 0\right), \\ \lfloor z \rfloor + 1 & \text{else,} \end{cases}$

Variance $\sigma^2 = \frac{k(1-\theta)}{\theta^2} = \frac{\mu}{\theta} = \mu + \frac{\mu^2}{k} > \mu$,

Coefficient of variation $\eta = \sqrt{\frac{1}{\mu} + \frac{1}{k}}$,

Coefficient of skewness $g_1 = \frac{2-\theta}{\sqrt{k(1-\theta)}} = \frac{1}{\sqrt{\mu}}\frac{2-\theta}{\sqrt{\theta}} > \frac{1}{\sqrt{\mu}}$.

The variance is always greater than the mean (as the variance equation above shows) and the distribution is therefore always **overdispersed**. Consequently, the reciprocal of k can be seen as a measure of the excess variance (i.e., clumping). That is, variance becomes ever more inflated as k decreases. For a given mean it is always more skewed than the Poisson distribution (for which $g_1 = \frac{1}{\sqrt{\mu}}$, see Section 6.2.6). For large sample sizes the negative binomial distribution approaches the Poisson distribution as $\theta \to 1$ and as $n(1-\theta) \to \mu$. See Problem 6.13 for the role of the summary statistics for the geometric distribution.

The CDF is the probability of getting the kth successes *on or before* the nth trial; that is, $B^*(n; k, \theta) = \sum_{i=k}^{n} b^*(i; k, \theta)$. Sometimes we want to know the probability of getting k successes in n *or more* trials; that is, $\mathcal{B}^*(n; k, \theta) = \sum_{i=n}^{\infty} b^*(i; k, \theta)$. We may relate this to the CDF of the binomial distribution by noting that "The probability with which more than n trials will be required for k successes is exactly the probability of fewer than k successes in n ordinary binomial observations" (Stuart *et al.* 1999 [304]). That gives us these useful expressions

$$\mathcal{B}^*(n+1; k, \theta) = B(k-1; n, \theta), \tag{6.17}$$
$$B^*(n; k, \theta) = 1 - \mathcal{B}^*(n+1; k, \theta). \tag{6.18}$$

A special case that may be used in studies of unbiasedness is $B^*(2k-1; k, \frac{1}{2}) = \frac{1}{2}$ (see Problem 6.14).

Two alternative forms of the pmf are sometimes reported. In the first, attention is focused not on n, but on the number of trials *in excess* of k required to produce the kth success. Call this number s. Then, by substituting $s = n - k$, the pmf becomes $b_s^*(s; k, \theta) = \binom{s+k-1}{k-1}\theta^k(1-\theta)^s$, for $s = 0, 1, 2, \ldots$ (the s subscript denotes that this alternative form of the distribution is being used, where the first of the function's argument is s, not n). From the properties of the binomial coefficient this pmf can also be written as $b_s^*(s;k,\theta) = \binom{s+k-1}{s}\theta^k(1-\theta)^s$, also for $s = 0, 1, 2, \ldots$. Of interest to microbiologists is the fact that this form can also be derived from the Poisson distribution, with $\mu = rt$, as discussed in Section 6.2.6. One then obtains the negative binomial distribution if r is assumed to follow a gamma distribution (Hilborn & Mangel, 1997, p. 84 [149]). By this means, overdispersion of microorganisms can be accounted for.[5]

The second alternative form follows by generalizing the distribution for application to biological populations, noting that k need not be an integer (e.g., Elliott 1983 [81], Hilborn & Mangel 1997 [149]). It is best parameterized by μ and k and so the pmf becomes $b_s^*(s;k,\mu) = \binom{s+k-1}{s}(\frac{k}{\mu+k})^k(\frac{\mu}{\mu+k})^s$, for $s = 0, 1, 2, \ldots$ (see Problem 6.15). Direct calculation of the binomial terms for noninteger k can be carried out using gamma functions (being the generalization of factorials with non-integer arguments). For integer k we may use the recursive relation $b_s^*(s;k,\mu) = \binom{s+k-1}{s}(\frac{\mu}{\mu+k})b_s^*(s-1;k,\mu)$, where $b_s^*(0;k,\mu) = (\frac{k}{\mu+k})^k$ (see Problem 6.16).

6.2.8 Hypergeometric distribution

This distribution arises if sampling of a discrete variable is without replacement. It is essentially an elaboration of the binomial distribution for the case where only two outcomes are possible.[6] We want the distribution for obtaining k successes in n random trials by sampling without replacement from a population of size N in which there are m "successes." The pmf is

$$h(k;n,N,m) = \frac{\binom{m}{k}\binom{N-m}{n-k}}{\binom{N}{n}}, \quad (6.19)$$

for $k = 0, 1, 2, \ldots, n$; $k \leq m$; and $n - k \leq N - m$. Summary statistics are:

Mean $\mu = \frac{mn}{N}$,

Mode No closed form, but for large N (as is usually the case) the mode $\rightarrow \frac{m(n+1)}{N} - 1$ and thus is very close to the mode of the binomial distribution,

Variance $\sigma^2 = \frac{nm(N-m)(N-n)}{N^2(N-1)}$,

Coefficient of variation $\eta = \sqrt{\frac{(N-m)(N-n)}{nm(N-1)}}$,

Coefficient of skewness $g_1 = \frac{(N-2m)(N-2n)}{N-2}\sqrt{\frac{N-1}{nm(N-m)(N-n)}}$.

A recursive algorithm for the calculation of this pmf is given in the answer to Problem 6.17.

---

[5] Note too that in QRA dose-response analysis it is common to assume that r follows a beta distribution, in which case a more complex hypergeometric distribution is obtained (Teunis & Havelaar 2000 [307])—a Kummer confluent hypergeometric function, often approximated by the "Beta-Poisson" dose-response model, as we shall see in the appendix of Chapter 9.
[6] For multiple outcomes it can be generalised to the **multivariate hypergeometric distribution**.

6.2.9 Multinomial distribution

This distribution is a generalization of the binomial, where each trial has more than two possible outcomes and the probabilities of the respective outcomes are the same for each trial. This distribution arises when constructing MPN tables for microbiological enumerations (see Chapter 10). We denote the number of possible outcomes as m and their respective probabilities on any one trial are $(p_1, p_2, \ldots, p_m)$, where $\sum_{i=1}^{m} p_i = 1$). We are interested in the probability of getting k_1 outcomes of the first kind, k_2 outcomes of the second kind, and k_m outcomes of the last kind, where $\sum_{i=1}^{m} k_i = n$. The pmf is

$$m(k_1, k_2, \ldots, k_m; n, p_1, p_2, \ldots, p_m) = \binom{n}{k_1 k_2, \ldots, k_m} p_1^{k_1} p_2^{k_2} \ldots p_m^{k_m}, \quad (6.20)$$

for $k_i = 0, 1, 2, \ldots, n$ for each i. The term in parentheses is the multinomial coefficient ($= \frac{n!}{k_1! k_2! \ldots k_m!}$). Summary statistics are plentiful (Stuart & Ord 1994, p. 260 [303]).

6.2.10 Student's t-distribution

The pdf for a continuous random statistic t with f degrees of freedom is given by

$$p(t) = \frac{1}{\sqrt{f} \, \mathrm{B}(\frac{1}{2}, \frac{f}{2})} \left(1 + \frac{t^2}{f}\right)^{-\frac{1}{2}(f+1)}, \quad (6.21)$$

for $-\infty < t < \infty$, $f > 0$. This distribution is symmetrical about $t = 0$ (and so $g_1 = 0$), where its mean, median and mode are also located. Its variance is $\frac{f}{f-2}$ and so is undefined for $f \leq 2$ (Casella & Berger 2002 [45]). Percentiles are written as $t_{1-\alpha, f}$; the area to the right of this value is α.

For $f < 1000$ its CDF is given by $1 - \frac{1}{2} I_\xi(\frac{1}{2}, \frac{f}{2})$, where $\xi = \frac{t^2}{f+t^2}$ (Johnson & Kotz 1970 [169]), where $I_\xi(x, y)$ is the incomplete beta function ratio. For $f > 1000$ the normal CDF suffices (the t-distribution approaches the unit normal distribution as $f \to \infty$).

The CDF may also be calculated independent of the incomplete beta function ratio from the following simplification of Owen's integration-by-parts algorithm for the noncentral t-distribution presented in Section 6.3.4 (in which the "noncentrality parameter" is set to zero).

For $f = 1$

$$\Pr(t \leq t) = \frac{1}{2} + \frac{\tan^{-1}(t)}{\pi}.$$

For f even and > 1

$$\Pr(t \leq t) = \frac{1}{2} \left[1 + A\sqrt{B} \left(1 + m_2 + m_4 + \ldots + m_{f-2}\right)\right].$$

For f odd and > 1

$$\Pr(t \leq t) = \frac{1}{2} + \frac{\tan^{-1}(A) + AB(1 + m_3 + m_5 + \ldots + m_{f-2})}{\pi},$$

where $A = \frac{t}{\sqrt{f}}$, $B = \frac{f}{f+t^2}$, and $m_k = \frac{k-1}{k} B m_{k-2}$ (with $m_0 = m_1 = 1$). This algorithm is simple, fast, and theoretically exact, so it is subject only to computer roundoff error affecting the last significant decimal figure. It is similar to that presented by Cooper (1985a [57]).

6.2.11 Chi-square distribution

The pdf for a continuous random statistic x with f degrees of freedom is given by

$$p(x) = \frac{1}{2^{\frac{f}{2}} \Gamma(\frac{f}{2})} x^{\frac{f-2}{2}} e^{-\frac{x}{2}}, \tag{6.22}$$

for $x > 0$, $f > 0$. Percentiles of this distribution are written as $\chi^2_{1-\alpha,f}$; the area to the right of this value is α. Summary statistics are

Mean $\mu = f$,

Mode $\tilde{f} = \begin{cases} f - 2 & \text{for } f > 2, \\ 0 & \text{else,} \end{cases}$

Variance $\sigma^2 = 2f$,

Coefficient of variation $\eta = \sqrt{\frac{2}{f}}$,

Coefficient of skewness $g_1 = 2\sqrt{\frac{2}{f}} > 0$.

Being a special case of the gamma distribution (with $\alpha = \frac{f}{2}$ and $\beta = 2$), its CDF is just an incomplete gamma function ratio, $P(\frac{f}{2}, \frac{x}{2})$. If $\frac{f}{2}$ is an integer it may be more simply evaluated as $1 - \exp(-\frac{x}{2}) \sum_{i=0}^{r-1} (\frac{x}{2})^i / i!$, where $r = \frac{f}{2}$.

6.2.12 F-distribution

The pdf for a continuous random statistic x with f_1 and f_2 degrees of freedom is given by

$$p(x) = \frac{f_1^{\frac{f_1}{2}} f_2^{\frac{f_2}{2}}}{B(\frac{f_1}{2}, \frac{f_2}{2})} \frac{x^{\frac{f_1}{2}-1}}{(f_2 + f_1 x)^{\frac{f_1+f_2}{2}}}, \tag{6.23}$$

for $x > 0$, $f_1 > 0$, and $f_2 > 0$. This distribution has both "numerator" and "denominator" degrees of freedom (f_1 and f_2) because it describes the distribution of a ratio (of variances). Percentiles are written as $F_{1-\alpha,f_1,f_2}$; the area to the right of this value is α. Its summary statistics are

Mean $\mu = \frac{f_2}{f_2-2}$ for $f_2 > 2$,

Mode $\frac{f_1-2}{f_1}\frac{f_2}{f_2+2}$ for $f_1 > 2$,

Variance $\sigma^2 = \frac{2f_2^2(f_1+f_2-2)}{f_1(f_2-2)^2(f_2-4)}$ for $f_2 > 4$,

Coefficient of variation $\eta = \sqrt{\frac{2(f_1+f_2-2)}{f_1(f_2-4)}}$ for $f_2 > 4$,

Coefficient of skewness $g_1 = \frac{2f_1+f_2-2}{f_2-6}\sqrt{\frac{8(f_2-4)}{f_1(f_1+f_2-2)}} > 0$ for $f_2 > 6$.

The CDF may be calculated as $\Pr(x \le X) = I_{X_1}(\frac{f_1}{2}, \frac{f_2}{2})$, where $X_1 = \frac{f_1 X}{f_2+f_1 X}$ (Johnson & Kotz 1970, p. 78 [169]), where I is the incomplete beta function ratio. Percentiles of Students's t-distribution can be calculated from the F-distribution with $f_1 = 1$, using $t_{1-\alpha,f} = \sqrt{F_{1-2\alpha,1,f}}$.

Problems

Distribution calculation problems The first two have significant implications for setting microbiological standards (and for any other right-skewed water quality variables), representing a case for clear understanding of mathematical/statistical aspects in doing so. The rest of the Problems are more general, having to do with properties of distributions and associated calculations.

6.1 Show that the CDF of the three-parameter lognormal distribution (Equation 6.3) is given by $F(x; \mu_y, \sigma_y, \tau) = G\left(\frac{\ln(x-\tau)-\mu_y}{\sigma_y}\right)$, where $G(x)$ is the CDF of the unit normal distribution.

6.2 Safe shellfish harvesting criteria (USDHHS 1993 [322]) require that the median concentration of fecal coliforms in the shellfish-growing water should not exceed 14 per 100 mL, nor should the 90%ile exceed 43 per 100 mL (based on 15 samples, a "systematic random sampling strategy" and a particular microbiological test). Assuming that fecal coliforms follow the normal distribution and are completely borderline for compliance with this standard, what is that distribution's coefficient of variation? What is it if the distribution is the two-parameter lognormal? Discuss the implications.

6.3 Generalize the results obtained in Problem 6.2 to the ratio of *any* two percentiles of normal and (two-parameter) lognormal distributions. Use the results to calculate the ratio of the 99%ile to the 95%ile for normal and two-parameter lognormal distributions with coefficient of variation = 0.5 and also 1.0.

6.4 Show that the gamma distribution CDF, $F(x; \alpha, \beta)$, is the incomplete gamma function ratio $P(\alpha, x/\beta)$, where $P(a, x) = \frac{1}{\Gamma(a)}\int_0^x e^{-\xi}\xi^{a-1}\,d\xi$.

6.5 Show that the beta distribution pmf, $F(x; \alpha, \beta)$, can be evaluated from $\exp\{\ln[\Gamma(\alpha+\beta)] - \ln[\Gamma(\alpha)] - \ln[\Gamma(\beta)] + (\alpha-1)\ln(x) + (\beta-1)\ln(1-x)\}$.

6.6 Show that the beta distribution CDF, $F(x; \alpha, \beta)$, is the incomplete beta function ratio $I_x(\alpha, \beta)$, where $I_x(\alpha, \beta) = \frac{1}{B(\alpha,\beta)} \int_0^x \xi^{\alpha-1}(1-\xi)^{\beta-1} \, d\xi : \alpha, \beta > 0$.

6.7 Show for the binomial distribution that $B(k; n, \theta) = \mathcal{B}(n-k; n, \theta)$ only if the distribution is symmetric—$B(k; n, \theta)$ is the probability of getting up to k "successes" in n random dichotomous trials each of which has the probability θ of success, and $\mathcal{B}(k; n, \theta)$ is the probability of getting k or more successes in those n trials.

6.8 Compare the normal approximation to the binomial pmf (see page 143) to the exact binomial value (use $n = 1000$, $\theta = 0.1$ and $e = 90$). Also, compare the Poisson approximation to the binomial pmf (also on page 143) to the exact binomial value (use $n = 1000$, $\theta = 0.01$ and $e = 3$).

6.9 Derive the recursive algorithm given in Section 6.2.5 for the calculation of binomial probabilities. Hint: First demonstrate that $b(k+1; n, \theta) = \frac{\theta(n-k)}{(k+1)(1-\theta)} b(k; n, \theta)$. What is the advantage of this algorithm over the direct algorithm presented in Problem 6.10?

6.10 Show that the binomial pmf $b(k; n, \theta)$ can be calculated directly from $\exp[\ln \Gamma(n+1) - \ln \Gamma(k+1) - \ln \Gamma(n-k+1) + k \ln(\theta) + (n-k) \ln(1-\theta)]$. Can the CDF $[B(n; k, \theta)]$ be calculated using this formula?

6.11 Derive a recursive algorithm for Poisson probabilities by noting that $p(k+1, \mu) = \frac{\mu}{k+1} p(k; \mu)$.

6.12 Show that the negative binomial and binomial distributions' pmfs are related by $b^*(n; k, \theta) = \frac{k}{n} b(k; n, \theta)$.

6.13 What role does the parameter θ play in the geometric distribution? Can this be generalized to the negative binomial distribution?

6.14 Show that, for the negative binomial distribution, $B^*(2k-1; k, \frac{1}{2}) = \frac{1}{2}$.

6.15 Show that the negative binomial pmf $b^*(n; k, \theta) = \binom{n-1}{k-1} \theta^k (1-\theta)^{n-k}$ for $n = k, k+1, k+2, \ldots$ can also be expressed as $b_s^*(s; k, \mu) = \binom{s+k-1}{s} (\frac{k}{\mu+k})^k (\frac{\mu}{\mu+k})^s$ for $s = 0, 1, 2, \ldots$, where μ is the distribution's mean.

6.16 Derive a recursive algorithm for the negative binomial distribution.

6.17 Derive a recursive algorithm for the hypergeometric distribution.

6.3 CALCULATING CONFIDENCE AND TOLERANCE LIMITS ON NORMAL PERCENTILES

6.3.1 Two-sided equi-tailed confidence limits on normal percentiles

If we know the mean (μ) and standard deviation (σ) of a normal population then, as shown in Section 2.5.1 we can calculate the Pth percentile of that distribution simply from $X_p = \mu + (Z_p)\sigma$, where Z_p is the $100p\%$ile (i.e., the $P\%$ile) of the unit normal distribution. But when we don't know these population parameters (and we seldom do) the best we can do is estimate them from a random sample $X_1, X_2, \ldots, X_n$ using the sample mean $\bar{X}$ and standard deviation S in place of μ and σ. Were we to then use $X_p = \bar{X} + (Z_p)S$ as the estimate of our percentile we would obtain a different value for X_p each time an estimate is made—and recall that classical confidence intervals are based on the idea of repetitive sampling. About half of these estimates would be above the true (but unknown) value, the rest would be below it. To accommodate this variability we construct a two-sided confidence interval with lower limit defined as $\bar{X} + k_{\text{LCL}}S$ and upper limit defined as $\bar{X} + k_{\text{UCL}}S$, where $k_{\text{LCL}} < Z_p$ and $k_{\text{UCL}} > Z_p$.[7] We can calculate these k values by requiring that the areas in each tail of that distribution (the area to the left of k_{LCL} and the area to the right of k_{UCL}) are a mere $\frac{\alpha}{2}$.

Let's first focus on the lower tail. In this case we require that the probability is $\frac{\alpha}{2}$ that the lower confidence limit ($\bar{X} + k_{\text{LCL}}S$) will be no greater than the Pth percentile. We can write this as the following probability statement: $\Pr(\bar{X} + k_{\text{LCL}}S \leq \mu + (Z_p)\sigma) = \frac{\alpha}{2}$. Now we need to find the distribution followed by the term in parentheses. To do so we multiply both sides of the inequality by $\frac{\sqrt{n}}{S}$ (multiplying both sides of an inequality by the same positive term does not change its probability) to obtain

$$\Pr\left(\frac{\frac{\bar{X}-\mu}{\sigma}\sqrt{n} - Z_p\sqrt{n}}{\frac{S}{\sigma}} \leq -k_{\text{LCL}}\sqrt{n}\right) = \frac{\alpha}{2}.$$

The first term on the numerator is the normal deviate [see Equation (2.15)], and it follows the unit normal distribution.[8] The term on the denominator is distributed as $\chi_f/\sqrt{f}$, with $f = n - 1$ degrees of freedom.[9] Standard statistical theory shows that the term in brackets therefore follows the "noncentral t-distribution" (denoted as the t'-distribution) with "noncentrality parameter" $-k_{\text{LCL}}\sqrt{n}$ (Johnson & Kotz 1970, p. 201 [169]). So we write this equation as $\Pr(t' \leq -t| - \gamma) = \frac{\alpha}{2}$, where the critical value is $t = k_{\text{LCL}}\sqrt{n}$ and the (positive) noncentrality parameter is $\gamma = Z_p\sqrt{n}$. Now Owen (1968 [251]) has shown that $\Pr(t' \leq t|\gamma) = 1 - \Pr(t' \leq -t| - \gamma)$, so we may

---

[7] Were the distribution of these estimates around the true percentile to be symmetric we could have defined the confidence limits as Z_p plus or minus a single constant.

[8] From the central limit theorem, the normal deviate $\frac{\bar{X}-\mu}{\sigma}$ follows a normal distribution with mean zero and standard error $\frac{1}{\sqrt{n}}$, so $\frac{\bar{X}-\mu}{\sigma}\sqrt{n}$ is distributed as the unit normal.

[9] Recall from Section 2.7.2 that the standardized variance of normal data ($\frac{fS^2}{\sigma^2}$) follows the chi-square distribution, so the ratio $\frac{S}{\sigma}$ is distributed as $\chi/\sqrt{f}$.

finally write our equation as $\Pr(t' \leq t|\gamma) = 1 - \frac{\alpha}{2}$. We therefore can find k_{LCL} by finding the value of t that satisfies this equation, by iteration, and then calculating $k_{\text{LCL}} = \frac{t}{\sqrt{n}}$.

For the upper confidence limit we need to satisfy the equation $\Pr(\bar{X} + k_{\text{UCL}} S \geq \mu + (Z_p)\sigma) = \frac{\alpha}{2}$. By an entirely similar argument, noting that for continuous variables x and y we have $\Pr(x < y) = \Pr(x \leq y)$ and using the probability rule for mutually exclusive events [Equation (2.3)], we obtain t as the solution of $\Pr(t' \leq t|\gamma) = \frac{\alpha}{2}$ and so calculate $k_{\text{UCL}} = \frac{t}{\sqrt{n}}$.

Table A.4.1 gives values of k_{LCL} and k_{UCL} for four percentiles.

A note on the noncentral t'-distribution. The Student's t-distribution is a special case of the t'-distribution, occurring when $\gamma = 0$ (Figure 6.1). Also, the unit normal distribution is a special case of Student's t-distribution for infinite degrees of freedom, in which case variance is known. Note that whereas the unit normal and Student's t-distributions are symmetric, the noncentral t-distribution is slightly asymmetric.[10] A method for the calculation of its CDF is given in Section 6.3.4.

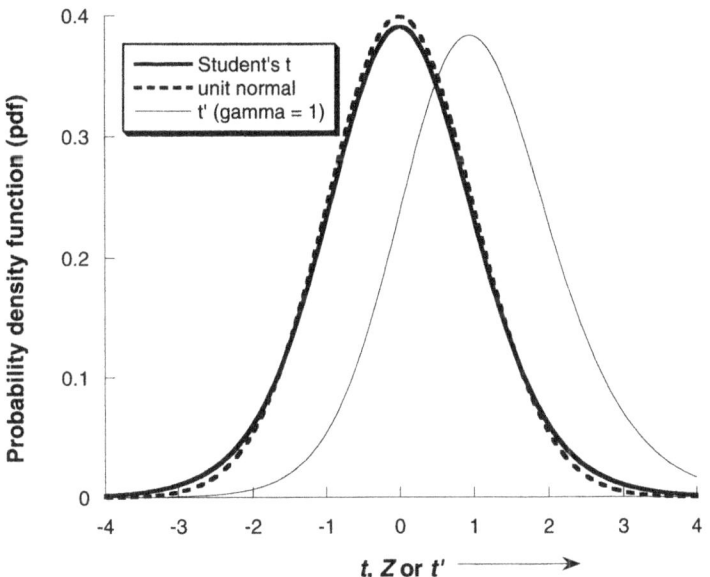

Fig. 6.1 Probability density functions for three sampling distributions: unit normal, Student's t, and noncentral t (for which the variate is t'). Both the t and t' distributions have $f = 12$ degrees of freedom, and t' has noncentrality parameter $\gamma = 1$.

---

[10] The pdf for t' was calculated using Equation (8) of Johnson & Kotz (1970, p. 205 [169]), truncating its infinite sum to 60 terms.

6.3.2 One-sided confidence (tolerance) limits on normal percentiles

The derivation of lower and upper one-sided confidence limits follows the same reasoning as for two-sided confidence limits, except that $\frac{\alpha}{2}$ is replaced by α. In the case of tolerance limits, the probability statement made for the lower limit is that the probability is $1 - \alpha$ that at least a proportion q of the population lies *above* $\bar{X} + kS$. The probability statement made for the upper tolerance limit is that the probability is $1 - \alpha$ that at least a proportion q of the population lies *below* $\bar{X} + kS$. Table A.4.2 gives k for a high (95%) level of confidence for those six percentiles. Results agree with Table 1 in Aldenberg & Jaworska (2000 [2]), except for minor discrepancies for $n = 500$ for the 99%ile and 98%ile cases. The table also gives values of k for median and low (5%) levels of confidence. Aldenberg & Slob (1993 [3]) present equivalent factors for a logistic distribution, sometimes used in ecotoxicity standards.

6.3.3 Two-sided tolerance limits on percentiles

Once again we have a random sample $X_1, X_2, \ldots, X_n$ from a normal distribution with mean μ and variance σ^2. We wish to compute the value of k such that with probability $1 - \alpha$ at least a proportion q of the sampled distribution lies between $\bar{X} - kS$ and $\bar{X} + kS$. This problem was solved by Wald & Wolfowitz (1946 [332]; see also Odeh & Owen 1980 [246]). In brief, the procedure is to solve the following equation for k

$$\frac{2\sqrt{n}}{\sqrt{2\pi}} \int_0^\infty \Pr\left(\chi_f^2 \geq \frac{fr^2}{k^2}\right) e^{-\frac{1}{2}n\bar{X}^2} \, d\bar{X} = 1 - \alpha, \qquad (6.24)$$

where r is a variable to be calculated by satisfying $G(\bar{X} + r) - G(\bar{X} - r) = q$. Table A.4.3 gives the k factors for three confidence levels, each with four coverages. It was calculated by integrating Equation (6.24) using Simpson's rule with increments of $\sqrt{n}\bar{X}$ of 0.05 (as reported by Odeh & Owen 1980 [246]) with an upper limit of 10 (beyond which the integrand effectively vanishes). The equation for r was solved by a Newton–Raphson technique. The results in Table A.4.3 agree with parallel cases in Tables 3.4 (for 90%, 95% and 99% coverages) of Odeh & Owen (1980 [246]).

Note that Wald and Wolfowitz also reported an approximate procedure in which $k = r\sqrt{\frac{f}{\chi_{f,1-\alpha}^2}}$ and r is calculated by solving the equation $G(\frac{1}{\sqrt{n}} + r) - G(\frac{1}{\sqrt{n}} - r) = q$. The results of this approximate procedure have been tabled in a number of places (e.g., Bowker & Lieberman 1972 [31]). They differ from the exact values in Table A.4.3 by at most 1.8%—generally by much less.

Finally, note that this tolerance interval is of the *Control Center* type; that is, it controls the proportion of the distribution between its (equal-area) tails. There is another type, *Control Both Tails* (Odeh & Owen 1980 [246]) in which each tail area is controlled separately. These intervals are generally slightly narrower than those of the *Control Center* type.

6.3.4 Calculating noncentral t-probabilities

Integral form

$$\Pr(t' \leq t|\gamma) = \frac{\sqrt{2\pi}}{\Gamma(\frac{1}{2}f)2^{\frac{1}{2}(f-2)}} \int_0^\infty G\left(\frac{tx}{\sqrt{f}} - \gamma\right) x^{f-1} G'(x)\, dx,$$

for $f > 0$, where $G(x)$ are the DF and pdf for the unit normal distribution, and γ is the noncentrality parameter. An algorithm for $G(x)$ has been given in Equation (2.5.1) of Chapter 2. Other, equivalent, integral forms of Equation (6.3.4) may be stated. This is the one used by Owen (1965 [250]) to obtain his remarkable algorithm.

Algorithm Owen (1965 [250]) obtained this algorithm by integrating repeatedly by parts, resulting in the following.

For $f = 1$

$$\Pr(t' \leq t|\gamma) = G(-\gamma\sqrt{B}) + 2T(\gamma\sqrt{B}, A).$$

For f even and > 1

$$\Pr(t' \leq t|\gamma) = G(-\gamma) + \sqrt{2\pi}(M_0 + M_2 + \cdots + M_{f-2}).$$

For f odd and > 1

$$\Pr(t' \leq t|\gamma) = G(-\gamma\sqrt{B}) + 2T(\gamma\sqrt{B}, A) \\ + 2(M_1 + M_3 + \cdots + M_{f-2}),$$

where $A = \frac{t}{\sqrt{f}}$ and $B = \frac{f}{f+t^2}$ and where

$$T(h, a) = \frac{1}{2\pi} \int_0^a \frac{\exp\left[-\frac{1}{2}h^2(1+x^2)\right]}{1+x^2}\, dx, \qquad (6.25)$$

is a function related to the unit normal distribution, as discussed by Owen (1956 [248]), and

$$\begin{aligned} M_0 &= A\sqrt{B}\, G'(\gamma\sqrt{B})\, G(\gamma A\sqrt{B}), \\ M_1 &= B\left[\gamma A M_0 + \frac{A}{\sqrt{2\pi}} G'(\gamma)\right], \\ M_2 &= \frac{1}{2} B\left[\gamma A M_1 + M_0\right], \\ &\vdots \\ M_k &= \frac{k-1}{k} B\left[a_k \gamma A M_{k-1} + M_{k-2}\right], \end{aligned}$$

where

$$a_k = \frac{1}{(k-2)a_{k-1}} \quad \text{for } k \geq 3, \text{ with } a_2 = 1.\text{[11]}$$

This is an extremely accurate algorithm, being theoretically exact for even f; for odd f accuracy is limited only be the accuracy of the evaluation of $T(h,a)$. However, care is needed for large γ—computer arithmetic precision errors can cause $G'(\gamma)$, and possibly $G'(\gamma\sqrt{B})$, to be zero, and so M_0 is zero also, in which case all subsequent calculations are in error. In that case one can use the approximate result given by Abramowitz & Stegun (1972 [1])

$$\Pr(t' \leq t|\gamma) \approx G\left[\frac{t\left(1 - \frac{1}{4f}\right) - \gamma}{\sqrt{1 + \frac{t^2}{2f}}}\right].$$

The algorithm is similar to that presented by Cooper (1985b [58]). The $T(h,a)$ integral admits analytical expressions in only a few special cases,[12] and so it must be evaluated numerically. To do that Simpson's rule has been used, with 10,000 segments, in which case results agree to 16 decimal places with those obtained using Mathematica® (Wolfram 1999 [344]). It is more accurate that than those obtained using the TFN program (Young & Minder 1985 [347]) where agreement is down to the eighth decimal place only (superior accuracy can be desirable if repetitive calculations are to be performed).

This distribution has a number of special cases discussed in Owen (1965, 1968 [250, 251]) and Odeh & Owen (1980 [246]); that is,

$$\begin{align}
\Pr(t' \leq t|\gamma) &= 1 - \Pr(t' \leq -t|-\gamma), \\
\Pr(t' \leq 0|\gamma) &= G(-\gamma), \\
\Pr(t'_{f=1} \leq 1|\gamma) &= 1 - P^2, \text{ where } P = G(\gamma/\sqrt{2}), \\
\Pr(t'_{f=1} \leq t|0) &= \frac{1}{2} + \frac{\tan^{-1}(t)}{\pi}.
\end{align}$$

Problems

6.18 Show that the ordinate for the HDR (highest density region) on Figure 3.5 is 0.0170.

6.19 How would you construct an algorithm to find the HDR of an asymmetric unimodal distribution?

---

[11] So $a_3 = 1, a_4 = \frac{1}{2}, a_5 = \frac{2}{3}, a_6 = \frac{3}{8}, a_7 = \frac{8}{15}, a_8 = \frac{15}{48}, \ldots$
[12] From Owen (1956 [248]): $T(h,0) = 0; T(0,a) = \frac{1}{2\pi}\tan^{-1}(a)$ [and so $T(0,\infty) = \frac{1}{4}$]; $T(h,1) = \frac{1}{2}G(h)[1 - G(h)]; T(h,\infty) = \frac{1}{2}[1 - G(h)]$ if $h \geq 0, T(h,\infty) = \frac{1}{2}G(h)$ if $h \leq 0; T(\infty,a) = 0$.

6.4 DEVELOPMENT OF THE DETECTION FORMULAE

The development of decision rules for these tests is well described in standard texts. These rules are of course based on *sample* statistics and give rise to the detection regions described in the main text. Development of detection probability formulae is not so well known, and it is now given. Sampling is assumed to be from normal populations in all cases. Note that, in contrast to decision rules, detection curves are based on *population* statistics.

6.4.1 One-sided test, unknown variance

Consider the one-sample test of the superiority hypothesis **H**: $\mu \geq \mu_0$. We would commit a Type II error by failing to reject this hypothesis when in fact it was false (i.e., when **K** was true). From the decision rule given in Tables 5.3 and 5.4 (i.e., reject **H** if the test statistic $T = \sqrt{n}\frac{\bar{X}-\mu_0}{S} < -t_{1-\alpha,f}$, where $f = n - 1$) we see that the Type II error would occur if (merely by bad luck) we did not get a value of the test statistic less than $T < -t_{1-\alpha,f}$ when in fact μ was less than μ_0. The probability of this error happening is therefore $\beta = \Pr(T \geq -t_{1-\alpha,f} | \mu < \mu_0)$.

We now need a means of computing this probability. To do so, we define the ratio $V = \frac{\bar{X}-\mu}{\sigma/\sqrt{n}}$, where V is the ratio of the difference between the sample and population means to the standard error of the population mean. Then, noting from Table 5.4 that $\gamma = \Delta\sqrt{n} = \frac{\sqrt{n}(\mu-\mu_0)}{\sigma}$, we have $T = \frac{V+\gamma}{W}$, where $W = \frac{S}{\sigma}$ is the ratio of sample to population standard deviations. From standard statistical theory (e.g., Freund 1992 [101]), V follows the unit normal distribution[13] for *any* value of μ. Also, W follows the chi-distribution, $W \sim \frac{\chi_f}{\sqrt{f}}$ (i.e., W^2 is distributed as $\frac{\chi^2}{f}$, with f degrees of freedom). Under these conditions, T follows the "noncentral t-distribution"— denoted as the t'-distribution—with "noncentrality parameter" γ (Johnson & Kotz 1970, p. 201 [169]), as discussed in Section 6.3.1.

The preceding argument shows that the Type II error risk, occurring "under **K**" (i.e., when **H** is false), follows the t'-distribution (because T follows that distribution). In that case we have $\Delta < 0$ and so $\gamma < 0$. Therefore $\beta = \Pr(t' \geq -t_{1-\alpha,f}|\gamma)$, for $\gamma < 0$. Accordingly, Power $= 1 - \beta = \Pr(t' < -t_{1-\alpha,f}|\gamma)$, with $\gamma < 0$, and it follows that the Type I error risk is $\Pr(t' < -t_{1-\alpha,f}|\gamma)$, with $\gamma \geq 0$. Therefore the detection probability is just $\Pr(t' < -t_{1-\alpha,f}|\gamma)$, with no restriction on the sign of γ.

Note that once the significance level (typically 5%) and sample size are set, then $t_{1-\alpha,f}$ is a constant. However, γ varies with μ and thus power varies with μ also. Accordingly, power can be computed as a function of either γ or the effect size Δ (because $\gamma = \Delta\sqrt{n}$). It is conventional to report it as a function of Δ.

Development of the detection formulae for the two-sample case is straightforward; all that need be noted is that the definitions of f and γ change to $f = 2(n-1)$ and $\gamma = \Delta\sqrt{\frac{n}{2}}$.

---
[13]That is, $V \sim n(0, 1)$, a normal distribution with zero mean and unit variance.

6.4.2 One-sided test, known variance

Consider again the one-sample test of the superiority hypothesis **H**: $\mu \geq \mu_0$. Following the reasoning for the unknown-variance case, $\beta = \Pr(T \geq -Z_{1-\alpha}|\mu < \mu_0)$. To discover the form of this probability, we define V as before, noting that $S = \sigma$ (because the variance is known) and thus $W = 1$. Again, V is distributed as the unit normal, for any value of μ. Therefore, noting that $T = V + \Delta\sqrt{n} = V + \gamma$, then Power = $1 - \beta = \Pr(V + \gamma < -Z_{1-\alpha}) = \Pr(V < -Z_{1-\alpha} - \gamma) = G(-Z_{1-\alpha} - \gamma)$ for $\gamma < 0$. The Type I error risk is $G(-Z_{1-\alpha} - \gamma)$, with $\gamma \geq 0$. Accordingly, the detection probability is just $G(-Z_{1-\alpha} - \gamma)$, with no restriction on the sign of γ.

The variance-known formulae may also be derived from the variance-unknown case by noting that as the degrees of freedom become very large then $\Pr(t' < t_{1-\alpha,f}|\gamma) \to G(Z_{1-\alpha} - \gamma)$.[14] That is, the asymptote of the noncentral t-distribution as $f \to \infty$ is the unit normal distribution whose mean is shifted to the right by the noncentrality parameter. For example, consider Figure 6.1 as $f \to \infty$ while maintaining $\gamma = 1$ (i.e., considering ever-smaller differences between means). The asymptotic form of the t'-distribution is the dashed line shifted to the right by γ (in that case by one unit on the abscissa). Again, extension to the two-sample case is straightforward.

6.4.3 Point-null hypothesis test, unknown variance

Consider the one-sample test of $\mathbf{H_0}$: $\mu = \mu_0$. We would commit a Type II error by failing to reject this hypothesis when in fact it was false (i.e., when **K** was true). From the decision rule given in Tables 5.4 and 5.7 (i.e., reject $\mathbf{H_0}$ if $|T| = \sqrt{n}\frac{|\bar{X} - \mu_0|}{S_e} > t_{1-\alpha/2,f}$) we see that the Type II error would occur if (merely by bad luck) we did not get a value of $|T|$ greater than $t_{1-\alpha/2,f}$ when in fact $\mu \neq \mu_0$. The probability of this error happening is therefore $\beta = \Pr(-t_{1-\alpha/2,f} \leq T \leq t_{1-\alpha/2,f}|\mu \neq \mu_0)$.

We now need a means of computing this probability. To do so we define the ratios V and T as before (Section 6.4.1) so that T follows the t'-distribution. Therefore the Type II error risk, occurring "under **K**" (i.e., when $\mathbf{H_0}$ is false), follows the t'-distribution (because T follows that distribution when $\mathbf{H_0}$ is false). And so, when $\gamma \neq 0$, $\beta = \Pr(-t_{1-\alpha/2,f} \leq t' \leq t_{1-\alpha/2,f})$, so that Power = $1 - \beta = \Pr(t' < -t_{1-\alpha/2,f}) + \Pr(t' > t_{1-\alpha/2,f}) = \Pr(t' < -t_{1-\alpha/2,f}) + [1 - \Pr(t' \leq t_{1-\alpha/2,f})]$. Noting that $\Pr(x \leq y) = \Pr(x < y)$ for continuous variables, we can write Power = $\Pr(t' < -t_{1-\alpha/2,f}|\gamma) + [1 - \Pr(t' < t_{1-\alpha/2,f}|\gamma)]$. Owen (1968 [251]) has shown that $\Pr(t' \leq t|\gamma) = 1 - \Pr(t' \leq -t|-\gamma)$, so we may finally write the power equation as $\Pr(t' < -t_{1-\alpha/2,f}|\gamma) + \Pr(t' < -t_{1-\alpha/2,f})|-\gamma)$, as in Table 5.7. By also using this equation when $\mathbf{H_0}$ is true ($\gamma = 0$), we have the power formula as the detection probability.

Note that once the significance level (typically 5%) and sample size are set then $t_{1-\alpha/2,f}$ is a constant. However, γ varies with μ and thus power varies with μ also. Accordingly, power can be computed as a function either of γ or of the effect size Δ

---

[14] From formulae presented by Owen (1968, Section 15 [251]) we have γ and 1 as the asymptotic mean and variance of the noncentral t-distribution as $f \to \infty$.

(because $\gamma = \Delta\sqrt{n}$). It is conventional to report it as a function of Δ. Extension to the two-sample case is straightforward, as it was for one-sided tests.

6.4.4 Point-null hypothesis test, known variance

Consider again the one-sample test of the point-null hypothesis $\mathbf{H_0}$: $\mu = \mu_0$. Following the reasoning for the unknown-variance case, $\beta = \Pr(-Z_{1-\alpha/2} \leq T \leq Z_{1-\alpha/2}|\mu \neq \mu_0)$. To discover the form of this probability we define V as before, noting that $S = \sigma$ (because the variance is known) and so $W = 1$. Again, V is distributed as the unit normal, for any value of μ, as is T. So, noting that $T = V + \Delta\sqrt{n} = V + \gamma$, when $\gamma \neq 0$ we have Power $= 1-\beta = \Pr(V+\gamma < -Z_{1-\alpha/2}) + \Pr(V+\gamma > Z_{1-\alpha/2})$, which simplifies to Power $= G(-Z_{1-\alpha/2} - \gamma) + G(-Z_{1-\alpha/2} + \gamma)$. By also using this equation when $\mathbf{H_0}$ is true ($\gamma = 0$), we have the power formula as the detection probability.

The variance-known formulae may also be derived from the variance-unknown case by noting that as the degrees of freedom become very large then $\Pr(t' < t_{1-\alpha/2,f}|\gamma) \to G(Z_{1-\alpha/2}-\gamma)$. That is, the asymptote of the noncentral t-distribution as $f \to \infty$ is a simple unit normal distribution whose mean is shifted to the right by the noncentrality parameter. Again, see Figure 6.1. The two-sample case is a trivial extension of the above.

6.4.5 Equivalence hypothesis, known variance

When the variance is known, interval tests for the means of normal populations have desirable properties.[15] Such tests will be the most powerful of all possible alternatives, over all possible differences. These properties mean that the tests will be "of size α." Consequently, at the at the edge of the equivalence interval the test's probability of rejecting a true hypothesis is guaranteed to *equal* the test's a priori significance level (α).[16]

Consider a single-sample case. The hypotheses are: $\mathbf{H_e}$: $\mu_L \leq \mu \leq \mu_U$ and $\mathbf{K_e}$: $\mu < \mu_L$ or $\mu > \mu_U$. Given data $X_1, X_2, \ldots, X_n$, the decision rule for the test of $\mathbf{H_e}$ is

$$\text{Reject } \mathbf{H_e} \text{ if } \bar{X} < C_L \text{ or } \bar{X} > C_U,$$

where C_L and C_U are critical values of $\bar{X}$ chosen so that

$$\int_{C_L}^{C_U} g\left(\bar{X}; \mu_L\right) \, d\bar{X} = \int_{C_L}^{C_U} g\left(\bar{X}; \mu_U\right) \, d\bar{X} = 1-\alpha, \qquad (6.26)$$

where g is the probability density function of $\bar{X}$ when μ is the parameter (Mood & Graybill 1963, p. 296 [230]). Equation (6.26) guarantees that the chance of falsely rejecting $\mathbf{H_e}$ at either edge of the equivalence interval is α. The power of this test

---

[15] There is a UMP test of hypothesis $\mathbf{H_i}$ and a UMPU test of hypothesis $\mathbf{H_e}$ (Wellek 2003, p. 38 [340]).
[16] The tests would be of "level" α if that probability was *no greater* than α—Casella & Berger 2002 [45].

(denoted by Θ) is given by the complement of the type II error (the probability of getting $C_L \leq \bar{X} \leq C_U$, when in fact $\mu < \mu_L$ or $\mu > \mu_U$), that is

$$\Theta = 1 - \int_{C_L}^{C_U} g(\bar{X}; \mu) \, d\bar{X} \qquad \text{for } \mu < \mu_L \text{ or } \mu > \mu_U. \tag{6.27}$$

Solving for the critical values We calculate C_L and C_U by noting that $\bar{X} \sim \mathrm{n}\left(\mu, \frac{\sigma^2}{n}\right)$ and so $g(\bar{X}, \mu) = \exp\left[-\frac{1}{2}\left(\frac{\bar{X}-\mu}{\sigma/\sqrt{n}}\right)^2\right] \Big/ (\sqrt{2\pi}\sigma/\sqrt{n})$. Equations (6.26) and (6.27) then become

$$G\left(\frac{C_U - \mu_L}{\sigma/\sqrt{n}}\right) - G\left(\frac{C_L - \mu_L}{\sigma/\sqrt{n}}\right) = 1 - \alpha, \tag{6.28}$$

and

$$G\left(\frac{C_U - \mu_U}{\sigma/\sqrt{n}}\right) - G\left(\frac{C_L - \mu_U}{\sigma/\sqrt{n}}\right) = 1 - \alpha, \tag{6.29}$$

where $G(x)$ is the cumulative probability distribution function for the unit normal distribution; that is,

$$G(x) = \int_{-\infty}^{x} \frac{1}{\sqrt{2\pi}} e^{-\frac{\xi^2}{2}} \, d\xi,$$

and ξ is a dummy variable of integration. To solve Equations (6.28) and (6.29) we first define dimensionless test statistic values as

$$c_L = \frac{C_L - \mu_0}{\sigma/\sqrt{n}} \quad \text{and} \quad c_U = \frac{C_U - \mu_0}{\sigma/\sqrt{n}},$$

where

$$\mu_0 = \frac{\mu_L + \mu_U}{2},$$

lies at the middle of the equivalence interval, so that the test's decision rule is now stated in terms of dimensionless quantities, as

$$\text{Reject } \mathbf{H_e} \text{ if } T < c_L \text{ or } T > c_U \qquad \text{where } T = \frac{\bar{X} - \mu_0}{\sigma/\sqrt{n}}.$$

Also, we define the effect size as

$$\Delta = \frac{\mu - \mu_0}{\sigma}.$$

Defining the "noncentrality parameter" as

$$\gamma = \frac{\mu - \mu_0}{\sigma/\sqrt{n}} = \sqrt{n}\Delta,$$

we can calculate c_L and c_U by requiring that

$$G(c_U - \gamma_L) - G(c_L - \gamma_L) = 1 - \alpha, \quad (6.30)$$

and

$$G(c_U - \gamma_U) - G(c_L - \gamma_U) = 1 - \alpha. \quad (6.31)$$

Now, it may simply be shown that $\gamma_U = -\gamma_L$ and, less simply, that $c_U = -c_L (= c_e)$.[17] Therefore, using the property of the unit normal distribution that $G(x) = 1 - G(-x)$ [Equation (2.17)], we can collapse Equations (6.30) and (6.31) to a single equation, requiring that

$$G(c_e + \gamma_U) - G(-c_e + \gamma_U) = 1 - \alpha, \quad (6.32)$$

and so the decision rule is simply

Reject $\mathbf{H_e}$ if $|T| > c_e$,

where c_e is the value that satisfies Equation (6.32) (see Table 5.9). This value has been solved using standard Newton–Raphson numerical root-finding techniques.[18]

Detection probability From Equation (6.27) the power of the test is

$$\Theta = 1 - \left[G\left(\frac{C_U - \mu}{\sigma/\sqrt{n}}\right) - G\left(\frac{C_L - \mu}{\sigma/\sqrt{n}}\right) \right] \quad \text{for } \mu < \mu_L \text{ or } \mu > \mu_U,$$

which may be simplified to the form shown of the detection probability on Table 5.9

$$D = \Theta = G(-c_e + \gamma) + G(-c_e - \gamma), \quad (6.33)$$

removing the restriction on the range of μ. Derivation of decision rules and power functions for the two-sample case is a simple extension of the above developments, noting that the difference is sample means is distributed as $d \sim n(\delta, 2\sigma^2/n)$, whereas $\bar{X} \sim n(\mu, \sigma^2/n)$.

6.4.6 Inequivalence hypothesis, known variance

For the single sample case the hypotheses are: $\mathbf{H_i}$: $\mu < \mu_L$ or $\mu > \mu_U$ and $\mathbf{K_i}$: $\mu_L \leq \mu \leq \mu_U$. Given data $X_1, X_2, \ldots, X_n$, the decision rule for the test of $\mathbf{H_i}$ is

Reject $\mathbf{H_i}$ if $C_L < \bar{X} < C_U$,

---

[17] Using Equation (2.17) we can write Equation (6.30) as $G(-c_L + \gamma_L) - G(-c_U + \gamma_L) = 1 - \alpha$. Then, noting that $\gamma_L = -\gamma_U$, and comparing this equation with Equation 6.31, we have $-c_L = c_U$ and $-c_U = c_L$.

[18] That is, define $F(c_e) = G(c_e + \gamma_U) - G(-c_e + \gamma_U) - (1 - \alpha)$. Given the unit normal distribution, $F'(c_e) = \{\exp[-\frac{1}{2}(c_e + \gamma_U)^2] + \exp[-\frac{1}{2}(-c_e + \gamma_U)^2]\}/\sqrt{2\pi}$, and the kth approximation to c_e is given by $(c_e)_k = (c_e)_{k-1} - F[(c_e)_{k-1}]/F'[(c_e)_{k-1}]$, with a starting value derived by taking $C_U = \delta_U$ and so $(c_e)_{k=0} = \gamma_U$.

where C_L and C_U are critical values of $\bar{X}$ chosen so that

$$\int_{C_L}^{C_U} g(\bar{X}; \mu_L) \, d\bar{X} = \int_{C_L}^{C_U} g(\bar{X}; \mu_U) \, d\bar{X} = \alpha,$$

guaranteeing that the test will only reject $\mathbf{K_i}$ at a rate of α at the edges of the equivalence interval. The power of this test is

$$\Theta = 1 - \left[\int_{-\infty}^{C_L} g(\bar{X}; \mu) \, d\bar{X} + \int_{C_U}^{\infty} g(\bar{X}; \mu) \, d\bar{X} \right] \quad \text{for } \mu_L \leq \mu \leq \mu_U. \quad (6.34)$$

Solving for the critical values To obtain the critical values we proceed as above, obtaining

$$G(c_i + \gamma_U) - G(-c_i + \gamma_U) = \alpha, \quad (6.35)$$

and the decision rule is simply

$$\text{Reject } \mathbf{H_i} \text{ if } |T| \leq c_i,$$

where c_i is the value that satisfies Equation (6.35) (see Table 5.9). This value is also solved in the calculator by using standard Newton–Raphson numerical root-finding techniques.[19]

Detection probability From Equation (6.34) the power of the test may be simplified to

$$\Theta = G(c_i - \gamma) - G(-c_i - \gamma) \quad \text{for } \mu_L \leq \mu \leq \mu_U.$$

The detection probability is therefore

$$D = 1 - \Theta = G(-c_i + \gamma) + G(-c_i - \gamma), \quad (6.36)$$

as shown on Table 5.9.

Once again, derivation of decision rules and power functions for the two-sample case is a simple extension of the above developments, noting that the difference is sample means is distributed as $d \sim n(\delta, 2\sigma^2/n)$, whereas $\bar{X} \sim n(\mu, \sigma^2/n)$.

6.4.7 (In)equivalence hypotheses, unknown variance

When the variance is unknown, interval tests for the means of normal populations will have neither the UMP nor UMPU property. That means that no single test will be the most powerful in all situations.

The starting point for these tests is the inequivalence hypothesis for the two sample case using the two one-sided test ("TOST") for two samples, as proposed by

---

[19]Redefining $F(c_i)$ as $F(c_i) = G(c_i + \gamma_U) - G(-c_i + \gamma_U) - \alpha$.

Schuirmann (1987 [283]). While having its detractors (Berger & Hsu 1996 [18]) this test does possess generally desirable power over a wide range of effect sizes (Schuirmann 1996 [284]). It is also simple to perform, requiring only the use of standard tables of the Student t-distribution. Its essence is that the $\mathbf{H_i}$ hypothesis is rejected at level α if and only if a symmetrical $1 - 2\alpha$ confidence interval for the difference in means is completely contained in the equivalence interval.[20] The power function for this test was developed by Phillips (1990 [258]), based on the incomplete noncentral t-distribution function (Q_f) derived by Owen (1965 [250]), for which an accurate algorithm is given in Section 6.4.9. The single sample case uses this procedure also, substituting the appropriate standard error.

For tests of the equivalence hypothesis $\mathbf{H_e}$ we simply invert the Schuirmann procedure. That is, we can accept the equivalence hypothesis at level α if any part of a symmetrical $1 - 2\alpha$ confidence interval for the difference in means extends into the equivalence interval, for either the single-sample or two-sample cases. Its power function was given by McBride (1999 [205]).

The manner in which these confidence intervals correspond to equivalence tests is shown diagrammatically in Fig 5.10.

These tests can also be constructed from the "Union–Intersection" and "Intersection–Union" approaches (Casella & Berger 2002 [45]).

6.4.8 Detection regions for equivalence tests and the Power Approach

Consider the same-size, two-sample variance-unknown cases in Tables 5.8–5.10. To make matters simple, and without loss of generality, we will take an equivalence interval symmetric about zero so that $\delta_U = -\delta_L$; as a consequence, $\delta_0 = 0$, and $\gamma_U = -\gamma_L = \frac{\delta_U}{S_{e,\bar{x}-\bar{y}}}$. In testing the equivalence hypothesis, equivalence is inferred when we accept $\mathbf{H_e}$. Now from Table 5.9, when the variance is unknown we *reject* $\mathbf{H_e}$ when $T_l < -t_{1-\alpha,f}$ or $T_r > t_{1-\alpha,f}$. So we would *accept* $\mathbf{H_e}$ when $T_l \geq -t_{1-\alpha,f}$ and $T_r \leq t_{1-\alpha,f}$. When the variance is known, we accept $\mathbf{H_e}$ when $|T| \leq c_e$, where c_e satisfies $G(c_e + \gamma_U) - G(-c_e + \gamma_U) = 1 - \alpha$. Therefore the boundaries between the detection and nondetection regions for tests of the equivalence hypothesis are defined by

$$S_{e,\bar{x}-\bar{y}} = \begin{cases} \frac{\delta_L - d}{t_{1-\alpha,f}} \text{ for } d \leq \delta_L \text{ and } \frac{d-\delta_U}{t_{1-\alpha,f}} \text{ for } d \geq \delta_U & \text{unknown variance,} \\ \frac{d}{c_e} & \text{known variance.} \end{cases}$$

Similarly, in testing the inequivalence hypothesis with unknown variance equivalence is inferred, by rejecting $\mathbf{H_i}$, if $T_l > t_{1-\alpha,f}$ and $T_r < -t_{1-\alpha,f}$. When the variance is known, equivalence is inferred if $|T| < c_i$ where c_i satisfies $G(c_i +$

---

[20] The procedure uses a $1 - 2\alpha$ confidence interval in an α level test (*not* a $1 - \alpha$ interval) because it is an amalgam of two comparisons, each using an α-level one-sided interval—hence the name "two one-sided test". The $1 - 2\alpha$ interval must be symmetric about the difference in sample means for this to work (Berger & Hsu 1996 [18]).

$\gamma_U) - G(-c_i + \gamma_U) = \alpha.$[21] In this case the boundaries between the detection and nondetection regions are defined by

$$S_{e,\bar{x}-\bar{y}} = \begin{cases} \frac{d-\delta_L}{t_{1-\alpha,f}} \text{ for } \delta_L \leq d \leq 0 \text{ and } \frac{\delta_U-d}{t_{1-\alpha,f}} \text{ for } 0 \leq d \leq \delta_U & \text{unknown variance,} \\ \frac{d}{c_i} & \text{known variance.} \end{cases}$$

Figure 5.12 shows examples of the shape of these regions. In Figure 5.13 we superpose the detection region for the point-null hypothesis t-test (variance-unknown) with power to detect any difference beyond the equivalence interval constrained to be no less than 80%. This is known as the "Power Approach" (Schuirmann 1987 [283]; Chow & Liu 1992 [46]). From Tables 5.4 and 5.7 it is a simple matter to show that the sloping sides of this region are given by

$$S_{e,\bar{x}-\bar{y}} = \frac{d}{t_{1-\alpha/2,f}}. \tag{6.37}$$

The top of this region is derived by requiring that two conditions are satisfied (Chow & Liu 1992 [46]): (a) $\mathbf{H_0}$ is rejected (i.e., we have "detection"), and (b) there is sufficient power ($\Theta_{\min}$) to detect differences beyond the equivalence interval. This means that we require that

$$\Pr(|T| > t_{1-\alpha/2,f}, \text{ given } |\delta| = \delta_U) \geq \Theta_{\min}.$$

(The usual—arbitrary—convention is to take $\Theta_{\min} = 0.8$.) This probability statement can be simplified by noting that T follows the noncentral t-distribution with noncentrality parameter γ, and so we require that

$$\Pr(t' \leq t_{1-\alpha/2,f} | \gamma = \gamma_U) < \beta_{\max},$$

where t' is the noncentral t-variate and $\beta_{\max} (= 1 - \Theta_{\min})$ is the maximum permissible Type II error rate. From Table 5.10 the noncentrality parameter is defined as $\gamma_U = \frac{\delta_U}{\sigma}\sqrt{\frac{n}{2}}$ (noting that $\delta_0 = 0$) and this reveals a conceptual difficulty with the Power Approach (Schuirmann 1987 [283]). That is, γ_U contains an unknown parameter (σ), but we need to know its value in order to calculate the t'-probability. Perhaps the best we can do is to replace σ by our best estimate of its value (the pooled samples' standard deviation, S_p) and so use $\hat{\gamma}_U$ $(= \frac{\delta_U}{S_p}\sqrt{\frac{n}{2}})$ instead of γ_U. Also, we note that while accurate calculation of the noncentral t-probability can be achieved using formal algorithms (see Section 6.4.9), or by interpolation in published tables (e.g., Owen 1962 [249]; Pearson & Hartley 1972 [256]), the resulting procedure is somewhat less than elegant. This clumsiness may be obviated by using an approximation based on the central t-distribution (Chow & Liu 1992 [46]). That is

---

[21] Both c_i and c_e are decreasing functions of γ_U, and $c_i < \gamma_U < c_e$, always. For example at $\alpha = 0.05$ and $\gamma_U = 1.1$ we have $c_i = 0.11470$ (as reported in Table 4.1 by Wellek 2003 [340]) and $c_e = 2.74544$. For $\alpha = 0.05$ and $\gamma_U = 500$ we have $c_i = 498.355$ and $c_e = 501.645$.

$$\Pr(t' \leq t_{1-\alpha/2,f}|\hat{\gamma}_U) \approx \Pr(t \leq t_{1-\alpha/2,f} - \hat{\gamma}_U),$$

where t is distributed as central (Student's) t. (By referring to Figure 6.1 we can see the reasonableness of this approximation, given that t' is typically only slightly skewed.[22]) We therefore obtain a requirement that $\Pr(t \leq t_{1-\alpha/2,f} - \hat{\gamma}) < \beta_{\max}$. Finally, using $\hat{\gamma}_U = \delta_U/S_{e,\bar{x}-\bar{y}}$ (as may be simply derived from Table 5.10, we can invert the probabilities in this inequality to require that

$$t_{1-\alpha/2,f} - \frac{\delta_U}{S_{e,\bar{x}-\bar{y}}} < -t_{1-\beta_{\max},f}.$$

Therefore the detection region's flat top is located at

$$S_{e,\bar{x}-\bar{y}} = \frac{\delta_U}{t_{1-\alpha/2,f} + t_{1-\beta_{\max},f}}. \tag{6.38}$$

For example, in Figure 5.13 we have $f = 18$, $\alpha = 0.05$, $\beta_{\max} = 0.20$ and $\delta_U = 20$. Then, using Table A.2.2, $t_{1-\alpha/2,f} = 2.1009$ and $t_{1-\beta_{\max},f} = 0.8620$. The flat top is therefore located at a standard error value of $S_{e,\bar{x}-\bar{y}} = 20/(2.101+0.862) = 6.7501$.

Finally, by equating Equations (6.37) and (6.38) we obtain the width of the flat top as

$$d = \pm \delta_U \frac{t_{1-\alpha/2,f}}{t_{1-\alpha/2,f} + t_{\Theta_{\min}}},$$

which, for power greater than 50%, is always less than the width of the equivalence interval.

Schuirmann (1987 [283]) demonstrates that the Power Approach test does not maintain a true level of significance below α. For infinite degrees of freedom, this level approaches β—an undesirable property.

6.4.9 Calculating the incomplete noncentral t-probabilities

Integral form

$$Q_f(t,\gamma;0,R) = \frac{\sqrt{2\pi}}{\Gamma\left(\frac{1}{2}f\right) 2^{\frac{1}{2}(f-2)}} \int_0^R G\left(\frac{t\gamma}{\sqrt{f}} - \gamma\right) x^{f-1} G'(x)\,dx,$$

for $f, R > 0$ where A and B are defined as previously.

Algorithm Owen (1965 [250]) obtained this algorithm by integrating repeatedly by parts, resulting in the following.

---

[22] The skewness coefficient for the t'-distribution is rather complex.

f odd : $Q_f(t, \gamma; 0, R) =$
$$G(R) - 2T\left(R, \frac{AR - \gamma}{R}\right) - 2T\left(\gamma\sqrt{B}, \frac{\gamma AB - R}{B\gamma}\right)$$
$$+ 2T(\gamma\sqrt{B}, A) + \begin{Bmatrix} 0 \text{ if } \gamma < 0 \\ -1 \text{ if } \gamma \geq 0 \end{Bmatrix}$$
$$+ 2\left(M_1^* + H_1 + M_3^* + H_3 + \cdots + M_{f-2}^* + H_{f-2}\right),$$

where $T(h, a)$ is defined by Equation (6.25),

f even : $Q_f(t, \gamma; 0, R) = G(-\gamma)$
$$+ \sqrt{2\pi}\left(M_0^* + H_0 + M_2^* + H_2 + \cdots + M_{f-2}^* + H_{f-2}\right),$$

and where

$$M_0^* = A\sqrt{B}G'(\gamma\sqrt{B})\left[G(\gamma A\sqrt{B}) - G\left(\frac{\gamma AB - R}{\sqrt{B}}\right)\right],$$
$$M_1^* = B\left[\gamma A M_0^* + A\frac{G'(\gamma)}{\sqrt{2\pi}}\right] - L_0,$$
$$M_2^* = \frac{1}{2}B\left[\gamma A M_1^* + M_0^*\right] - L_1,$$
$$\vdots$$
$$M_k^* = \frac{k-1}{k}B\left[a_k\gamma A M_{k-1}^* + M_{k-2}^*\right] - L_{k-1}.$$

(A more complex, but equivalent, form for M_1^* was given by Owen[23]). The a_k terms are defined as for the noncentral t-distribution, and

$$\begin{aligned} L_0 &= ABG'(R)G'(AR - \gamma), \\ L_{k-1} &= a_{k+2}RL_{k-2} \quad \text{for } k \geq 2, \\ H_0 &= -G'(R)G(AR - \gamma), \\ H_k &= a_{k+2}RH_{k-1} \quad \text{for } k \geq 1, \\ a_{k+2} &= \left(\frac{k-1}{k}\right)a_k, \text{ for } k \geq 3, \text{ with } a_1 = a_2 = 1. \end{aligned}$$

---

[23] L_0 was not defined and the L_k recursion was started from $L_1 = \frac{1}{2}ABRG'(R)G'(AR - \gamma)$.

Problems

6.20 Show that as the width of the equivalence interval $\to 0$, the decision rule presented in Table 5.9 for the test of the equivalence hypothesis with known variance becomes identical to the rule for the point-null hypothesis test with known variance (in Table 5.7). Does this convergence of decision rules also occur for the tests presented in those tables when the population variance is unknown?

6.21 Discuss the behavior of tests of the inequivalence hypothesis as the width of the equivalence interval $\to 0$.

6.22 Show that the detection probability for the inferiority hypothesis, with variance unknown, is $\Pr(t' < -t_{1-\alpha,f}| - \gamma)$, where t' is the noncentral t-variate, α is the significance level, f is degrees of freedom, and γ is defined on Table 5.4.

6.23 Show that the top of the detection region for the inequivalence hypothesis (variance unknown) in Figure 5.12 is where the standard error equals $20/t_{1-\alpha,f}$.

6.24 What happens to the shape of the "chimney on the teepee" in Figure 5.12 as the standard error on the y-axis increases?

Part II

Problems and Solutions

'There is no problem that cannot be solved.'
—François Viète (1591), quoted by Hollingdale (1989 [155])

'Models, of course, are never true, but fortunately it is only necessary that they be useful.'
—George Box (1979 [32])

7
Formulating water quality standards

Many statistical issues in water quality management arise from the requirements for water management agencies to set standards and to assess compliance with them. Before considering details of those issues, it is helpful to first set them in the context of general practice adopted in many countries.

7.1 SETTING THE SCENE

Water quality standards, by definition, spring from a statutory instrument (such as an Act of Parliament in systems of government based on the British model). Whenever possible, they should be enforceable; compliance should be mandatory. They can place limits on water quality variables for wastewater effluents, for stormwater, for drinking waters, for unit processes, and for receiving waters. Some can only be stated in narrative form, because the science relating to them is too imprecise. Rules or advice stating how compliance is to be assessed should be given in the standard or in accompanying documents, with "rules" applying to numerically explicit standards and "advice" referring to narrative standards. Their typical modes of promulgation are as follows.

Standards for wastewater effluent discharges are usually stated in a "consent" (or "permit"). They include limits on some combination of rate of wastewater flow, as well as concentration or loads of contaminants. A variety of statistics are used to state the limits (means, medians, percentiles, minima, and maxima). Sampling may be continuous, discrete or in proportion to flow. A permissive burden of proof is often taken, but sometimes data are taken at "face-value."

172 FORMULATING WATER QUALITY STANDARDS

Some standards may also be stated by means of a "general authorization" in a water quality management plan, especially if they are held to cause only a minor impact (but see Problem 7.1).

Standards for stormwater discharges can also be stated in permits (including "general authorizations"), or by an authorizing procedure in a water quality management plan. (Issues involved with setting stormwater standards are discussed in Problem 7.2.)

Standards for drinking water for water treatment and distribution systems are usually promulgated by a health-oriented central government agency. These standards contain limits on concentrations of chemicals (a continuous variable) or on the presence/absence of potentially harmful microorganisms (a discrete variable). Because the health of water-drinkers is at risk in a poorly performing system, compliance assessment should always take the precautionary approach, minimizing the "consumer's risk."

Standards for unit processes may be imposed to control the output from individual components, so that the combined output of all processes is satisfactory. This may be particularly appropriate for a drinking-water treatment plant, in which a precautionary or face-value stance may be taken. (See Problem 7.3.)

Standards for receiving waters may be stated in a statute *and* in a water management plan (as discussed further below). They may refer not only to physical/chemical contaminant levels, but also to impacts on aquatic biological communities. A precautionary or face-value stance may be taken. The "raw score" approach of the US Environmental Protection Agency for listing TMDLs is an example of the face-value stance (Smith *et al.* 2003 [290]).[1]

7.1.1 Explicit and narrative statements

In some cases the standards in statutory instruments can be quite explicit, e.g., in the bathing waters Directive issued by the European Union (Pedini 1976 [257]), or in drinking water standards issued by a government health agency (e.g., MoH 2000 [229]). However, in other cases the expression of standards in statutory instruments is of a rather general form (e.g., "the water shall not be rendered unsuitable for bathing by the presence of contaminants" in the New Zealand Resource Management Act 1991), leaving the development of a more explicit form to be effected through water management plans for individual regions and/or individual consent application hearings—a two-stage approach. In the USA, narrative standards are often included in the TMDL process (Freedman *et al.* 2003 [100]). This more flexible region-by-region development of standards allows for regional variability. It is often aided by the availability of "guidelines" or "criteria", either national (e.g., USEPA 1986 [324];

---
[1] Total maximum daily loads (TMDLs) set pollutant loading targets for streams that are not meeting water quality standards with regard to technology-based controls on point sources (under Section 303 of the Clean Water Act).

USEPA 1999 [325]) or international (e.g., WHO 2003 [342]). These are usually advisory only, although in the absence of formal standards they are often taken as "default standards".

An advantage of this two-stage region-by-region approach is that it allows for locally-important issues and environmental concerns to be catered for at that level. Some may see a disadvantage in that it can mean that environmental standards may differ somewhat between regions.

General discussion of the broader field of setting environmental standards is given by Barnett & O'Hagan (1977 [12]). Here we note the key general issues that arise in setting water quality standards—in statutory instruments, in plans or in consents.

7.2 KEY QUESTIONS

- **What is the standard aiming to achieve?** A standard should include a general statement of intent, usually having to do with maintaining stated water uses (e.g., managing a water body to minimize health risks for swimmers). Some uses can (and should) be passive—preserving pristine environments and maintaining habitats that foster biodiversity are uses that need to be valued alongside more active uses. In doing so, note that merely aiming for "high-quality water" is not always appropriate—nor even necessarily desirable. For example, if the presence of aquatic wildfowl is to be maintained the water will be contaminated by bird feces and may not be considered to be of "high quality" with respect to its appearance and degree of fecal contamination. There is obvious potential for conflicting aspirations for active and passive water uses, so public input into the setting of standards is a wise course of action. After all water quality should be managed for the "public good"—both for the present and for future generations.

- **Has the full body of science knowledge been considered?** Formulation of a standard is greatly facilitated by convening a group of appropriate experts and policy development advisers, to debate and formulate its contents so that it is scientifically sound. It can be beneficial for a member of that group to have awareness of statistical issues, to address issues of sampling implications, methods for and feasibility of compliance assessment.

- **Have the appropriate variables been selected? Is their assay feasible, and reasonably affordable?** Here again, an expert group can best resolve this issue, although for public bodies (e.g., a local authority operating a sewerage system) some community input may be desirable in deciding what is reasonably affordable.

- **Is there consistency between the standard's formulation and its implementation?** Some water-quality microbiological standards have been based on a study where samples were taken only at adult chest depth (Cabelli 1989 [38]). If a standard uses relationships in that study but requires sampling in knee-depth

174 FORMULATING WATER QUALITY STANDARDS

water (for reasons of sampler-safety), there may be an important inconsistency, because microbiological quality is generally worse in shallow at-beach water and so the standard will be stricter than might be supposed.

- **Should effluent standards be just that—limits on the effluent's constituents only?** Or should they also include some standards on the receiving waters? (See Problems 7.4 and 7.5.)

- **Does the standards' form of expression make compliance feasible?** Water quality management is seriously compromised if unattainable targets are set—for example requiring that a water treatment plant filter produces water with turbidity less than 0.01 NTU for at least 95% of the time—even attaining 0.1 NTU for at least 95% of the time may be difficult with some technologies.

- **Are the standards measurable?** It is generally desirable that standards be measurable. This is particularly so for drinking-water and wastewater treatment plants—compliance assessment is otherwise problematical. However, *some* receiving water standards are necessarily of a narrative nature—for example, requiring that there be "no significant adverse effects on aquatic life" (New Zealand Resource Management Act 1991).

- **Are the standards enforceable?** Standards for drinking-water and wastewater treatment plants should be enforceable, because they directly impact on ecological and human health. Statistical issues are an integral part of these issues (e.g., Problem 7.6) as we now discuss.

7.3 ENFORCEABILITY

Enforceability means that if standards are breached, penalties may be imposed (e.g., fines or enforcement notices requiring improved treatment technology to be installed or an upgrading of a drinking water distribution system). Therefore the funders, designers, and operators of treatment plants need to know levels of quality are to be required of them and how those requirements are to be assessed. In contrast, standards for receiving waters may not be enforceable in the sense of imposing penalties on particular water users—it can be difficult to identify any particular cause of a breach. Nevertheless, if a breach of those standards is caused by an effluent discharge, that discharge may well be in breach of its effluent consent, in which case penalties may indeed apply. Conversely, if the receiving water standard is breached but all treatment plants comply with their effluent limits, the appropriate resource management agency (assuming that there is one) should check on the reasons why and, if held to be necessary, formulate policies to address the cause or ameliorate the effects. It may be held that the receiving water standard is too strict.[2] Alternatively effluent consents

---

[2]There are many examples where environmental standards have been relaxed over time—for example, the USEPA "criteria" for ammonia in freshwater for 1986 versus 1999 (USEPA 1986 [324]; USEPA 1999 [325]). Of course, "who decides?" and "why?" are crucial issues when making such decisions and, recognizing that, some jurisdictions allow for challenges of proposed changes through the legal system.

may need revision (i.e., they are too generous). These matters involve the ability to detect effects, as well as setting standards assessing compliance with them.

7.4 REQUIREMENTS FOR ENFORCEABILITY

Ideally, enforceable standards should state the following (recognizing that they seldom do contain all of them):

- The variable or variables for which limits will be defined.

- Requirements for the sampling and analytical methods to be used to make the measurements (these need not be standardized; but methods should be comparable in terms of their ability to detect contaminants).

- Limits on contaminants or unit processes, including the statistic(s) to be compared to those limits (means, medians, percentiles, duration, and frequency of exceedances), including an explanation for the choice of compliance statistics (percentiles, averages, maxima—see Problems 7.7–7.9 for a discussion of these issues).

- How compliance with these limits is to be assessed, by defining either:

    - the time periods over which compliance is to be assessed, including specification of whether those periods are adjacent or running (see Problems 7.10 and 7.11);
    - the sampling regime (e.g., grab versus continuous sampling, systematic-random versus expected-adverse-condition sampling);
    - the sampling frequency;
    - the burden of proof being adopted;
    - the statistical methodology to be used, and how that accounts for the burden of proof;

- or, more flexibly, by defining:

    - a standard for statistical verification of a compliance methodology (this is the "statistically verifiable ideal standard" of Barnett & O'Hagan (1997 [12]), leaving the form of the compliance rule to be negotiated for each and every case.[3] (See Problem 7.12 for the kind of thinking called for here.)

These issues cannot not resolved using statistical methods alone (as discussed in Section 1.3). Indeed some do not need input from statistical analyses at all. For

---

[3]These authors noted that this approach had yet (as at 1997) to be implemented. However, the flexibility it introduces could be very useful in situations where a statutory instrument (such as a general Directive) is otherwise overly explicit.

example: Have the appropriate variables been selected? Has the full body of science been utilized? But many issues do benefit from the results of statistical investigations, in one form or another. For example, consider the question of the appropriate burden of proof. This invokes public policy issues about risk-acceptance and cost-acceptance, ideally being resolved in dialogue with the affected communities and environmental professionals. The extent of this dialogue can be strongly influenced by the available statutory instruments.[4] But the dialogue itself benefits greatly by being advised of the results of statistical analyses, because they can demonstrate the possible *explicit* forms of consent under discussion, and their practicality and cost implications. In particular, as noted in Section 5.3.1, a precautionary proof of safety stance is more onerous than a permissive proof of hazard stance and the consequences of this (more intensive sampling) are not obvious to many. Another is, What is the appropriate length of a compliance assessment period—instantaneous, a week, a month, a season, a year?

None of the issues are *solely* statistical in nature, not even "how many samples should be taken?" (See Problem 7.13.) As noted by Shabman & Smith (2003 [287]), concerning the appropriate error risks (burden of proof) to be adopted when listing and "delisting" TMDLs:

> The public and political leaders may shrink from making such decisions, instead claiming that these decisions are in the realm of science. They are not decisions made by science, but they are decisions to be informed by science. Too often what looks like science—choosing an error rate for a statistical test for a water assessment—is an unrecognized policy decision. It is time to make this clear.

So the dominant statistical issue in setting standards concerns the need for standards to be written such that that they are statistically coherent and enforceable, and that they properly reflect agreed public policy so far as that is possible. Statistical investigations have an important role to play in that process.

Problems

7.1 Discuss the issues arising when "minor" wastewater discharges are granted a "general authorization" (e.g., in a watershed management plan).

7.2 Should stormwater standards be stated in permits or given a more general authorization, such as in a water quality management plan?

7.3 Pathogenic microorganisms such as oocysts of *Cryptosporidium parvum* may pass through a poorly performing filtration unit in a water treatment plant. Continuous monitoring of such microorganisms is therefore desirable, but it is not always feasible—even frequent assays may be impractical and/or excessively costly. Accordingly, limits may be placed on a suitable surrogate in the filter's product water that that can be monitored continuously (or, at least, frequently). Turbidity is commonly used for this purpose, as it is easily measured continuously. Consider a case where

---

[4]For example, New Zealand's Resource Management Act (1991) has promoted community dialogue before the resource management agency decides on standards, either in plans or in consents, not least because the decision on that standard is open to appeal to an independent Court (the Environment Court).

a new standard has been proposed that a treatment plant would be noncomplying if *any* of its filters produces water with turbidity in excess of 0.3 NTU turbidity units for more than 5% of any one day (i.e., 72 minutes). What is the impact of this rule on the supplier's risk for plants with many filters versus those with few? Can you devise a fairer standard? Would it be better to place a standard only on the combined output of the filters?

7.4 Discuss the merits of including receiving water standards in effluent consents.

7.5 Consider the case where an effluent consent contains only receiving water standards, not effluent standards. For example, "...the consent holder shall not cause the dissolved oxygen in the stream anywhere downstream of the point of discharge to fall below 80% saturation." Is this good practice?

7.6 Fines or enforcement notices are seldom imposed for effluent consents that exceed a maximum limit (a 100%ile). Why is this?

7.7 Discuss the merits and disadvantages of setting maximum standards for effluents.

7.8 Discuss the merits and disadvantages of setting maximum standards for drinking waters.

7.9 Discuss the merits and disadvantages of setting maximum standards for receiving waters.

7.10 Discuss the merits of having "running" compliance assessment periods, versus adjacent periods.

7.11 A precautionary proof of safety compliance rule can be based on one-sided confidence limits. So can permissive proof of hazard rule. How can this be done? How do these approaches compare with a face-value stance?

7.12 Discuss the notion that considerable care is needed if null hypothesis tests are to be used for drinking-water standards, and that it would be best if they were not used at all in this case. What type of test should be used?

7.13 When is the question "How many samples should be taken to assess compliance with a water quality standard?" *solely* a statistical issue?

8
Percentile standards (and the Reverend Bayes)

In Chapter 2 we noted that water quality management is increasingly using percentile standards, both for effluent quality and receiving water quality. Here we elaborate on these issues and show how a Bayesian analysis can be helpful in determining appropriate compliance rules.

8.1 TWO FORMS OF PERCENTILE STANDARDS

Percentile standards may be based either on percentiles *of samples* or on percentiles *of time*. An example of the former is the influential NSSP shellfish water standard (USDHHS 1993 [322]) requiring that, for a particular MPN setup, no more than 10% of samples should exceed 43 fecal coliforms per 100 mL of water. This is the face value stance to the burden of proof, discussed in Section 1.3—because it is based on a percentage of samples, not on a percentile of time. In that case there is not much statistical analysis to do, other than to compute percentiles of samples. Even so, the need to be aware that there is more than one way—and no one correct way—to estimate percentiles is emphasized again (as stated in Section 2.13.4).

On the other hand, some effluent consents, drinking-water standards, and receiving water standards require that some limit not be exceeded for a stated percentage of time. After all, it is the percentage of time that aquatic microorganisms and water users are exposed to excessive contamination that is the central issue, rather than the percentage of samples containing that contamination. But few water quality variables can be measured continuously, so in deriving compliance rules for these standards we must make allowances, one way or another, for sampling error—taking a precautionary or a permissive approach, or possibly treating data at face value. Note

180 PERCENTILE STANDARDS (AND THE REVEREND BAYES)

too that when a percentile standard is being set for an existing discharge, with the intent of merely continuing its authorization, one *should* take a permissive approach. For example, say that an existing wastewater treatment plant's discharge has displayed a long-term 95%ile concentration of 40 parts of ammoniacal-nitrogen per million, and that percentile limit is set in a consent (or permit) for future operation of the plant. If compliance is to be assessed every six months, then, if the treatment plant continues to produce a similar effluent quality, it will be found to be in breach of its consent *about half the time* (see Problem 8.1).

In this chapter we concentrate mostly on nonparametric approaches to percentile estimation (for the first form of percentile standards) and, in a more lengthy discussion, on allowing for sampling error (for the second form of those standards). The latter uses both classical and Bayesian methods. Parametric methods are discussed by Gilbert (1987 [118]) and by Smith *et al.* (2003 [290]), particularly using one-sided confidence (or tolerance) intervals on percentiles, as discussed in Section 3.1.3.

8.2 CALCULATING NONPARAMETRIC SAMPLE PERCENTILES

We have a set of data: $X_1, X_2, \ldots, X_n$. If the distribution from which these observations have been drawn is unknown, nonparametric percentiles can be calculated using the following three-step procedure:

1. The data are put into ascending order: $Y_1, Y_2, \ldots, Y_n$.

2. The "rank" of the required percentile is calculated using one of a number of possible formulae.

3. An interpolation is made between the adjacent data (this is necessary because the rank is seldom an integer).

If the computed rank is less than 1, or is greater than the number of samples (n), the percentile cannot be calculated. Five procedures for computing the rank of the required sample percentile are in common use in the water industry (Ellis 1989 [84]). From highest to lowest value of the rank these are:

$$\begin{aligned} \text{Weibull:} \quad r &= p(n+1), \\ \text{Tukey:} \quad r &= \frac{1}{3} + p\left(n + \frac{1}{3}\right), \\ \text{Blom:} \quad r &= \frac{3}{8} + p\left(n + \frac{1}{4}\right), \\ \text{Hazen:} \quad r &= \frac{1}{2} + pn, \end{aligned} \tag{8.1}$$

where p is the proportion corresponding to the required percentile (so $p = 0.95$ for a 95%ile), and r is its rank. In addition, we have the formula used in a spreadsheet (Microsoft©Excel 2000) and in some other software packages (Mathsoft 1997 [203]):

Excel: $r = 1 + p(n-1)$.

The percentile, X_p, is usually calculated from linear interpolation formula

$$X_p = (1 - r_{\text{frac}}) Y_{r_{\text{int}}} + r_{\text{frac}} Y_{r_{\text{int}}+1}, \tag{8.2}$$

where $Y_1, Y_2, \ldots, Y_n$ are the n data (X_i) arranged in ascending order, and the subscripts "frac" and "int" denote the integer and fractional parts of r. Simple algebra, using the requirement that $1 \leq r \leq n$, reveals these minimum sample size requirements to enable the percentile calculation to be made:

$$\text{Weibull: } n \geq \frac{p}{q} \text{ and } n \geq \frac{q}{p},$$

$$\text{Tukey: } n \geq \frac{1+p}{3q} \text{ and } n \geq \frac{1+q}{3p},$$

$$\text{Blom: } n \geq \frac{3+2p}{8q} \text{ and } n \geq \frac{3+2q}{8p},$$

$$\text{Hazen: } n \geq \frac{1}{2q} \text{ and } n \geq \frac{1}{2p},$$

$$\text{Excel: } n \geq 1,$$

where $q = 1 - p$ is the exceedance probability, and, if necessary, n is rounded up to the next highest integer. The last equation demonstrates the extraordinary result that *any* percentile can be calculated from a single sample using the Excel formula, making this formula an unfortunate choice.

8.2.1 Comparing the alternatives

If we wish to estimate the 95%ile using the Weibull, Tukey, Blom, Hazen and Excel formulae, the above equations show that we will need at least $n = 19, 13, 13, 10,$ or 1 data, respectively. If in fact we had $n = 19$ data, the ranks would be $r_{\text{Weibull}} = 19.00$, $r_{\text{Tukey}} = 18.70$, $r_{\text{Blom}} = 18.66$, $r_{\text{Hazen}} = 18.55$, and $r_{\text{Excel}} = 18.10$. Further, let's imagine that the two largest of these 19 data are $Y_{18} = 63$ and $Y_{19} = 170$. In that case, using Equation (8.2), $(X_{p=0.95})_{\text{Weibull}} = 0.00 \times 63 + 1.00 \times 170 = 170.00$, $(X_{p=0.95})_{\text{Tukey}} = 0.30 \times 63 + 0.70 \times 170 = 137.90$, $(X_{p=0.95})_{\text{Blom}} = 0.34 \times 63 + 0.66 \times 170 = 133.62$, $(X_{p=0.95})_{\text{Hazen}} = 0.45 \times 63 + 0.55 \times 170 = 121.85$, $(X_{p=0.95})_{\text{Excel}} = 0.90 \times 63 + 0.10 \times 170 = 73.70$.

The highest result (Weibull) is over twice as large as the smallest result (Excel), demonstrating the need for caution in estimating percentiles, especially from small sets of right-skewed data. There is even some danger of this happening for larger sample sizes if the data are drawn from right-skewed distributions (e.g., of microbiological concentrations in bathing beach water). For example, if we had $n = 100$ data, the ranks are: $r_{\text{Weibull}} = 95.95$, $r_{\text{Tukey}} = 95.65$, $r_{\text{Blom}} = 95.6125$, $r_{\text{Hazen}} = 95.5$, and

$r_{Excel} = 95.05$. If the 95th and 96th ranked data are very different, then the difference between the Weibull and Excel estimators will produce remarkably different results.

> **There is no one "correct" way to estimate percentiles. The different estimators may give *very* estimates for datasets drawn from skewed distributions.**

8.2.2 Median versus geometric mean

In Section 2.2.5 (on page 22) we noted that the sample median and geometric mean can be quite dissimilar. For example, in reviewing the performance of a number of waste stabilization ponds, Hickey *et al.* (1989 [148]) reported a geometric mean for ammoniacal nitrogen as 3.1 g m^{-3}, whereas the median concentration is 7.0 g m^{-3}. In contrast, the median chloride concentration (68 g m^{-3}) was less than the geometric mean chloride concentration (92 g m^{-3}). This calls for care when translating standards based on the geometric mean into percentile standards based on the median. Merely using the historical geometric mean concentration as the median concentration for a percentile standard can bias the seriously compliance rule.

> **Do not automatically use historical geometric mean (or arithmetic mean) values in a new median percentile standard. Always check the comparability between the geometric mean and median (and between the arithmetic mean and median).**

8.3 PERCENTILES VERSUS MAXIMA

Deciding on whether to use maxima (100%iles) or high percentiles (e.g., a 95%ile) in standards can greatly benefit from careful discussion between water managers and users. The merits of each option has a lot to do with the nature of the water in question—drinking water, wastewater or environmental water—as discussed in Section 2.13.4 and Problems 7.7–7.9. One of the major features of concern is that setting a maximum standard can discourage sampling, because the more samples we have the more likely it is that a maximum will be exceeded. This is indicated in Figure 8.1.

For low numbers of samples (but greater than 10—the Hazen estimator requires at least 10 observations to be able to estimate a 95%ile) the maximum and 95%ile can coincide. But as more observations are made it becomes increasingly likely that these two statistics will diverge, with the observed 95%ile declining but the observed maximum increasing, as shown on Figure 8.2. (See also Problem 8.3.)

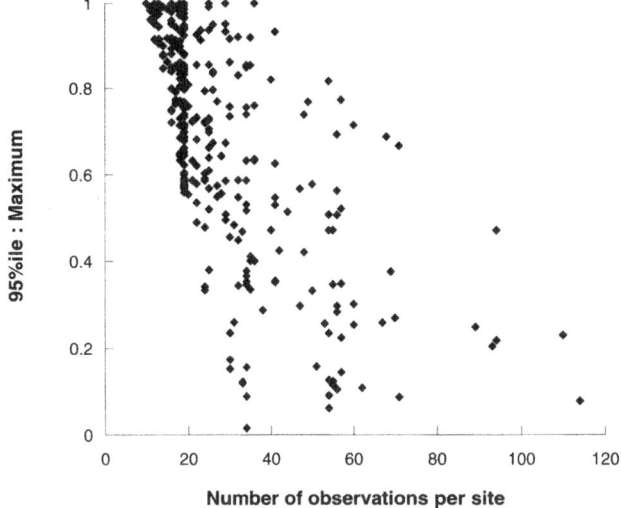

Fig. 8.1 Ratio of 95%ile to maximum for *E. coli* concentrations at 317 stream and river sites, using the Hazen rank estimator [Equation (8.1)] to estimate the 95%ile (data from Larned *et al.* 2004 [178]).

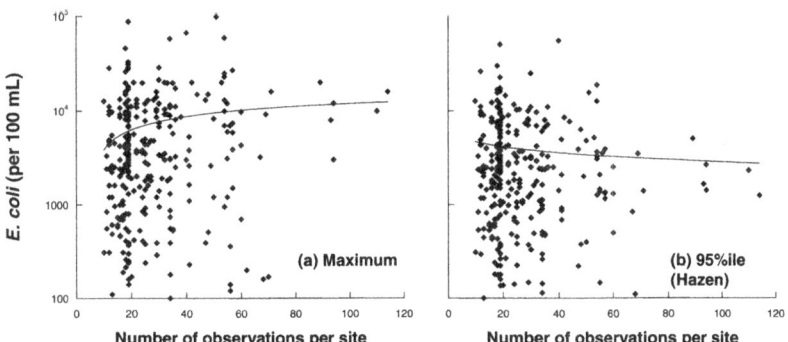

Fig. 8.2 Maximum and 95%ile *E. coli* at 317 sites (lines are logarithmic regressions, chosen to reflect central tendency of the data, and the data are from Larned *et al.* 2004 [178]).

8.4 PERCENTILE COMPLIANCE RULES

While 95%iles are the most common high percentile used in water standards, we'll first consider 98%iles, to avoid any confusion with 95% confidence or credible intervals. Much of the following material is based on McBride & Ellis (2001 [209]); some similar material also appears in Smith *et al.* (2001 [289]).

8.4.1 Developing rules using the Classical approach

Consider a consent (permit) requiring, among other things, that the total inorganic nitrogen concentration ("TIN") in the effluent from a coastal sewage treatment plant should be below 10 g m^{-3} for at least 98% of a three-month summer period, based on two samplings every outgoing tide. Compliance with this 98%ile standard is therefore to be assessed using about $n = 350$ samples. Let's say that it has been agreed that a proof of hazard approach is to be adopted, so we will assume that the effluent is in compliance with its standard unless data are convincing to the contrary. Accordingly, the "producer's risk" (also called the "supplier's risk") is to be kept below a small number. We'll denote that number by α, and use e to denote the number of exceedances of the 10 g (TIN) m^{-3} percentile limit in a three-month period in which 350 samples are collected.

This problem can now be treated as a one-sided test of an inferiority hypothesis. That is (using the nomenclature in Chapters 4 and 5), we test the hypothesis **H**: $x \leq X$, where x is the probability of exceedance for any single random sample, and $X = 0.02$. Assuming only that we have random sampling, we can use the following formula to identify the maximum permissible number of exceedances under this proof of hazard approach while still reaching a verdict of "compliance":

$$p_h = \Pr(e > e_h | n, x = X) \leq \alpha, \tag{8.3}$$

where e_h is the maximum permissible number of exceedances among n random samples in a proof of hazard approach.[1] Note that the conditional probability in Equation (8.3) includes $x = X$, whereas **H** posits that $x \leq X$. This is because one-sided hypotheses are tested assuming that they are *only just true*. Therefore, the left-hand side of this equation gives the probability of obtaining more than e_h exceedances for a "borderline" effluent—it was only just complying (i.e., $x = X = 0.02$). We select e_h as the lowest value of e that satisfies $p_h \leq \alpha$ (the equality $p_h = \alpha$ will seldom be attained because exceedance data are discrete, not continuous). In that case we reject **H** and so infer a breach of standard, there being only a small chance of getting more than e_h exceedances if in fact the effluent was borderline. Note that under this proof of hazard approach the proportion of permissible exceedances (i.e., $\frac{e_h}{n}$) will always exceed X (see Problem 8.4).

The required probability (p_h) can be calculated by noting that for random sampling e is distributed as cumulative binomial (regardless of the distribution of the underlying process), that is,

$$p_h = \sum_{e=e_h+1}^{n} \binom{n}{e} X^e (1-X)^{n-e} = 1 - B(e_h; n, X), \tag{8.4}$$

where $\binom{n}{e}$ is the binomial coefficient and $B(e_h; n, X)$ is the cumulative binomial probability of getting up to e_h exceedances in n random samples each of which

---

[1] McBride & Ellis (2001 [209]) use e_{bd} instead of e_h, the double subscript denoting "benefit of doubt" (see Section 1.3). Here we use the single subscript h, denoting the equivalent term "proof of hazard".

has a probability X of exceedance [see Equation (6.12)]. This probability can be calculated from the algorithm given in Section 6.2.5, or by using available calculators (e.g., Excel's BINOMDIST function). For example, taking $n = 350, X = 0.02$ and $\alpha = 0.05$ we obtain $p_h = 0.0515$ for $e_h = 11$ and $p_h = 0.0257$ for $e_h = 12$. Accordingly, from Equation (8.3), we must allow up to $e_h = 12$ exceedances of the 10 g m^{-3} limit in those 350 samples (i.e., about 3.4% of the samples). That is, we take e_h as the highest value of e that satisfies the requirement that $p_h \leq \alpha$.

However, if the consumer's risk is to be controlled, in a precautionary proof of safety approach, the tested hypothesis must now posit breach (i.e., $x \geq X = 0.02$). That is, we test a superiority hypothesis. Because we are demanding a high standard of proof before this hypothesis will be rejected, the proportion of allowable exceedances (i.e., $\frac{e_s}{n}$) will never exceed X (see Problem 8.5). The appropriate formula is

$$\begin{aligned} p_s &= \Pr(e \leq e_s | n, x = X) = \sum_{e=0}^{e_s} \binom{n}{e} X^e (1-X)^{n-e} \\ &= B(e_s; n, X) \leq \alpha, \end{aligned} \quad (8.5)$$

where e_s is the maximum permissible number of exceedances in a proof of safety approach.[2] It is the highest value of e satisfying $p \leq \alpha$, for a borderline effluent. Again taking $n = 350, X = 0.02$ and $\alpha = 0.05$, we obtain $p_s = 0.0285$ and $p_s = 0.0797$ for $e_s = 2$ and 3, respectively. Accordingly we can allow up to only $e_s = 2$ exceedances of the 10 g m^{-3} limit in those samples (i.e., about 0.6% of the samples).

Note that this proof of safety approach means that α has switched roles: Under the permissive proof of hazard approach it was the maximum acceptable *producer's* risk; under the precautionary proof of safety approach it becomes the maximum acceptable *consumer's* risk. Also, p_h is the p-value for the test of the inferiority hypothesis used in the proof of hazard approach, whereas p_s is the p-value for the test of the superiority hypothesis used in the proof of safety approach.

These rather straightforward calculation procedures can be used to devise compliance rules for percentile standards, as we shall see. But first, let's consider an alternative approach.

8.4.2 Developing rules using the Bayesian approach

The classical approach developed so far has two worrisome aspects:

- The compliance rule is developed *assuming borderline compliance*.

- No account whatsoever is taken of any prior knowledge or opinion as to the likelihood of compliance or the extent of any noncompliance.

---
[2]McBride & Ellis (2001 [209]) use e_{fs} instead of e_s, the double subscript denoting "fail safe" (see Section 1.3). Here we use the single subscript s, denoting the equivalent term "proof of safety."

A Bayesian approach allows us to move away from these concerns. That is, we will consider all possible states of exceedances ($0 \leq x \leq 1$) and we will also incorporate various forms of prior "information." As a result we can talk of the "confidence of compliance", which is the probability that $x \leq X$ given our data—the number of exceedances (e_h or e_s) in n observations. Furthermore, we consider only those data, not all data more extreme than e_h or e_s as we did in Equations (8.4) and (8.5). To quantify the confidence of compliance, we can appeal to Bayes' rule to obtain probabilities for a range of x values, given the a number of data n displaying e exceedances. However, we must first note that the form of Bayes' rules we have thus far seen [Equations (2.13) and (2.14)] are stated in terms of discrete events (A and B) rather than continuous variables (such as x). Accordingly, we need the rule to be re-stated in terms of a continuous parameter (x) and discrete data (e and n).[3] That form is (using Lee 1997 [179], Press 2003 [264])

$$h(x|e,n) = g(x) \left[\frac{L(e|n,x)}{\int_0^1 L(e|n,x)g(x)} \right], \qquad (8.6)$$

where h and g are probability densities, L is the "likelihood function" and the term in brackets is the "standardized likelihood." Note that this equation is of the general form noted in Section 4.9, that is, posterior $\propto$ prior $\times$ likelihood. Furthermore, using the standardized likelihood turns the proportionality into equality.

Considering Equation (8.6) in more detail, $g(x)$ is the unconditional *prior* probability density function for the exceedance probability, reflecting prior knowledge or belief about x prior to collecting n new data. Also, $h(x|e,n)$ is the conditional *posterior* probability density function for the exceedance probability, given e exceedances in n new random samples. This updating of probability densities, from prior to posterior, is accomplished through the likelihood function, $L(e|n,x)$. This function reflects the information contained in the data, and in the data only, as explained below.

But first, let's understand the form of Equation (8.6), by comparing it with a simple rearrangement of Bayes' rule for k multiple mutually exclusive events [Equation (2.14)]:

$$\Pr(B_j|A) = \Pr(B_j) \frac{\Pr(A|B_j)}{\sum_{i=1}^{k} \Pr(A|B_i)\Pr(B_i)}. \qquad (8.7)$$

Comparing Equation (8.6) with Equation (8.7) we see that

- the probabilities become probability densities (so h will have to be integrated over x to obtain a probability);

- x takes the role of B_j;

- e takes the role of A;

- n is an additional component to all the conditional probability densities;

---

[3]There are also form of Bayes' rule for continuous data and a discrete parameter and also for the case where both the data and the parameter are continuous (Press 2003 [264]).

- the finite sum on the denominator of Equation (8.7) becomes an integral over the infinitely many possible values of x in its domain ($0 \leq x \leq 1$).

For a set of n samples the likelihood function $L(e|n,x)$ is the probability mass function for a particular value of e, with x (rather than e) being regarded as the function's parameter. Furthermore, we know that under random sampling the distribution of exceedances is binomial. So L is a statement (not of probability, but of likelihood) only about the data obtained. It particular, it is the binomial term being summed in Equations (8.4) and (8.5), for any value of x (not just for X as in those equations), that is,

$$L(e|n,x) = \binom{n}{e} x^e (1-x)^{n-e}.$$

(Note that the sum of $L(e|n,x)$ over all possible values of e is unity, but the integral of L over all x [in Equation (8.6)] is not—hence it is a likelihood, not a density.) Therefore, noting that the binomial coefficients cancel in the standardized likelihood function, the posterior probability density is given by

$$h(x|e,x) = g(x) \left[\frac{x^e (1-x)^{n-e}}{\int_0^1 x^e (1-x)^{n-e}\,dx} \right] g(x). \tag{8.8}$$

The probability of the exceedance rate being no greater than X, which we call "Confidence of Compliance," denoted as CC), is calculated by integrating this posterior density from $x = 0$ to $x = X$, so that

$$CC = F(x \leq X|e,n) = \int_0^X h(x|e,n)\,dx, \tag{8.9}$$

where F is a cumulative probability. We also define the "Confidence of Failure" as

$$CF = 1 - CC = F(x > X|e,n) = \int_X^1 h(x|e,n)\,dx.$$

Choosing a prior distribution The choice of the prior density $g(x)$ may be approached in two general ways. In either case it is highly desirable to use a "conjugate distribution." This guarantees that the prior and posterior probability densities belong to the same family of distributions, making the calculations required by Equations (8.8) and (8.9) much simpler. Therefore, noting that the likelihood function follows a binomial distribution, we choose $g(x)$ to follow the versatile two-parameter beta distribution, because these distributions are conjugate to a binomial likelihood, (Lee 1997 [179]). We write that distribution as $\text{Be}(\alpha, \beta)$, where α and β are the distribution's parameters (see Section 6.2.4 for mathematical details of that distribution).

In the first approach we can use a "reference prior", reflecting the views of someone with no particular *a priori* beliefs, as will be shown in Figure 8.3. Noting that an optimal reference prior does not exist (see Lee 1997 [179]), perhaps the most obvious selection is the uniform prior, also known as "Bayes' postulate," in which $g(x) = 1$ for all possible values of the exceedance rate (i.e., $0 \leq x \leq 1$).[4] This choice implies that all values of x are equally likely and is a special case of the beta distribution; that is, $\text{Be}(1,1)$, as shown in the Figure (and also in Figure 2.10). However, the U-shaped prior derived from "Jeffreys' rule" (Section 3.3 of Lee 1997 [179])—$\text{Be}(\frac{1}{2}, \frac{1}{2})$—can be argued to be a better choice.[5] It implies that extreme values of the exceedance rate (i.e., $x \to 0$ or $x \to 1$) are more likely than intermediate values, as shown in Figure 8.3.[6] (See Section 8.4.2 to see why this apparently undesirable behavior at $z \to 1$ does not pose a problem.)

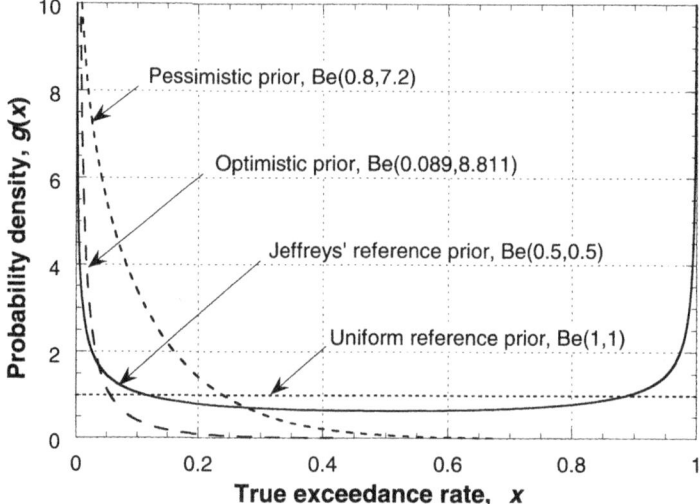

Fig. 8.3 Four prior densities compared; each curve encloses a unit area.

It bears repeating material on page 58, that such choices of prior probabilities densities are often stated to be "uninformative," whereas the best that may be said of them is that they may be called "impartial". Even that descriptor is problematical; after all, the flat $\text{Be}(1,1)$ and U-shaped $\text{Be}(\frac{1}{2}, \frac{1}{2})$ are supplying different prior "information."

The alternative approach arises if we have more particular views about the prior distribution of exceedances, and so supply a more "informative" prior distribution. For example, it may be held to be very unlikely that x would be too much above zero, and we may want to so specify our own prior distribution. So, based on past experience, we could take the expected value of the true exceedance rate as $\bar{X} = 0.1$ and its variance as $S^2 = 0.01$ (corresponding to meeting a 90%ile standard). Or

---

[4] Lee notes that Bayes himself was not entirely convinced of the validity of this postulate, and that this was partly responsible for the fact that his paper was not published in his lifetime.
[5] It is invariant under any change of scale of measurement.
[6] Reprinted from *Water Research* **35**(5), McBride & Ellis, Confidence of Compliance: A Bayesian approach for percentile standards, 1117–1124, 2001, with permission from Elsevier.

we could hold that the true exceedance rate is $\bar{X} = 0.01$ with variance $S^2 = 0.001$ (corresponding to meeting a 99%ile standard). We can then calculate approximate values of α and β from the known expressions for the population mean and variance of the beta distribution (see Section 6.2.4), that is, $\mu = \frac{\alpha}{(\alpha+\beta)}$ and $\sigma^2 = \frac{\alpha\beta}{(\alpha+\beta)^2(\alpha+\beta+1)}$, from which, by making the approximations $\mu \approx \bar{X}$ and $\sigma^2 \approx S^2$ we obtain

$$\alpha_{\text{prior}} = \bar{X}\left[\frac{\bar{X}(1-\bar{X})}{S^2} - 1\right] \text{ and } \beta_{\text{prior}} = \alpha_{\text{prior}}\left[\frac{1}{\bar{X}} - 1\right]$$

We therefore obtain $\text{Be}(0.8, 7.2)$ and $\text{Be}(0.089, 8.811)$ as our prior distributions for these two cases. These "hockey stick"-shaped densities are also shown in Figure 8.3, denoted as the pessimistic prior and the optimistic prior, respectively.

Calculating posterior probabilities Because h is a beta density, it follows that F is an incomplete beta function ratio, written as $I_X(\alpha, \beta)$. Inserting the beta pdf (see Section 6.2.4) into Equations (8.8) and (8.9), we obtain $\alpha_{\text{posterior}} = \alpha_{\text{prior}} + e$ and $\beta_{\text{posterior}} = \beta_{\text{prior}} + n - e$ (as shown by Lee (1997 [179])). Accordingly, the posterior distribution function is given by

$$CC = F(x \leq X|e, n) = I_X(\alpha_{\text{prior}} + e, \beta_{\text{prior}} + n - e), \qquad (8.10)$$

which can be calculated using numerical methods (e.g., Press *et al.* 1992 [265]).

Confidence of Compliance graphs Figure 8.4 displays the sensitivity of Confidence of Compliance (and Confidence of Failure) to the choice of prior distribution for a 95%ile, using both reference priors and both informative (optimistic and pessimistic) priors.[6] This graph, and those that follow, include marker lines at $CC = 5\%$ and 95%, for reasons to be explained. Three particular features should be noted from this graph:

1. The uniform prior gives the most pessimistic CC curves.

2. CC curves for Jeffreys' prior sit roughly in the middle between those for pessimistic and optimistic priors.

3. Despite Jeffreys' prior being U-shaped, allowing high (and low) exceedance rates to be more likely than moderate values, the influence of the prior near $x = 1$ is quickly extinguished.

As a consequence, Jeffreys' emerges as a suitable reference prior.[7]

Confidence of Compliance graphs for the 80th, 90th, 95th, 98th and 99th percentiles are shown on Figure 8.5, all using Jeffreys' prior.[6] These graphs can be used to devise proof of hazard and proof of safety compliance rules, as in Section 8.4.3.

---

[7] A further possibility—Haldane's $\text{Be}(0, 0)$ prior—acts rather similarly but has the unfortunate property that $CC = 100\%$ when no exceedances occur [because $I_X(0, n) = 1$].

190 PERCENTILE STANDARDS (AND THE REVEREND BAYES)

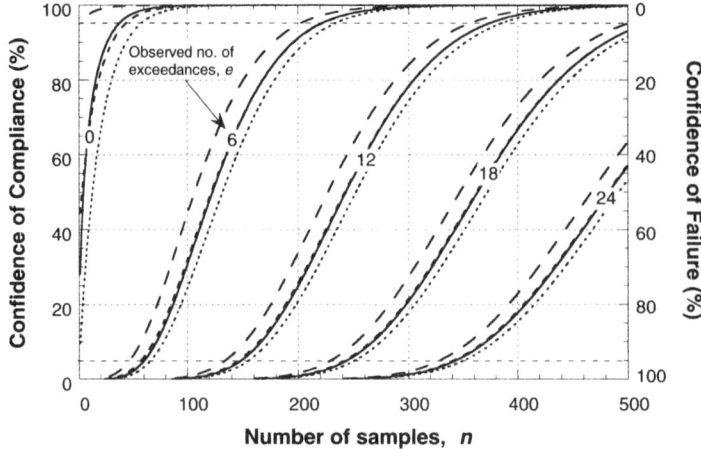

Fig. 8.4 "Confidence of Compliance" with a 95%ile: sensitivity to prior distributions using four priors: —— Jeffreys' Be($\frac{1}{2}, \frac{1}{2}$), – – optimistic Be(0.089, 8.811), - - - pessimistic Be(0.8, 7.2), - - - - uniform Be(1, 1).

8.4.3 Comparing the Classical and Bayesian results

The Confidence of Compliance graphs can be used when considering sampling requirements for compliance rules seeking to control either the consumer's risk or the supplier's risk. Tables 8.1 and 8.2 demonstrate the results for a 95%ile, with maximum risks constrained to be less than 5%.[6] Results for the consumer's risk using the Bayesian approach are obtained by reading n values for which CC first exceeds 95% (i.e., for which CF first falls below 5%). For Jeffrey's prior and a 95%ile this may be done using the top dashed line on Figures 8.5(c) and 8.5(d). Results for the supplier's risk using the Bayesian approach are obtained by reading n values for which CF first exceeds 95% (i.e., for which CC first falls below 5%), using the bottom dashed line in the figures. The classical results are obtained from Equations (8.4) and (8.5) as the n values where the appropriate p-values first falls below 5%.

In these tables we see that the uniform prior produces the most onerous sampling requirements of the four Bayesian options, with the classical approach even more severe. To explain this, note that the the Bayesian method would have predicted $CC = p_h$ if, and only if, we had chosen Be(1, 0) as the reference prior (see Problem 8.6).[8] Because Be(1,0) concentrates all prior probability at the most extreme exceedance (its pdf is everywhere zero except at $x = 1$), we see that the classical approach is the most pessimistic of all cases. This is entirely attributable to the two "worrisome" aspects we observed at the start of Section 8.4.2.

---

[8] Such correspondences are known to occur for particular choices of priors in one-sided tests (Edwards *et al.* 1963 [79], Pratt 1965 [262], DeGroot 1973 [67], Casella & Berger 1987 [44], Lee 1997 [179])—but not always with this dramatic result!

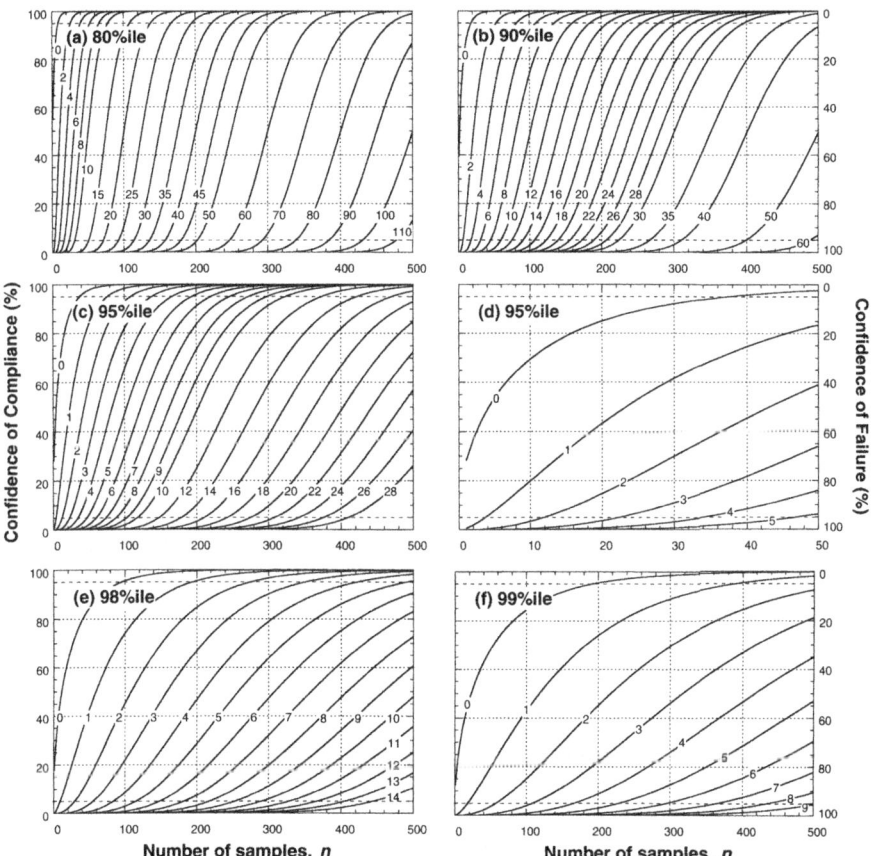

Fig. 8.5 Confidence of Compliance for five percentiles, using Jeffreys' prior. Numbers on the graphs are the observed number of exceedances. Note the condensed range on the horizontal axis for the 95%ile in part (d).

Examples Table 8.1 shows that using the Bayesian approach with Jeffreys' prior, the minimum number of complying samples that need to be collected to give 95% confidence of meeting a 95 percentile standard is 38, compared with the classical approach's requirement of 59 samples. However, if one transgression of the percentile limits occurs at least 77 samples are needed (compared with 93). Therefore, using this reference prior, substantial reduction in sampling effort can be expected while still meeting the same confidence criteria.[9] Returning to the example of the coastal sewage TIN discharge 98%ile standard we note from Figure 8.5(e) that for $n = 350$ samples the Confidence of Failure first exceeds 95% when $n = 12$ exceedances are allowed (in agreement with the classical approach). However, the Confidence of

---

[9]This reduction has been incorporated into New Zealand's drinking-water standards, MoH 2000 [229].

Table 8.1 Numbers of samples and maximum permissible exceedances needed to keep the consumer's risk below 5% when assessing compliance with a 95%ile standard

Maximum permissible exceedances	Classical approach	Numbers of samples			
		Bayesian approach using the following priors			
		Uniform	Jeffreys'	Optimistic	Pessimistic
0	59–92[a]	58–91[a]	38–76[a]	2–53[a]	44–79[a]
1	93–123	92–122	77–108	54–87	80–110
2	124–152	123–151	109–138	88–118	111–139
3	153–180	152–179	139–166	119–147	140–167
4	181–207	180–206	167–193	148–174	168–195
5	208–233	207–232	194–220	175–201	196–221
6	234–259	233–258	221–246	202–228	222–247
7	260–285	259–284	247–272	229–254	248–273
8	286–310	285–309	273–298	255–279	274–298
9	311–335	310–334	299–323	280–304	299–323
10	336–360	335–359	324–348	305–329	324–348

[a] It is not possible to keep the consumer's risk below 5% if less than 59, 58, 38, 2, or 44 samples are to hand (classical approach and using the four Bayesian priors, respectively).

Compliance first exceeds 95% when 3 exceedances are observed (versus 2 in the classical approach). So again, a more reasonable prior assumption (than given by the $Be(1,0)$ distribution implicit in the classical approach) results in less onerous compliance requirements.

Finally, note that most wastewater discharge issues include multiple determinands (e.g., biochemical oxygen demand, suspended solids, forms of nitrogen and phosphorus). Applying 95%ile standards to each in turn may cause the error rate to be higher than desired. Simulation studies, using randomization methods, are generally the most useful way to proceed to examine the complications that arise (Ellis 1986 [83]). In particular, this will account for the correlations that will occur between at least some of the determinands.

8.4.4 Conclusion

Bayesian approaches give direct answers to questions of Confidence of Compliance with percentile standards and thus enable the supplier's risk and the consumer's risk to be identified. This facility comes at the "cost" of having to state one's prior belief as to likely exceedance rates. However, the comparison between classical and Bayesian results demonstrates the strong similarity in their results if one uses a uniform reference prior distribution, and it also demonstrates that the classical approach in fact gives the most pessimistic results. Indeed, in making a confidence statement using that approach one has (perhaps unwittingly) adopted the most pessimistic prior. Using the proposed Bayesian technique (with Jeffreys' reference prior) makes compliance

Table 8.2 Numbers of samples and maximum permissible exceedances needed to keep the supplier's risk below 5% when assessing compliance with a 95%ile standard

Maximum permissible exceedances	Classical approach	Numbers of samples			
		Bayesian approach using the following priors			
		Uniform	Jeffreys'	Optimistic	Pessimistic
0	1[a]	b	b	b	b
1	2–7	1–6	1–3	1	1
2	8–16	7–15	4–11	2	2–7
3	17–28	16–27	12–22	3–10	8–18
4	29–40	28–39	23–34	11–21	19–30
5	41–53	40–52	35–46	22–33	31–43
6	54–67	53–66	47–60	34–46	44–57
7	68–81	67–80	61–74	47–60	58–71
8	82–95	81–94	75–88	61–74	72–85
9	96–110	95–109	89–102	75–89	86–100
10	111–125	110–124	103–117	90–103	101–115

[a] The risk is exactly 5% in this case.
[b] It is not possible to keep the supplier's risk below 5% in these cases.

rules less onerous, particularly for smaller numbers of samples, while still affording the desired degree of protection. Informative prior belief as to likely performance (good or bad) can easily be incorporated into the technique, as shown. At larger numbers of samples results for all techniques become more similar—the information in the data increasingly overwhelms that in the chosen prior distributions, a well known property of Bayesian analyses. These results have been obtained using a nonparametric approach; evidence suggests that little is to be gained in adopting a parametric approach (McBride 2003 [207]), except that compliance rules based on the proof of safety approach becomes less strict when many samples are taken.

Problems

8.1 In Section 8.1 we observed that requiring a wastewater treatment plant to meet its historical long-term 95%ile for ammoniacal-nitrogen would mean that the plant would be found to be in breach of its consent about half the time. Why is this? What would be the implications of requiring the plant to comply with its seasonal 95%ile?

8.2 Discuss the Hazen estimate of a 95%ile when 10 observations are available.

8.3 In Figure 8.2(b) the estimated 95%ile declines as the number of observations increases. Can you explain why this happens?

8.4 Explain why a proof of hazard approach means that the proportion of exceedances among samples is always higher than that contemplated for the whole population.

8.5 Explain why a proof of safety approach means that the proportion of exceedances among samples is always lower than that contemplated for the whole population.

8.6 Show that the p-value for the proof of hazard approach [Equation (8.4)] may also be written in terms of the incomplete beta function ratio, $p_h = I_X(e_h+1, n-e_h)$, and show that this is also the Confidence of Compliance that results when taking a $\mathrm{Be}(1,0)$ prior distribution.

9
Microbial water quality and human health

Fecal residues containing pathogenic microorganisms may enter water bodies by a number of means, including point sources of wastewater, stormwater outfalls, diffuse agricultural runoff, and direct deposition by animals (e.g., when crossing streams or grazing banks). These activities can be subject to control by water management agencies, through provisions in water and land management plans, or through consents. In formulating such controls, it can be beneficial to attempt to make quantitative estimates of health risks to people using the water for recreation (e.g., swimming, skiing), for drinking-water abstraction, or for consuming shellfish harvested from the water. Impacts on animal health can also be considered. This is the realm of quantitative microbial risk assessment (QMRA). It is useful in indicating the possible extent of any public health concerns. Note too that microbiological water quality guidelines (MfE/MoH 2003 [223], WHO 2003 [342]) make clear that complying with limits on traditional bacterial water quality "indicators" close to a wastewater outfall is not necessarily a guarantee of safety. In such cases a direct consideration of pathogens and their consequential risk to the health of water users is desirable. QMRA is used to do this. It can be especially useful in comparing the *relative benefits* of different wastewater management schemes. For example, when discharging a city's treated wastewater to the ocean, is it better to have a 2 km ocean outfall with upstream ultraviolet (UV) disinfection installed, or a 3 km outfall with no disinfection, given that these options may be cost-neutral? This is a situation currently (2005) being faced by Christchurch, New Zealand.

9.1 THE KEY ELEMENTS

A four-step procedure is used (Haas *et al.* 1999 [132]):

Hazard identification: What are the relevant pathogens?

Exposure assessment: What exposures might a population have when in contact with water?

Dose-response assessment: How much infection or illness would arise in that population if the water contains pathogens?

Risk characterization: Integrating the three preceding steps to indicate the extent of any public health concern.

Many aspects of the second and third steps are either uncertain or variable. Uncertainty is evident in dose-response curves, because they are developed from limited datasets describing clinical trials on a small number of subjects. An example of variability is the length of time different people at a beach stay in the water. To account for these features may have to use a Monte Carlo approach (Section 4.8), in which we place statistical distributions on the uncertain and variable factors and perform repeated sampling from those distributions to build a risk profile. For example, consider Christchurch City's proposal for an ocean outfall of treated wastewater. For each scenario of wastewater treatment, we could conduct the "thought experiment" of exposing a population of, say, 1000 people at a recreational site (a coastal beach) on 1000 separate occasions, giving a million exposures for each scenario, so that the variability and uncertainty is fully captured. (See Problem 9.1 to see why this isn't the same as exposing one person one million times.) A flow diagram for each of the QRA calculation sequences is given in Figure 9.1.

Some elaboration of the four steps is appropriate, taking, as an example, the discharge of treated sewage to the ocean via a submarine outfall.

9.1.1 Hazard identification

To identify potentially waterborne pathogens of public health significance patterns of community disease burden and wastewater characteristics should be considered. In countries with temperate climates, such as New Zealand, this indicates that the major pathogens of concern include the following.

Human adenoviruses, very infective, can cause respiratory illness, may be unusually resistant to disinfection, and present in substantial numbers in raw sewage (Crabtree *et al.* 1997 [62]).

Hepatitis A virus, extremely infectious and serious, but at lower concentrations in raw sewage.

THE KEY ELEMENTS   197

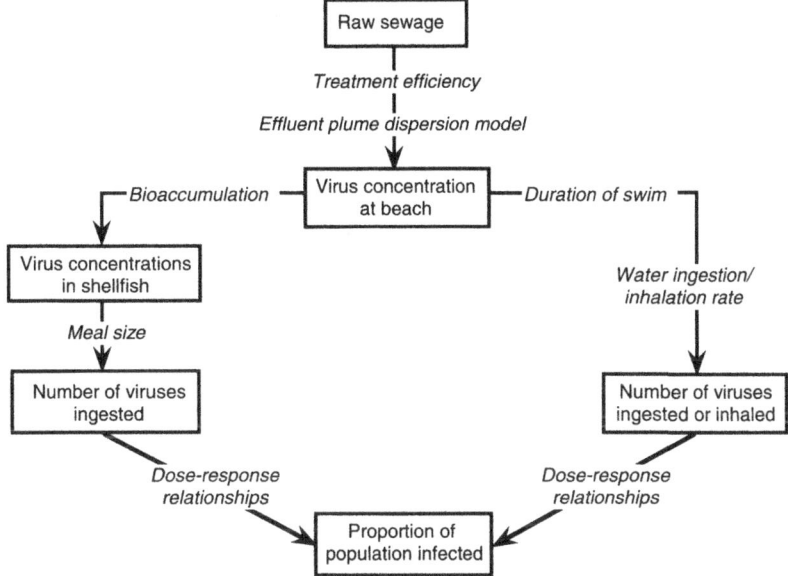

Fig. 9.1 General flow diagram for QRA calculation sequence for swimmers and for consumers of raw shellfish.

Human enteroviruses, generally less infective, less resistant to disinfection, and present in somewhat lower concentrations in sewage than adenoviruses, but may give rise to more serious illness.

Cryptosporidium **oocysts and** *Giardia* **cysts,** particularly a concern in drinking water (Craun *et al.* 1998 [63]) and swimming water (Baker *et al.* 1998 [9]). A waterborne outbreak in Milwaukee in 1993 is thought to have affected some 403, 000 people (MacKenzie *et al.* 1994 [194]) and has been claimed to have caused 50–100 fatalities (Hoxie *et al.* 1994 [156], Morris *et al.* 1996 [231]). Immune-compromised people are particularly at risk from this microorganism; its presence in drinking water has been associated with an outbreak of cryptosporidiosis in Las Vegas with at least 20 deaths (Goldstein *et al.* 1996 [120]).

Campylobacter, particularly for freshwater. Campylobacteriosis is the major component of New Zealand's "notifiable" gastrointestinal illness burden (Till & McBride 2004 [316]), with a current reported rate in excess of 350 per 100, 000 people per annum. *Campylobacter* are often detected in environmental water samples. Tends to occur "sporadically", rather than in outbreaks, so may be overlooked.

Salmonella, rotaviruses (Gerba *et al.* 1996 [112], noroviruses (formerly "Norwalk-like viruses") and pathogenic *E. coli* may also need to be considered. In any event, competent health authorities should be contacted while deciding on the microbial hazards of concern.

Usually, unless contamination is extreme, the pathogens of concern give rise to mild short-term illness, but there is no room for complacency—for example, see the concerns with *Cryptosporidium* raised above. Also, mild illnesses can give rise to serious sequelae. For example, a small percentage of cases of campylobacteriosis can result in the much more serious Guillain–Barré syndrome (McDowell & McElvaine 1997 [217]). For the USA, that proportion has been estimated as about 0.1% (Nachamkin *et al.* 1998 [237]). A Danish epidemiological study has shown that there is an enhanced risk of mortality within one year of infection by such "mild" pathogens, when compared to a control group who were not infected (Helms *et al.* 2003 [145]). And one must always be cognizant of the possibility of more serious pathogens being present (e.g., in the case of a disease outbreak in the contributing community), and of emerging pathogens (e.g., SARS).

9.1.2 Exposure assessment

Water contact For swimming and other recreational water activities (skiing, windsurfing, yachting, diving) estimates are needed of the amount of water ingested, inhaled or in contact with body surfaces. The total volume inhaled or ingested is obtained as the product of the time in contact with the water and the rate of ingestion or inhalation. Schernewski & Jülich (2001 [280], citing Johl *et al.* 1995 [165]) have noted that "10 ml to 100 ml water are incorporated during bathing." So the distribution of incorporation rate could be taken as a simple triangular distribution with minimum, median, and maximum $= 10, 50$ and 100 mL per hour. (See Problem 9.2 for a discussion of why it may be better to interpret this median as the mode.) In the absence of more information, this seems the most plausible distribution.

Pathogens in the water To obtain a dose from water contact, the estimates of exposure to water must be married with the pathogen concentrations in that water at the point of contact (for ingestion or inhalation) or with the contact time with the water (for skin/ear/nose/throat infections). It is usually impractical to adequately measure these concentrations at the sites of recreational activity in coastal areas. Even in rivers there are substantial difficulties, because the concentrations of individual pathogens can change with time, even when bacterial indicators are relatively constant. Accordingly the concentrations may be *derived* from distributions of pathogen concentrations in treatment plant effluents, or, failing that, by using literature values for concentrations in raw sewage and for the removal efficiency of the wastewater treatment processes. These effluent concentrations may then be translated into concentrations at the site of recreational use by using mathematical hydrodynamic dispersion models, taking into account time-varying factors such as wind direction and tidal state.

Three caveats must be made here.

First, care is needed in fitting distributions to concentrations that are assayed using serial-dilution multiple tube fermentation MPN techniques, particularly if there only a small number of tubes—see Chapter 10. Essentially this is because the MPNs constitute an unusual (discrete) number system—no two MPNs have the same "occurrence probability," and most numbers are actually impossible. Therefore a binning system may be required, as discussed in Chapter 10.

Second, virus assays for environmental waters are usually only dichotomous—that is, present/absent in a particular volume of water. They are often absent from the water. In that case an "added zeroes" distribution may be needed in order to assess the dose on each Monte Carlo trial, with elicited expert opinion (from aquatic virologists), as presented in Section 9.3.1.

Third, if the model requires raw sewage influent virus distributions, one can expect them to be always present, to at least some degree. If few data are available (e.g., for the Christchurch effluent), one may have to appeal again expert opinion in constructing a distribution. In that case the opinion given is based on experience with similar influents elsewhere and may be expressed as a minimum, medium (not as the mode), and maximum. In that case, an empirical "hockey stick" distribution may be constructed, as also shown in Section 9.3.2. Alternatively, a beta distribution could be used.

Shellfish meal size To calculate risks associated with consumption of (raw) shellfish taken from areas potentially affected by the wastewater discharge, we need the above environmental information *and* both the size of a meal and the degree of bioaccumulation of the pathogens in the shellfish flesh before harvesting. A first guess at the distribution of meal size can be taken from the literature; for example, Rose & Sobsey (1993 [273]) estimated the likely average serving for the USA to be 60–240 g. This estimated range can be improved if data are available from nutrition surveys. For example, Russell *et al.* (1999 [277]) report national nutrition data that can be best fitted to a loglogistic model, as shown in Figure 9.2. (See Section 9.3.3 for a discussion of this distribution.) The fit to the data (obtained using the fitting procedures embedded in @RISK, Palisade Corporation 2000 [252]) is very satisfactory. Or is it? In company with the normal distribution, and many others, this distribution has no upper bound, so the pdf (the thick curve in Figure 9.2) is nonzero to infinity. Put another way, the cumulative probability function (the thin curve in the Figure) never quite reaches one. But the data have a maximum meal size of 790 g. Under repeated sampling in Monte Carlo analysis, one can obtain implausibly large meal sizes, considerably beyond 790 g, albeit rarely.[1] Yet it can be the extremes of the distributions that can cause a rare-but-large risk to be predicted. This can have serious consequences, especially if resource management statutes consider environmental effects to include both (a) any effect of high probability or (b) any effect of low probability with a potentially high impact.[2] This problem can be avoided by truncating the distribution at plausible values (truncation is provided for in @RISK).

> **When using distributions with no upper bound, consider truncating the upper limit at a plausible value.**

---

[1] In one set of 1000 samplings from this distribution a maximum of 1748 g was obtained.
[2] As in Section 3 of New Zealand's Resource Management Act 1991.

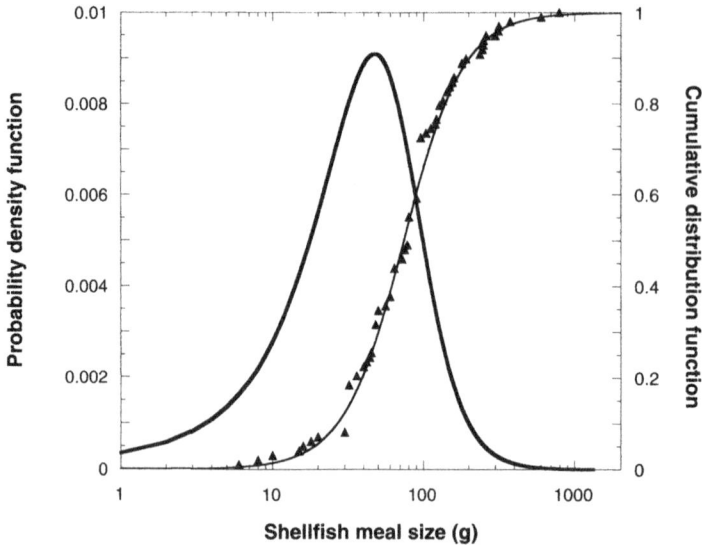

Fig. 9.2 Loglogistic fit for shellfish meal size (data from Russell *et al.* 1999, [277]). The loglogistic shape, scale and location parameters are $\alpha = 2.2046$, $\beta = 75.072$, and $\gamma = -0.903$ (mathematical details for this distribution are given in Section 9.3.3).

Bioaccumulation factors ("BAF") have been reported by Burkhardt & Calci (2000 [36]), as shown in Table 9.1.

Table 9.1 Coastal shellfish bioaccumulation factors for two seasons (USA)

Indicator organism	February–early October			Late October–January		
	n	BAF	S_e	n	BAF	S_e
Fecal coliforms	15	4.4	0.5	8	4.2	0.5
FRNA bacteriophage	15	2.9	0.5	8	49.9	7.4
C. perfringens spores	15	58.1	7.8	8	62.0	12.2

Source: Burkhardt & Calci (2000 [36]) (n = number of data; S_e = standard error). The pathogen dose ingested from consumption of a given mass of shellfish is taken as BAF × the number of pathogens present in the equivalent volume of seawater; for example, the virus dose ingested from consumption of 100 g of shellfish is taken as BAF × the concentration of viruses expressed as number per 100 mL of seawater. Note that this assumes that this concentration has persisted long enough for that degree of accumulation to occur.

The BAF parameters for FRNA bacteriophage in Table 9.1 are suitable estimates for viral pathogens, with fecal coliform and *Clostridium perfringens* spores factors sufficing for bacteria and cysts. A precautionary approach is then to describe the distribution of the BAF using the seasonal maximum. For example, in modelling viruses we can take a normal distribution with mean = 49.9 and a standard deviation of 20.9 (i.e., $S_e\sqrt{n}$).

9.1.3 Dose-response

Data from clinical trials on human volunteers are available, against which dose-response models have been fitted, using infection (not illness) as the endpoint (Teunis et al. 1996 [309], Haas et al. 1999 [132]). These models describe the probability of infection (P_{inf}) as a function of the *mean* ingested or inhaled dose (d) for a cohort. The trials use a number of cohorts, each given challenges from a preparation for which the *mean* dose is estimated. Therefore, d is not a *particular* dose. The distribution of pathogens in each preparation is assumed to be random, that is, Poisson (see Chapter 6.2.6). The proportions of each cohort becoming infected are plotted against the mean doses and an appropriate curve is fitted, as presented below (Figure 9.3). Mathematical development of the various dose-response curves are given in Section 9.3.4, including cases where the pathogen distribution is not random and is overdispersed.

When applying dose-response curves to water or shellfish flesh, it is usually assumed that the pathogens are distributed randomly in that medium also and thus follow the Poisson distribution. The resulting dose-response relations for viruses and protozoa are commonly of a simple single-parameter exponential type, in which there is a constant probability ($0 < r < 1$) of a single pathogen causing infection in any individual—the "pathogen survival probability." That is,

$$P_{\text{inf}}(d) = 1 - e^{-rd}. \tag{9.1}$$

The median infective dose, obtained by inserting $P_{\text{inf}} = \frac{1}{2}$ into Equation (9.1), is given by

$$d_{50} = -\frac{\ln(\frac{1}{2})}{r} \approx \frac{0.693}{r}. \tag{9.2}$$

Note that infective doses are defined over a population, not for an individual. Also, once one pathogen is present there is *some* risk, no matter how small. Therefore, as noted by Gale (2003 [105]),

> **The term "the infective dose" is meaningless;**
> **the "minimum infective dose" is one pathogen.**

For bacteria, such as *Campylobacter*, r is usually not constant between individuals (nor is it for rotavirus). Consequently, the exponential form rises too quickly toward 1 and so r is replaced by a two-parameter beta distribution, with parameters α and β. Once integrated this gives the infection probability as the "beta-Poisson" model

$$P_{\text{inf}}(d) \approx 1 - \left(1 + \frac{d}{\beta}\right)^{-\alpha} \quad \text{provided that } \beta \gg 1 \text{ and } \alpha \ll \beta, \tag{9.3}$$

where α is a shape parameter and β governs the curve's shift along the x (mean dose) axis (altering β causes the curve to shift left or right, without changing its shape).

If both $\beta \gg 1$ and $\alpha \ll \beta$, Equation (9.3) is a good approximation to the exact solution (a Kummer hypergeometric equation—see Section 9.3.4). In practice the requirement that $\beta \gg 1$ can be ignored when computing the curve, but care is needed when confidence or credible intervals are computed around that curve—they should be computed for the Kummer function (Teunis & Havelaar 2000 [307]). The median infective dose is

$$d_{50} = \beta \left(2^{\frac{1}{\alpha}} - 1 \right), \qquad (9.4)$$

so that the beta-Poisson model can be characterized by α and d_{50}, the latter perhaps being a more intuitively appealing parameter than β.

Values for the parameters r, α, and β are available for a range of pathogens (e.g., Haas *et al.* 1999 [132]). Figure 9.3 shows the nature of these models for three viruses and the bacterium *Campylobacter jejuni*, and it also shows their median infective doses. Note that while the dose-response models for viruses rise inexorably toward a unit probability, the beta-Poisson model for *Campylobacter* (once viewed on a linear dose scale) initially rises steeply but then becomes almost flat at $P_{\text{inf}} \approx \frac{1}{2}$. This reflects the fact that some people are remarkably resistant to infection by this bacterium. Teunis & Havelaar (2000 [307]) also derive 95% Bayesian credible intervals for the beta-Poisson equation (and for the exact Kummer curve), using uniform log-transformed prior distributions for α and β.

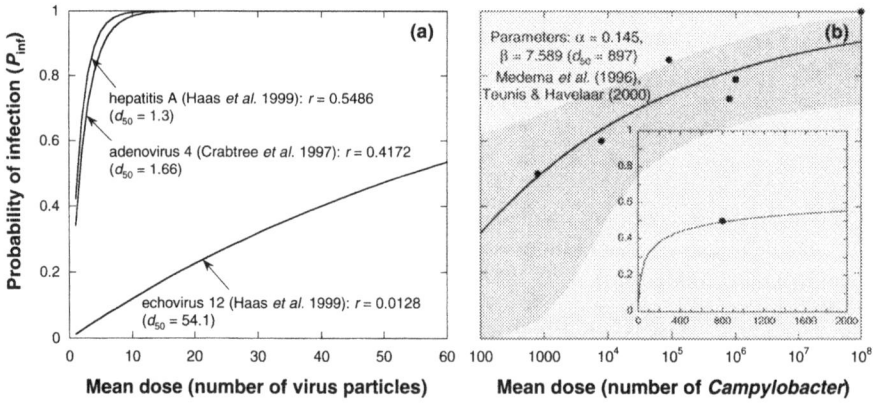

Fig. 9.3 Dose-response models for (a) three viruses (linear scale) and (b) the bacterium *Campylobacter jejuni* (log scale). Echovirus 12 is taken as a representative enterovirus—no other enterovirus dose-response data are available. The insert on the *C. jejuni* graph shows the dose-response model for low concentrations on a linear scale. Dots on (b) are the clinical trial data (*C. jejuni*, Strain A3249, Black *et al.* 1988 [26]). Shaded region is for the exact (Kummer) solution, copied from Figure 2 of Teunis & Havelaar (2000 [307]).

Haas *et al.* (Chapter 7, 1999 [132]) show how confidence limits on the dose-response curve can be generated using Excel's "Solver" add-in. More elaborate and accurate Bayesian procedures are described briefly by Teunis & Havelaar (2000 [307]).

9.1.4 Risk profiling

Given the variability and uncertainty in key components of our risk model (Figure 9.1), it is desirable to use a Monte Carlo approach to develop a risk profile.[3] Provided that a sufficient number of Monte Carlo trials are performed, these profiles give the percentage of time that the infection risk is no greater than a given value. They can be for a group of people going to the *same* location, or for each of them attending a *different* site. In the former case the risk profile is typically zero a majority of the time (because usually there is minimal pathogen contamination), but may rise steeply in the higher percentiles (because occasionally there may be significant contamination present and so would affect a proportion of people at that location). In the latter case the profile is much flatter, because some contamination can be expected at some locations at any time. An example profile for *Campylobacter* infection is given in Table 9.2. This was calculated from surveys of recreational use of inland waters and their *Campylobacter* contamination distributions in New Zealand (adenovirus infections exhibit a similar risk profile). (The actual risks will be lower than these figures suggest, because the *Campylobacter* concentrations tend to be higher when the flow is high and the water is turbid, in which case fewer people will use it for recreational pursuits.)

Single-site risk profiles have been used to compare the merits of various options for the discharge of Christchurch's treated wastewater via an ocean outfall (length of outfall, possible installation of UV disinfection prior to discharge). For example, the profile at the "Taylor's Mistake" beach for rotavirus health risk to consumers of raw shellfish for a 2-km outfall with UV in winter is zero up to 98%ile, reaching a maximum of 8 cases per 1000 people. Results from the computational hydrodynamic model and contaminant transport/inactivation model show that the plume from the offshore discharge seldom impacts the shoreline, which is why the risk profile is zero for the most part. For a 3-km outfall without UV the winter profile is zero until the 99%ile, reaching a maximum of 3 cases per 1000 people. Results such as these can be used to assess the relative impact of the outfall options.

Finally, note that it has been stressed in this chapter that the probabilities referred to apply to a cohort, rather than to an individual. Nevertheless, we can define an "individual's infection risk" (IIR), calculated as the total number of cases divided by the total number of exposures. It represents the risk to a swimmer or shellfish consumer on any day, having no prior knowledge of any beach contamination from

---

[3]In some cases a direct calculation using the *arithmetic* means of the variable and uncertain components can be used (Haas 1996 [130], Gale 2003 [105]). For this to be possible the dose-response should either be the simple exponential form, Equation (9.1), or be beta-Poisson at low doses, in which case Equation (9.3) is approximately linear; that is, $P_{\text{inf}} \approx \frac{\alpha d}{\beta}$.

Table 9.2 Calculated *Campylobacter* infection rates at New Zealand inland water locations, per 1000 people (numbers in parentheses are approximate 95% credible limits)

Percentile	Single site		Multiple sites	
Minimum	0	(0, 0)	25	(0, 104)
2.5th	0	(0, 0)	29	(0, 110)
5th	0	(0, 0)	29	(1, 117)
10th	0	(0, 0)	33	(1, 120)
20th	0	(0, 0)	36	(2, 125)
30th	0	(0, 0)	38	(2, 128)
40th	0	(0, 0)	39	(3, 131)
50th	0	(0, 0)	41	(3, 134)
55th	0	(0, 7)	42	(3, 135)
60th	1	(0, 26)	43	(4, 137)
65th	3	(0, 48)	44	(4, 138)
70th	9	(0, 150)	44	(4, 139)
75th	18	(0, 239)	45	(4, 141)
80th	26	(0, 291)	46	(5, 143)
85th	72	(1, 409)	47	(5, 146)
90th	131	(3, 498)	49	(6, 149)
95th	329	(21, 639)	52	(7, 152)
97.5th	435	(50, 681)	54	(7, 155)
Maximum	491	(88, 735)	62	(11, 167)

Source: McBride *et al.* 2002 [211]. The approximate credible intervals were calculated by adjusting the value of β in the beta-Poisson model [Equation (9.3)], thereby merely shifting the dose-response curve along the dose axis. This approximates the less regular shape of the boundaries of the credible regions (Figure 2 in Teunis & Havelaar [307]). For the upper limit $\beta = 0.11$ [and so, from Equation (9.4), $d_{50} = 13$] and the curve is shifted to the left. For the lower limit $\beta = 846$ (and so $d_{50} = 100,000$) and the curve is shifted to the right. (A more accurate procedure is to generate a large number of sets of dose-response parameters using bootstrap methods, and then sample from those sets on each Monte Carlo trial.)

the outfall. For example, simulating 1000 people attending a beach on each of 1000 days, the total number of cases is $1000m$, where m is the mean infection case rate over 1000 people and the total number of exposures is 10^6. So the IIR, expressed as a percentage, is $100 \times \frac{1000m}{10^6} = \frac{m}{10}$. In the Taylor's Mistake case the winter IIRs for the two options are 0.0044% and 0.0016%. Summer risks are negligible.

These results cover a range of concerns—from there being minimal predicted impact (Christchurch) to address a local issue, to there being a considerable concern for the potential for widespread *Campylobacter* infection from freshwater recreation.

9.2 RISK COMMUNICATION

Mere presentation of such risk profiles, unaccompanied by narrative and quantitative statements about uncertainty, can give rise to an unwarranted confidence of "certainty." So it can be worthwhile to list the uncertainties:

- Dose-response data are mostly restricted to healthy adults; thus small children, the elderly, and the immune-compromised are excluded.

- Pathogens present in sediments can be released to the water column if disturbed by swimmers or by livestock, causing elevated pathogen concentrations.

- Dose-response data can be very sparse in the region of most interest for health-risk studies. For example, the data for *Campylobacter* curve shown in Figure 9.3 are sparse in the region where the risk changes so much—between zero and about 400 bacteria. Indeed the lowest dose administered in the only relevant study (Black *et al.* [26]) was 800 bacteria and the median infectious dose for the fitted beta-Poisson (or hypergeometric) model is about 900 bacteria. Accordingly, credible intervals about the lower portions of the curve are rather wide.

- Dose-response studies are restricted to one, or at most a few, strains of each pathogen, but infectivity may vary between strains (e.g., *Cryptosporidium parvum*, Okhuysen *et al.* 1999, [247]).[4]

- The definition of the endpoint of dose-response studies can vary (e.g., the definition of "infection" in clinical trials sometimes includes stool tests, or stool tests and symptoms).[5]

- Numerative virus data, obtained using PCR techniques (polymerase chain reaction, Haas *et al.* 1999 [132]), does not necessarily mean that all the detected viruses will be infective (PCR techniques detect DNA, not viability).

- Shellfish bioaccumulation studies are sparse and exhibit variable accumulation factors (see Table 9.1).

These uncertainties do not necessarily mean that risk profiles will be underestimated. For example, a precautionary approach is usually taken in public health risk assessments, so that seasonal maxima of bioaccumulation factors are used. Also the analysis is based on infection rather than illness, so illness rates will be lower than those for infection (typically by 50% in the case of bacteria).

A risk analysis needs to be accompanied by quantitative measures of uncertainty (in particular, using credible intervals for dose-response curves and possibly bioaccumulation factors) and by narrative statements concerning the sources of uncertainty. Risk analysis professionals need to take care to always accompany analysis reports by such statements, also pointing out that results are most useful for *comparing* the effects of different options—that is, trading off the need for installing disinfection

---

[4]Clinical trials for different strains of *C. parvum* have received considerable attention (Messner *et al.* 2001 [221]; Teunis *et al.* 2002a, 2002b [310, 311]), with the result that the development of new USA drinking-water regulations (US Federal Register 40 CFR Parts 141 and 142, 2003) has revised the infectivity model (in the simple exponential dose-response model r was raised from its previous value of 0.0042 to 0.09, lowering the median infectious dose from 165 to 8).

[5]This is part of the reason for the revision of r in the *C. parvum* simple exponential model.

technology versus extending the length of a wastewater outfall pipe. The optimal form (or forms) of such a presentation is an issue yet to be satisfactorily addressed.

9.3 SOME USEFUL MATHEMATICAL DETAILS

Readers may skip this section without unduly compromising an understanding of the main issues.

9.3.1 Added-zeroes distributions for dichotomous data

If we are presented with virus data reported as present or absent in a volume V of environmental water, we need to be able to translate that into a dose contained in a volume v ingested or inhaled by a water user on any Monte Carlo trial. This can be done using the Poisson added-zeroes distribution (El-Sharaawi 1985 [86]) and elicitation (from environmental virologists) as to the typical and maximum concentrations of the virus in the volume V *if* the sample result shows the virus to be present. For example, in a freshwater study with 44% of $V = 1$ L samples being positive for the presence of human adenoviruses, virologists opined that the most likely number of those viruses among positive samples would be 5 and the maximum 100 (McBride et al. 2002 [211]). We can interpret these numbers as the mode and maximum of the mean (λ) of a Poisson distribution, and use the versatile beta distribution to describe its variation (by also setting $\lambda_{\min} = 1$, the necessary lower bound for positive samples). Defining P to be the proportion of positive results in all possible samples, then

$$P_{\text{none}} = \Pr(\text{obtaining } no \text{ viruses in volume } V) = (1 - P) + P e^{-\lambda}. \quad (9.5)$$

This equation adds the probability of getting an empty sample $(1-P)$ to the probability of getting a negative result from a contaminated sample $(P e^{-\lambda})$ (see Problem 9.4). Its complement is

$$P_{\text{some}} = \Pr(\text{obtaining } some \text{ viruses in volume } V) = P\left(1 - e^{-\lambda}\right). \quad (9.6)$$

[Equations (9.5) and (9.6) add to unity, as required.] We also have the probability of getting k viruses in a volume V as

$$P_k = \Pr(\text{obtaining } k \text{ viruses in volume } V) = P \frac{e^{-\lambda} \lambda^k}{k!} \text{ where } k > 0 \quad (9.7)$$

(see Problem 9.5 for the sum of P_k over all nonzero values of k). Therefore, the probability of getting c viruses in a different volume v is

$$P_c = \Pr(\text{obtaining } c \text{ viruses in volume } v) = P \frac{e^{-\lambda'} (\lambda')^c}{c!}, \quad (9.8)$$

where $\lambda' = \lambda \frac{v}{V}$ is the Poisson mean scaled by the volume ratio and c is an integer. Then we take a random sample from this distribution for the value of c on each trial. The dose is then obtained from a discrete two-point distribution with a zero dose assigned a probability P_{none} and a dose c assigned a probability P_{some}.[6]

9.3.2 Hockey stick distribution

Influent virus data can be expected to be right-skewed, along with most environmental microbiological data. Given estimates of the minimum, median and maximum of that distribution, it would be folly to use a single triangular distribution—the right tail would be "too fat" and the median would not be preserved (unless maximum − median = median − minimum). A simple solution is to join the median and maximum by a "hockey stick"—that is, the line **BCD** in Figure 9.4.

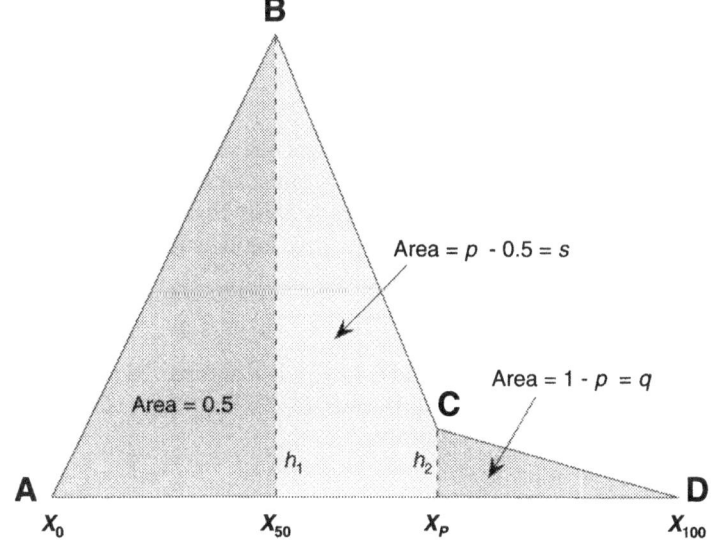

Fig. 9.4 "Hockey stick" empirical influent virus distribution (X_0 is the distribution's minimum, X_{50} is its median, X_{100} is its maximum and X_P is the Pth percentile: $p = P/100$). The value of P (>50) is to be supplied by the risk modeler.

To determine the position of the point **C** the we use the constraint that the total area under the distribution should be unity. To satisfy this requirement we find the value of the Pth percentile (X_P) given a user-supplied value of P. We also use linear interpolation between the distribution's breakpoints.[7] The algebra involves solving

---

[6] In @RISK this would be coded as c*RiskDiscrete({0,1},Pnone:Psome) (using named cells for c, P_{none} and P_{some}).

[7] Were we to join **BD** by a straight line, X_{50} would be the distribution's mode (not its median) and we would over-estimate the median and the frequency of most of the high values.

a quadratic, in the following calculation sequence (referring to the nomenclature on the Figure—see Problem 9.6):

$$h_1 = \frac{1}{X_{50} - X_0}, \quad (9.9)$$

$$X_P = \frac{1}{2}\left\{X_{50} + X_{100} + \frac{1}{h_1} - \sqrt{(X_{100} - X_{50})^2 + \frac{X_{50}(2-8q) + X_{100}(2-8s)}{h_1} + \frac{1}{h_1^2}}\right\}, (9.10)$$

$$h_2 = \frac{2q}{X_{100} - X_{50}}. \quad (9.11)$$

A pragmatic approach is to set $P = 95$—that is, the toe of the hockey stick is at the 95%ile. Random samples can be taken from this distribution (e.g., using @RISK's RiskGeneral function).

9.3.3 Loglogistic distribution

This distribution is sometimes used for human-related variables, obtaining a better fit to than the lognormal or gamma distributions can achieve. It can have pronounced right-skew (the ordinary logistic distribution is always symmetrical, and similar in shape to the normal distribution). The cumulative loglogistic distribution function is

$$F(x) = \frac{1}{1 + \xi^{-\alpha}},$$

where α (>0) is a shape parameter, and $\xi = \frac{x-\gamma}{\beta}$ is a normalized location parameter with scale and location parameters β (>0) and γ ($\leq x$). By differentiating this function we obtain the loglogistic density function

$$f(x) = \frac{\alpha \xi^{\alpha-1}}{\beta(1 + \xi^\alpha)^2}.$$

The skewness coefficient is complex, and is only defined for $\alpha > 3$—so it isn't defined for Figure 9.2. More usefully, its mode is

$$\tilde{f} = \begin{cases} \gamma + \beta \left(\frac{\alpha-1}{\alpha+1}\right)^{\frac{1}{\alpha}} & \text{for } \alpha > 1, \\ 0 & \text{for } \alpha \leq 1. \end{cases} \quad (9.12)$$

9.3.4 Dose-response curves

The manner in which dose-response curves are derived is instructive, because their mathematical forms follow from minimal plausible assumptions, and not from mere arbitrary curve fitting. So any claim that they invoke unnecessary mathematical sophistication rather misses the point that they embody simplicity, not complexity.

Simple exponential and beta-Poisson models Following Teunis *et al.* (1995 [308]) and Haas *et al.* (1999, Chapter 7 [132]), two probabilities are required:

- $P_1(i|d)$, the probability of ingesting precisely i pathogens from an exposure in which the *mean* dose is d;

- $P_2(k|i)$, the probability of k pathogens ($\leq i$) surviving to initiate infection.

With respect to the second probability, assume that infection occurs if only one pathogen arrives at an infection site in the human body—the "single-hit" hypothesis. Then, if the processes driving these probabilities are independent, the probability of infection for a given *mean* dose (d) is

$$P_{\mathrm{inf}}(d) = \sum_{k=1}^{\infty} \sum_{i=k}^{\infty} P_1(i|d) P_2(k|i), \qquad (9.13)$$

so we see that the infection probability is obtained by considering exposures to *all* possible doses (i)—i is summed from 1 to infinity—for a given mean dose (d). Therefore this probability applies to an *ensemble* of individuals, not to a *single* individual. Let's call such equations the *average dose-response relationship*.

If we assume that the distribution of the pathogens within a water volume is random then P_1 is a Poisson distribution, as presented in Section 6.2.6, that is,

$$P_1(i|d) = \frac{d^i}{i!} e^{-d}, \qquad (9.14)$$

and if we assume that each pathogen has an independent and identical survival probability (r) then P_2 is the binomial distribution,

$$P_2(k|i) = \frac{i!}{k!(i-k)!} (1-r)^{i-k} r^k. \qquad (9.15)$$

Substituting Equations (9.14) and (9.15) into Equation (9.13) we get

$$P_{\mathrm{inf}}(d) = \sum_{k=1}^{\infty} \sum_{i=k}^{\infty} \left[\frac{d^i}{i!} e^{-d} \right] \left[\frac{i!}{k!(i-k)!} (1-r)^{i-k} r^k \right]. \qquad (9.16)$$

Standard mathematical results, show that this rather complex expression can be re-expressed as the very much simpler exponential dose-response relationship (see Problem 9.7):

$$P_{\mathrm{inf}}(d) = 1 - e^{-rd}. \qquad (9.17)$$

If we relax the assumption that r is a constant, then Equation (9.15) must be replaced by a "mixture distribution," by integrating its right-hand side over a distribution of r, with probability density function $f(r)$ within the interval [0,1].[8] We therefore obtain

---
[8] Being a probability, the limits of integration cannot be beyond the [0,1] interval.

$$P_2(k|i) = \int_0^1 \left[\frac{i!}{k!(i-k)!}(1-r)^{i-k}r^k\right] f(r)\,\mathrm{d}r. \tag{9.18}$$

Using this equation is identical to applying the mixture operation directly to Equation (9.17), so that

$$\begin{aligned} P_{\mathrm{inf}}(d) &= \int_0^1 \left(1 - \mathrm{e}^{-rd}\right) f(r)\,\mathrm{d}r \\ &= 1 - \int_0^1 \mathrm{e}^{-rd} f(r)\,\mathrm{d}r. \end{aligned} \tag{9.19}$$

If $f(r)$ is taken to be the (extremely versatile) two-parameter β-distribution presented in Section 6.2.4, with parameters α and β, we obtain

$$P_{\mathrm{inf}}(d) = 1 - {}_1F_1(\alpha, \alpha+\beta, -d) \tag{9.20}$$

(see Problem 9.8), where ${}_1F_1$ is Kummer's confluent hypergeometric function. If $\beta \gg 1$ and $\alpha \ll \beta$ we can use the approximation that ${}_1F_1(\alpha, \alpha+\beta, -d) \approx (1+\frac{d}{\beta})^{-\alpha}$, so that we finally have the beta-Poisson equation

$$P_{\mathrm{inf}}(d) = 1 - \left(1 + \frac{d}{\beta}\right)^{-\alpha}. \tag{9.21}$$

This equation was first developed by Furumoto & Mickey (1967 [103]).

The title of the Teunis & Havelaar (2000 [307]) paper ("The beta Poisson dose-response model is not a single-hit model") draws attention to the fact that the confidence interval limits, or credible interval limits, of the beta-Poisson equation, being an approximation, may not be well-behaved at low doses, whereas the exact (Kummer) solution's limits are. Nevertheless, the beta-Poisson curve itself is generally a close approximation to Kummer's function.

9.3.5 Conditional dose-response relationships

Haas (2002 [131]) defines the conditional dose-response probability as the risk attendant on exposure to a *particular* dose (i). Accordingly we have

$$P_{\mathrm{inf}}^0(i) = \sum_{k=1}^{\infty} P_2(k|i),$$

where the 0 superscript denotes "conditional." If r is constant, then, noting that the sum of the right-hand side $= 1$ if the lower limit of summation is $k = 0$, along with using Equation (9.15), we can calculate

$$P_{\mathrm{inf}}^0(i) = 1 - \sum_{k=0}^{0} \frac{i!}{k!(i-k)!}(1-r)^{i-k}r^k = 1 - (1-r)^i, \tag{9.22}$$

(which is the complement of the probability that no pathogens will survive). Haas (2002 [131]) shows that this relationship holds also for the case where each pathogen in any host has a distribution of values, with the same distribution applying between hosts (in which case r is replaced by its average value $\bar{r}$). However, if there is host-to-host variability in that distribution (whether or not there is a distribution within each host) then, assuming that $\bar{r}$ follows a beta distribution, we end up with a beta-binomial curve of conditional dose-response, that is,

$$P_{\text{inf}}^0(i) = 1 - \frac{\text{B}(\alpha, \beta + i)}{\text{B}(\alpha, \beta)}, \qquad (9.23)$$

where $\text{B}(\alpha, \beta)$ is the beta function (see Problem 9.9).

These relationships can be used to determine the form of a dose-response relationship if the distribution of doses is not random (i.e., not Poisson). If it is Poisson, we obtain the simple exponential or beta-Poisson curves, see Haas (2002 [131]).

9.3.6 What if the distribution of pathogens is not Poisson?

If the distribution of pathogens is not random (i.e., Poisson, in which case the variance equals the mean), then it is highly likely that it will be *overdispersed*, with variance greater than the mean. Indeed, some hold that the overdispersion is the norm (ISO 2001 [163], Forster 2003 [98]). The negative binomial and "Poisson lognormal" are commonly used in such cases, especially for drinking water (e.g., Gale *et al.* 1997 [106]).

Haas *et al.* (1999, p. 272 [132]) and Teunis & Havelaar (2000 [307]) show that for constant r and pathogens distributed as negative binomial, the resulting infection probability for a population exposed to that distribution is also of a beta-Poisson form, that is,

$$P_{\text{inf}}^0(i) = 1 - \left(1 + \frac{rd}{k}\right)^{-k}, \qquad (9.24)$$

where k is the overdispersion parameter of the negative binomial distribution (see Section 6.2.7).

If r follows the beta distribution between hosts the resulting equation is written in terms of the Gauss hypergeometric function (Teunis & Havelaar 2000 [307], Haas 2002 [131])

$$P_{\text{inf}}^0(i) = \begin{cases} 1 - {}_2F_1\left(\alpha, k; \alpha + \beta; -\frac{d}{k}\right), \text{ or} \\ 1 - \left(1 + \frac{d}{k}\right)^{-k} {}_2F_1\left(\beta, k; \alpha + \beta; \frac{d}{d+k}\right). \end{cases} \qquad (9.25)$$

Equivalence between these two equations can be demonstrated (see Problem 9.10). Haas (2002 [131]) notes that these equations always produce a lower risk than the equivalent beta-Poisson model (with the same values of α and β). This seems to be true for *any* overdispersed distribution.

9.3.7 Comparing conditional and average-dose risks

As discussed thus far, an "individual" represents a cohort, not a single person, and so that "individual" is exposed to a mean dose not to a particular dose. That is why we have thus far used average dose-response relationships, such as simple exponential for constant survival probability [Equation (9.17)], or beta-Poisson for variable survival probability [Equation (9.21)]. This is certainly appropriate when analyzing dose-response data, because the data we fit to the models are for groups of people for whom we know only the mean dose.

However, when constructing risk models, it could be argued that that the "individual" is exposed to a particular dose (e.g., Hartnett *et al.* 2001 [140]). In that case the conditional dose-response relationships should be used in the model, that is, binomial [Equation (9.22)] or beta-binomial [Equation (9.23)]. Curves for constant and variable survival probabilities are shown in Figures 9.5 and 9.6. Each figure contains two cases: (a) a moderately infective pathogen and (b) a highly infective pathogen (as indexed by their d_{50} values).

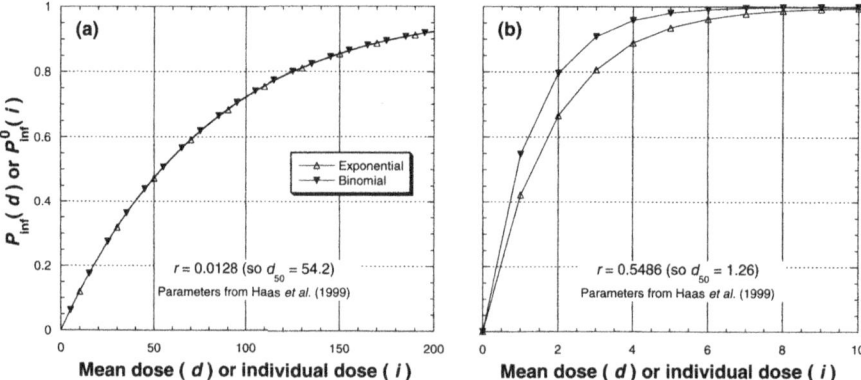

Fig. 9.5 Comparison of the simple exponential dose-response model for $P_{\inf}(d)$ [Equations (9.1) and (9.17)] with the binomial conditional dose response model for $P_{\inf}^0(i)$ [Equation (9.22)] for (a) echovirus 12 and (b) hepatitis A [d_{50} values are calculated from the simple exponential model, using Equation (9.2)].

The figures show that for moderately infective pathogens the two curves are very similar, so that risk calculations would be largely unaffected. But for pathogens with low median infective doses (e.g., hepatitis A and rotavirus) the choice does matter. Adopting the conditional dose-response curves for a single individual on each exposure occasion will result in higher risks being calculated. However taking just one individual per exposure event results in implausible jumps in risk profiles, particularly for very infectious pathogens (see Problem 9.1). Therefore the following prescriptions are made.

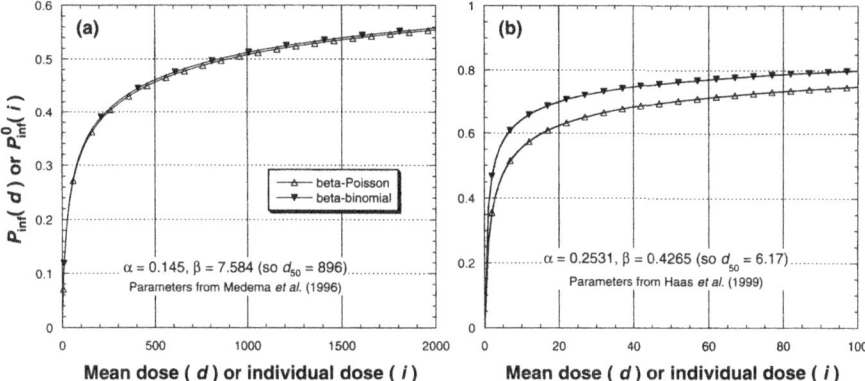

Fig. 9.6 Comparison of the beta-Poisson dose-response model for $P_{\text{inf}}(d)$ [Equations (9.3 and 9.21)] with the beta-binomial conditional dose response model for $P_{\text{inf}}^0(i)$ [Equation (9.23)] for (a) *Campylobacter* and (b) Rotavirus [d_{50} values are calculated from the beta-Poisson model, using Equation (9.4)].

> For very infectious pathogens, use conditional dose-response models for multiple individuals per exposure event; do not use models with single individuals per exposure event. For less infectious pathogens, use average dose models for a single individual per exposure event.

Problems

9.1 Discuss the difference in health risks predicted by QMRA based on a individual exposed one million times, versus 1000 individuals exposed 1000 times. *Hint*: contrast cases for exposure to a very infective versus a much less infective pathogen.

9.2 In Section 9.1.2 we noted that it may be better to use the mode rather than the median when constructing a distribution of ingestion or inhalation rates, given three statistics for these rates (in mL per hour): minimum = 10, median = 50, maximum = 100. Why is this?

The following problems use results given in Section 9.3.

9.3 What is the value of the mode of the shellfish meal size distribution shown in Figure 9.2?

9.4 What is the probability of getting a negative result when sampling a contaminated water in which pathogens are present and distributed randomly in a proportion P of samples?

9.5 What is the sum of Equation (9.7) over all nonzero values of the integer k?

9.6 Derive the hockey stick equations [Equations (9.9) – (9.11)].

9.7 Derive the simple exponential dose-response model for constant pathogen survival probability and random distribution of pathogens [Equation (9.17), which is also Equation (9.1)].

9.8 Derive the exact dose-response relationship for variable pathogen survival probability and a random distribution of pathogens [Equation (9.20)].

9.9 Derive the beta-binomial conditional dose-response relationship for variable pathogen survival probability [Equation (9.23)].

9.10 Show that the two forms shown in Equation (9.25) are identical.

10
MPNs and microbiology

Some microbiological water quality variables are assayed using multiple fermentation techniques—for example, the fecal indicator *E. coli* and the pathogens *Campylobacter* and *Salmonella*. To perform these assays, subsamples of the water are placed in sets of replicated tubes or wells containing compounds that will encourage the microorganism to grow in an incubator, if at least one cell is present. Or, in the case of viruses, PCR (polymerase chain reaction) techniques may be used to detect the presence of a virus in a well, via its DNA or RNA. Usually, there is more than one set. For example, there may be three sets of tubes each with five replicates: 5×100 mL, 5×10 mL, and 5×1 mL. This is a "decimal dilution series". Typically, up to four sets are used in such a series.

The pattern of positive results obtained is related to a "most probable number" (MPN) of microorganisms via a standard table (e.g., as contained in APHA 1998 [7]), and this is the final result of the test—perhaps accompanied by a confidence interval (also available in those tables). These tables have been worked out over many years, starting with McCrady (1915 & 1918 [213, 214]) and Greenwood & Yule (1917 [128]), and continuing to the present (e.g., Eisenhart & Wilson 1943 [80]; Cochran 1950 [48]; de Man 1975, 1977 & 1983 [68, 69, 70]; Best 1990 [24]) using a variety of approximate methods. These are all based on the assumption of random sampling from Poisson distributions (Tillett 1995 [317]). Interestingly, the early work of McCrady and Greenwood & Yule used an explicitly Bayesian approach (with a diffuse prior), the dominant statistical paradigm of those times.

Despite all this endeavor, there are some untidy aspects MPN tables and their usage:

216 MPNS AND MICROBIOLOGY

1. Numerical values of MPNs in standard tables may vary from one publication to another.

2. The width of "confidence intervals" about MPNs also varies between standard tables.

3. Fitting statistical distributions to MPNs (e.g., for Monte Carlo risk analyses) is not straightforward—there are "traps for the unwary."

10.1 MPNS IN STANDARD TABLES

Consider a pattern of 5-5-2 positive tubes in a setup with 5 tubes in each of three series of decimal dilutions, with volumes 10 mL, 1 mL and 0.1 mL. Two standard works have the equivalent MPN (per 100 mL of water) as 500 (APHA 1998 [7]) and as 540 (APHA 2001 [8]). Other examples may be cited. There are two reasons for such discrepancies: (i) most of the methods used to date for calculation of MPNs are approximate, and it is seldom clear exactly what approximations form the basis of particular tables; (ii) different and unstated rounding conventions appear to have been used (e.g., rounding 540 to 500, as noted above). These can simply be addressed and resolved:

1. by using the exact MPN methods described in Section 10.4,[1]

2. by authorities adopting the convention of stating the rounding criteria adopted in preparing their tables, so that readers are at least aware that the differences in MPN values between tables may be attributable to rounding conventions.

Having an exact method available also facilitates the production of "non-Standard" tables. For example, a 1×500 mL, 3×100 mL and 3×10 mL setup (making better use of a 1 L sample than a 3×100 mL, 3×10 mL and 3×1 mL setup for low concentrations). It is also better, sometimes much better, than approximate formulae, such as the disarmingly simple approximate formula proposed (without explanation) by Thomas (1942 [312]):

$$\text{MPN (per 100 mL)} = 100 \frac{P}{\sqrt{NT}}, \quad (10.1)$$

where P is the number of positive tubes, N is the volume in the negative tubes (mL), and T is the total volume in *all* the tubes. For example, consider the dilution series 3×10 mL, 3×1 mL, and 3×0.1 mL where a pattern of 3-3-1 positives is observed. Equation (10.1) gives MPN = $\frac{100 \times 7}{\sqrt{0.2 \times 33.3}} = 271.24$ per 100 mL. Direct enumeration of the exact equations (using the formulae in the Appendix) gives the result as 459.46 per 100 mL.

---
[1] Only approximate methods could be used in earlier times, because exact methods are computationally intensive.

Table 10.1 compares various standard tables' approximate results with the exact calculations, accompanied by their occurrence probabilities. Note that these probabilities vary considerably, and they are much smaller for the less probable patterns (e.g., a 1-2-3 pattern, not shown in the table, has an occurrence probability 6.14×10^{-8}). Tabulation of occurrence probabilities in standard tables helps analysts to discover laboratory methodological problems. For example, if there was an inhibiting material in the sampled water, its effect would be diluted in the second and third series, so a 1-2-3 pattern would be more likely to arise than its occurrence probability would suggest.

Table 10.1 Comparing approximate and exact MPN values for a 3×10 mL, 3×1 mL, 3×0.1 mL setup

Pattern	MPN per 100 mL, by reference				Exact[e]	Occurrence probability[f]
	a	b	c	d		
0-1-0	3	3.05	3.0	3.0	3.00	0.090
1-0-0	4	3.59	3.6	3.6	3.00	0.901
1-0-1	7	7.20	7.2	7.2	6.01	0.016
1-1-0	7	7.34	7.3	7.4	6.01	0.162
2-0-0	9	9.50	9.1	9.2	6.01	0.541
1-2-0	12	11.26	11	11	9.01	0.015
2-0-1	14	14.31	14	14	12.01	0.010
2-1-0	15	14.82	15	15	12.01	0.184
2-2-0	20	20.62	21	21	18.02	0.033
3-0-0	25	28.62	23	23	21.02	0.398
3-0-1	40	38.75	39	38	36.04	0.034
3-1-0	45	45.71	43	43	39.04	0.400
3-1-1	75	58.42	75	75	72.07	0.069
3-2-0	95	75.99	93	93	90.09	0.339
3-1-2	115	71.75	120	120	114.15	0.007
3-2-1	150	94.92	150	150	147.15	0.129
3-2-2	200	115.66	210	210	213.21	0.025
3-3-0	250	189.83	240	240	237.24	0.370
3-3-1	450	271.24	460	460	459.46	0.430
3-3-2	1100	438.40	1100	1100	1096.10	0.446

MPNs for many improbable patterns are not shown. [a]McCrady (1918 [214]). [b]Calculated from the equation of Thomas (1942 [312]), that is, Equation (10.1). [c]Woodward (1957 [345]). [d]de Man (1983 [70]). [e]Calculated from the value of n at the mode of Equation (10.10). [f]The mode of Equation (10.10).

From Table 10.1 we see that the exact computation tends to produce MPN values slightly lower than most other methods, and that the Thomas equation [Equation (10.1)] can produce considerably lower MPNs when a majority of tubes are positive.

10.2 MPN "CONFIDENCE INTERVALS"

A variety of methods have been used to compute these intervals, both frequentist and Bayesian, (as reviewed by McBride *et al.* (2003 [212]). For example, Woodward (1957 [345]) used an MPN-ordering procedure.[2] Others use lognormal approximations (Eisenhart & Wilson 1943 [80], Cochran 1950 [48])). Woodward's has been held to be most accurate of these methods (Loyer & Hamilton 1984 [193]). More recently the US Food and Drug Administration has endorsed a narrower set of confidence intervals,[3] being those published by de Man (1983 [70]). These are all frequentist methods, although earlier tables by de Man (1975, 1977 [68, 69]) were a form of Bayesian intervals, using a diffuse uniform prior.[4] These Bayesian intervals still appear in some standard tables for water and food samples (APHA 1998, 2001 [7, 8]). Given the manner in which confidence intervals are routinely interpreted (as discussed in Chapter 3), it could be argued that the standard tables *should* report Bayesian credible intervals.

Finally, Beliaeff & Mary (1993 [13]) have reported Bayesian credible intervals, using a diffuse prior, representing a computational implementation of equations developed by Greenwood & Yule (1917 [128]). They present both equi-tailed intervals and HPD (highest posterior density) region intervals, as discussed in Section 3.3.[5]

Comparisons between these intervals are given on Table 10.2. This shows the advantage of the Bayesian HPD intervals—they are the shortest of those presented. An interesting (and unresolved) issue is whether they should be shorter still. That is because the prior distribution used in their development states that before sample data are available it is believed that all microbial concentrations are equally likely. This implies that a water body is more likely to be grossly contaminated than it is to be healthy (there being a much larger range of concentrations implying contamination), even when historical sampling has routinely demonstrated a healthy state. To address this issue, one can adopt other more "informative" priors or adopt the "Empirical Bayes" approach (Reckhow 1996 [267], Carlin & Louis 2000 [40])), in which the data are used to guide the choice and parameter(s) of the prior distribution. Either approach will calculate shorter intervals. This is a fruitful research area. In conducting that research interaction and dialogue between statisticians and water managers is highly desirable, because choosing informative priors and using Empirical Bayes methods have subjective elements calling for professional judgments to be made.

10.3 SAMPLING AN MPN DISTRIBUTION

The ordinary number system has the (very desirable) property of unbiasedness. That is, if the true value of some quantity is x, where x can be any discrete or continuous

---
[2] Elaborated by Loyer & Hamilton (1984 [193]).
[3] http://vm.cfscan.fda.gov/ ebam/bam-a2.html.
[4] They can also be called "likelihood intervals".
[5] Roussanov *et al.* (1996 [274]) also report examples of these HPD intervals, along with an accompanying computer program, with essentially identical results.

Table 10.2 Comparing MPN confidence and credible limits for a 3 × 10 mL, 3 × 1 mL, 3 × 0.1 mL setup.

Pattern	95% confidence limits		95% credible limits		
	a	b	c	d	e
0-1-0	0.085–13	0.1–10	<1.0–17	0.7–17	0.1–15
1-0-0	0.085–20	0.2–17	<1.0–21	0.9–21	0.1–18
1-0-1	0.87–21	1.2–17	2–27	2.2–27	1.1–24
1-1-0	0.88–23	1.3–20	2–28	2.2–28	1.1–24
2-0-0	1.0–36	1.5–35	2–38	2.9–38	1.3–33
1-2-0	2.7–36	4–35	4–35	4.1–35	2.6–31
2-0-1	2.7–37	4–35	5–48	5.2–48	3.1–43
2-1-0	2.8–44	4–38	5–50	5.3–50	3.2–44
2-2-0	3.5–47	5–40	8–62	8.5–63	5.8–56
3-0-0	3.5–120	5–94	<10–130	8.7–130	3.8–108
3-0-1	6.9–130	9–104	10–180	15–180	8.1–150
3-1-0	7.1–210	9–181	10–210	17–210	8.4–180
3-1-1	14–230	17–199	20–280	28–280	16–250
3-2-0	15–380	18–360	30–380	33–390	18–340
3-1-2	30–380	30–360	40–350	44–350	29–320
3-2-1	30–440	30–380	50–500	56–510	34–450
3-2-2	35–470	30–400	80–640	86–640	59–580
3-3-0	36–1300	40–990	<100–1400	91–1400	40–1170
3-3-1	71–2400	90–1980	100–2400	180–2400	88–2070
3-3-2	150–4800	200–4000	300–4800	380–4800	190–4130

[a]Woodward (1957, [345]). [b]de Man (1983, [70]). [c]de Man (1979, [69]) [d]Beliaeff & Mary (1993, [13]) using a equi-tails regions with an area of 0.025 in each tail of the distribution. [e]Beliaeff & Mary (1993, [13]) using the HDR (Highest Density Region) with a total tail area = 0.05.

real number, then the the expected value from repetitive sampling from that distribution is x. This property is not shared by MPNs. For example, if the true value of a microorganism concentration is 28.8 per 100 mL, it is much more likely that the pattern obtained in a dilution series of 3 × 100 mL, 3 × 10 mL and 3 × 1 mL would be 3-3-0, corresponding to a MPN of 23.7 (with occurrence probability 0.370) than 28.8 (a 3-2-3 pattern, with occurrence probability 0.002). (See Table 10.3 for occurrence probabilities.) To make matters worse, many numbers are actually impossible as an MPN value (e.g., $28.7, 28.69, \ldots$). This pattern of varying occurrence probabilities guarantees that the PN distribution will be multi-modal—posing a considerable problem for distribution fitting.

This behavior poses difficulties for constructing frequency distributions. In particular, because of the varying occurrence probabilities, the MPN frequency distribution is very irregular and a fitting a unimodal distribution giving equal weight to all data is inappropriate. For example, fitting distributions to the top part of Figure 10.1, setting the "greater than" data to 111 per 100 mL gives rise to a geometric distribution [see

Table 10.3 Occurrence probabilities (P) for a 3 × 10 mL, 3 × 1 mL, 3 × 0.1 mL setup.

#	MPN	P	#	MPN	P	#	MPN	P
0-0-0	<0.3	–	0-3-1	1.2	6×10^{-6}	2-3-1	3.3	3×10^{-4}
1-0-0	0.3	0.901	0-2-2		2×10^{-6}	2-2-2		7×10^{-5}
0-1-0		0.090	1-0-3		6×10^{-7}	3-0-1	3.6	0.034
0-0-1		0.009	0-1-3		6×10^{-8}	3-1-0	3.9	0.400
2-0-0	0.6	0.541	0-3-2	1.5	9×10^{-8}	2-2-3		9×10^{-7}
1-1-0		0.162	0-2-3		9×10^{-9}	2-3-2	4.2	1×10^{-5}
1-0-1		0.016	2-2-0	1.8	0.033	2-3-3	5.1	2×10^{-7}
0-2-0		0.005	2-1-1		0.009	3-0-2	6	0.002
0-1-1		0.002	2-0-2		3×10^{-4}	3-1-1	7.2	0.069
0-0-2		5×10^{-4}	1-3-1		3×10^{-5}	3-2-0	9	0.339
1-2-0	0.9	0.015	1-2-2		9×10^{-6}	3-0-3	9.3	4×10^{-5}
1-1-1		0.004	1-1-3		3×10^{-7}	3-1-2	11.4	0.007
0-2-1		0.001	0-3-3		5×10^{-10}	3-2-1	14.7	0.129
0-3-0		2×10^{-4}	3-0-0	2.1	0.398	3-1-3	15.6	3×10^{-4}
1-0-2		1×10^{-4}	1-3-2		7×10^{-7}	3-2-2	21.3	0.025
0-1-2		1×10^{-5}	1-2-3		6×10^{-8}	3-3-0	23.7	0.370
0-0-3		2×10^{-7}	2-2-1	2.4	0.002	3-2-3	28.8	0.002
2-1-0	1.2	0.184	2-1-2		2×10^{-4}	3-3-1	45.9	0.430
2-0-1		0.018	2-0-3		2×10^{-6}	3-3-2	110	0.446
1-3-0		6×10^{-4}	2-3-0	2.7	0.003	3-3-3	>110	–
1-2-1		5×10^{-4}	1-3-3		6×10^{-9}			
1-1-2		5×10^{-5}	2-1-3	3.0	2×10^{-6}			

"#" denotes pattern of positive tubes.

Equation (6.16)] with parameter $\theta = 0.00657$.[6] This is a very long-tailed distribution and it may be better to use a binning procedure, as also given in the Figure. (Furthermore, fitting distributions to the "greater than" data is problematical.) That is:

1. "Bin" the data according to some rational criterion; that is, select the concentrations at the floor and roof of contiguous bins and count up all MPNs inside each bin, of whatever occurrence probability.

2. Fit a distribution to the frequencies over the bins.

Then in a Monte Carlo risk analysis procedure a two-step procedure can be used:

1. On any iteration, draw a random sample from that distribution to determine which bin is to be selected.

2. Make a random draw from that bin using a uniform distribution between its floor and roof, so returning a microorganism concentration.

---

[6]Candidate distributions were examined using the BestFit procedure in @RISK.

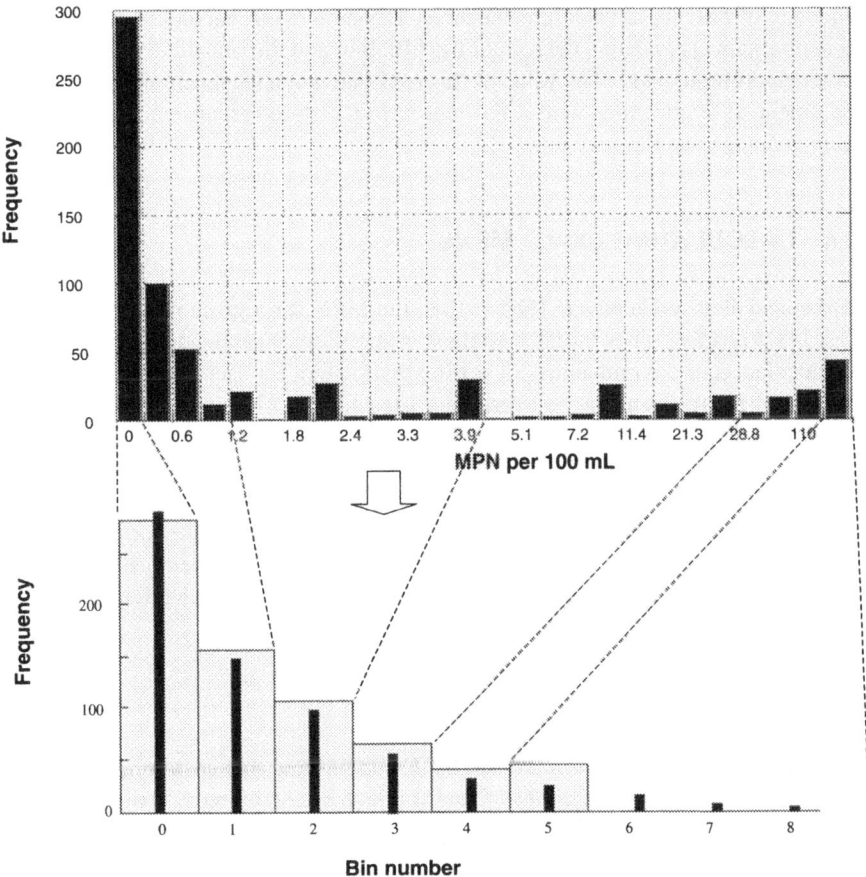

Fig. 10.1 Binning procedure for MPN data (data are *Campylobacter* concentrations (per 100 mL) in freshwaters (McBride *et al.* 2002 [211]). *Note*: x axis not to scale.

Example For a 3×10 mL, 3×1 mL, 3×0.1 mL setup given in Table 10.3 we can define bins such that internal bin should contain two MPNs with occurrence probabilities greater than 0.2. Bins must also be defined at the extremes of the data (no tubes positive or all tubes positive, corresponding to zero and >110 per 100 mL). This procedure results in internal bin boundaries at 0.3, 1, 4, 25, and 110 per 100 mL, as shown on Figure 10.1. The bin frequencies can be well described by geometric distribution [Equation (6.16)], with parameter $\theta = 0.425$. Note that this fit includes bins $6, 7, 8, \ldots$ whereas our maximum bin number is 5. But this merely means that whenever a random draw is made from this distribution, any bin number greater than 4 signals that the microorganism concentration is in the ">110" category and the draw is to be made from bin number 5. Because the sum of all the vertical bars in bins 5 and beyond matches the observed frequency in bin number 5 closely, the frequency of drawing a value greater than 110 per 100 mL should be satisfactory. All that remains is to define the roof of the fifth bin, and this should be elicited from aquatic microbiologists.[7] The value of this roof may not have much impact in a risk assessment, if it coincides with a flat region in the dose-response curve [e.g., for *Campylobacter*, see Figure 9.3(b)].

10.4 CALCULATING EXACT MPNS

Following Tillett & Coleman (1985 [318]), the MPN for a given pattern of results (e.g., 1-2-2 positive tubes out of a total set of $1 \times 3 \times 3$) can be derived by repetitive calculation of the probability of getting that pattern for each and every value of numbers of microorganisms between nominated limits. These are the "occurrence probabilities." The distribution of these probabilities over those microorganism numbers is smooth and unimodal, so the MPN is the number of them at the distribution's mode divided by the sample volume placed in the tubes. To develop the required formula we first consider each set of tubes separately and obtain the formula for the probability of getting any number of positive tubes in that set, it being assumed that one microorganism will always cause a tube to return a positive result after incubation. The procedure given below can be programmed relatively easily.

10.4.1 Considering each set separately

Denote the number of tubes in any set by R and assume that each tube in this set contains the same volume of sample. The formula for the required probability comes from "occupancy theory" (David & Barton 1962 [64]), who noted that it flows from Abraham De Moivre's 1718 treatise *Doctrine of Chances*).[8] Its development is as follows. If n microorganisms are distributed at random among the R replicate test tubes, the probability that r tubes will receive at least one microorganism is

$$\Pr(r|R,n) = \binom{R}{r} \sum_{i=0}^{r} \left[\binom{r}{i}(-1)^i \left(\frac{r-i}{R}\right)^n \right], \quad (10.2)$$

for $r(\leq n) = 1, 2, \ldots, R$, where $\binom{R}{r} = \frac{R!}{r!(R-r)!}$ is the binomial coefficient (the number of combinations of R objects taken r at a time), i is a dummy summation index, and the "|" symbol denotes "given." Three special cases (boundary conditions) must be stated separately (they will be needed later). First, if there are no microorganisms in the set (i.e., $n = 0$), it is certain that no tubes will be positive. Conversely, it is impossible for any tube to be positive if the microorganism is absent. Therefore

---

[7] The elicited maximum *Campylobacter* concentration in the New Zealand freshwater study was taken as 2000 per 100 mL (McBride *et al.* 2002 [211]).
[8] That is, if x balls are tossed at random at y cavities, what is the chance that a particular cavity will contain at least one ball?

$$\Pr(r|R, 0) = \begin{cases} 1 & \text{if } r = 0, \\ 0 & \text{if } r = 1, 2, \ldots, R. \end{cases} \quad (10.3)$$

Second, it is impossible for there to be no positive tubes if microorganisms are present, that is,

$$\Pr(0|R, n) = 0 \quad \text{for } n \geq 1. \quad (10.4)$$

Third, it is impossible for the number of positive tubes to exceed the number of microorganisms, that is,

$$\Pr(r|R, n) = 0 \quad \text{for } r(\leq R) > n. \quad (10.5)$$

For the remaining general case (i.e., $1 \leq r \leq R, n \geq 1$ and $r \leq n$) there are two means of calculating the probabilities given by Equation (10.2). The first is to use a recursion formula, that is,

$$\Pr(r|R, n) = \frac{r}{R} \Pr(r|R, n-1) + \frac{R+1-r}{R} \Pr(r-1|R, n-1), \quad (10.6)$$

for $r(\leq n) = 1, 2, \ldots, R$.[9] This equation can be calculated for any n using the boundary conditions given in Equation (10.3) and an upper bound given by Equation (10.5). The second approach is to evaluate Equation (10.2) directly, simplifying it (by canceling out the $r!$ term, and grouping constant terms outside the summation symbol) to

$$\Pr(r|R, n) = \frac{1}{R^n} \frac{R!}{(R-r)!} \sum_{i=0}^{r-1} (-1)^i \frac{(r-i)^n}{i!(r-i)!}, \quad (10.7)$$

for $r(\leq n) = 1, 2, \ldots, R$, again using Equations (10.3) and (10.5), as appropriate. See Problem 10.3 for a discussion of the merits of these approaches.

10.4.2 Combining the sets

Consider the series of three fermentation tubes in which we have:

- I tubes each of volume V_I of which i are positive (so $0 \leq i \leq I$);
- J tubes each of volume V_J of which j are positive (so $0 \leq j \leq J$);
- K tubes each of volume V_K of which k are positive (so $0 \leq k \leq K$).

The total volume in the tubes is

$$V = IV_I + JV_J + KV_K, \quad (10.8)$$

---

[9] See Problem 10.2 for a derivation of Equation (10.6).

and so for each randomly distributed microorganism present in the tubes:

- the probability that it is in an I-tube is $p_I = IV_I/V$;
- the probability that it is in a J-tube is $p_J = JV_J/V$;
- the probability that it is in a K-tube is $p_K = KV_K/V$.

These probabilities are the proportion of total sample volume in each set.

Now, consider n microorganisms randomly distributed among the tubes. The probability of obtaining n_I microorganisms in the I series, n_J microorganisms in the J series and n_K microorganisms in the K series is obtained from the multinomial distribution (see Section 6.2.9), that is,

$$\Pr(n_I, n_J, n_K|n) = \frac{n!}{n_I! n_J! n_K!} p_I^{n_I} p_J^{n_J} p_K^{n_K}, \tag{10.9}$$

where $n = n_I + n_J + n_K \geq i + j + k$ (equality would hold only if each positive tube contained only one microorganism). Now we need to consider all the possible ways of obtaining i positive tubes in series I, j positive tubes in series J, and k positive tubes in series K, over a range of values of n. So we sum the product of the multinomial probabilities with the result of the previous section—the probability that i tubes in the I series will receive at least one bacterium, and so on for series J and K. This summation is over n_I and $n_J (\leq n - n_I)$ only, because n_K is then fixed (at $n_K = n - n_I - n_J$):

$$\Pr(i,j,k|n) = \sum_{n_I=0}^{n} \sum_{n_J=0}^{n-n_I} \Pr(i|I, n_I) \Pr(j|J, n_J) \Pr(k|K, n_K) \Pr(n_I, n_J, n_K|n).$$
$$\tag{10.10}$$

See Problem 10.4 for the form of this equation with 1, 2 or 4 series.

Problems

10.1 Discuss how MPN values relate to the pattern of positive tubes in Table 10.3.

10.2 Derive the recursion Equation (10.6), for the probability of getting r positive tubes in a set of R tubes, given n microorganisms to randomly distribute in that set.

10.3 Discuss the relative merits of evaluating Equation (10.2) via the recurrence relation [Equation (10.6)] versus the direct equation [Equation (10.7)].

10.4 What are the occurrence formulae corresponding to Equation (10.10) for setups with 1, 2, or 4 dilution series?

11
Trends, impacts, concordance, detection limits

11.1 DETECTING AND ANALYZING TRENDS

Water quality trend analysis has increasingly been based on nonparametric methods, particularly those based on Kendall's tau, the Mann–Kendall test (Section 2.9) and the Sen slope estimator (Section 2.12). The attraction of these methods is their ability to handle non-normal data with censored, tied and missing values. A popular version of the test is the seasonal Kendall trend test, an extension of the Mann-Kendall test, that removes seasonal cycles. It is well described in detail in texts (e.g., Gilbert 1987 [118]; Loftis 1989 [188]) and software (IDT 1998 [160]), so only a brief description is given here.

11.1.1 Trend slope estimator

We use the example where we have 60 observations of total nitrogen concentrations from five years of monthly sampling at a river site (Smith *et al.* 1996 [288]). Denote the concentration in any month i and year j as $c_{i,j}$. The procedure is as follows:

1. Select a particular month i and compute all possible annual slopes between years for that month. There are 10 of these: $\frac{c_{i2}-c_{i1}}{1}$, $\frac{c_{i3}-c_{i1}}{2}$, $\frac{c_{i4}-c_{i1}}{3}$, $\frac{c_{i5}-c_{i1}}{4}$, $\frac{c_{i3}-c_{i2}}{1}$, $\frac{c_{i4}-c_{i2}}{2}$, $\frac{c_{i5}-c_{i2}}{3}$, $\frac{c_{i4}-c_{i3}}{1}$, $\frac{c_{i5}-c_{i3}}{2}$, $\frac{c_{i5}-c_{i4}}{1}$.

2. Repeat this procedure for all months (i.e., for $i = 1, 2, \ldots, 12$) for each of the five years, giving 120 slopes.

3. Rank these slopes and take the median as the trend slope estimate. This is the average value of the 60th and 61st slopes.

11.1.2 Seasonal Kendall trend test

The α-level procedure to test the hypothesis that there is no trend is as follows:

1. Select a particular month i and compute the signs of all possible differences in concentrations between years for that month. Cross-month differences are not considered, thus removing seasonality. The resulting 10 differences are: $\text{sgn}(c_{i2} - c_{i1})$, $\text{sgn}(c_{i3} - c_{i1})$, $\text{sgn}(c_{i4} - c_{i1})$, $\text{sgn}(c_{i5} - c_{i1})$, $\text{sgn}(c_{i3} - c_{i2})$, $\text{sgn}(c_{i4} - c_{i2})$, $\text{sgn}(c_{i5} - c_{i2})$, $\text{sgn}(c_{i4} - c_{i3})$, $\text{sgn}(c_{i5} - c_{i3})$, $\text{sgn}(c_{i5} - c_{i4})$.

2. Replace $\text{sgn}(x)$ by 1 if the sign is positive or by -1 if it is negative, and by 0 if there is no difference (i.e., there is a tie, so that $x = 0$). Then sum the 10 values and call that sum S_i.

3. Compute the variance of S_i, that is, $\frac{n_i(n_i-1)(2n_i+5) - \sum [t_{ip}(t_{ip}-1)(2t_{ip}+5)]}{18}$, where n_i is the number of data for month i, t_{ip} is the number of tied data in the pth tied group for the ith month, and the summation is over the number of tied groups for that month.[1]

4. Sum the S_i series over all months, that is, $S = \sum_{i=1}^{12} S_i$, so S is the sum of 120 terms.

5. Compute the variance of S from $\text{var}(S) = \sum_{i=1}^{12} \text{var}(S_i)$.

6. Compute the test static (Z_{sK}) from the large sample normal approximation (with continuity correction of one unit), that is,

$$Z_{sK} = \begin{cases} \dfrac{S-1}{\text{var}(S)} & \text{if } S > 0, \\ 0 & \text{if } S = 0, \\ \dfrac{S+1}{\text{var}(S)} & \text{if } S < 0. \end{cases} \quad (11.1)$$

7. Reject the hypothesis if $|Z_{sK}| > Z_{1-\alpha/2}$.

In our example, the lowest nonzero value of $|Z_{sK}|$ is 0.0707. This value occurs when there are no ties and the number of terms of opposite sign differs by 2, so that $S = \pm 2$ and $\text{var}(S) = \sum_{i=1}^{12} \frac{5 \times 4 \times 15}{18} = 200$. Equation (11.1) then gives $|Z_{sK}| = 1/\sqrt{200} = 0.0707$. The associated p-value is 0.9436.[2] Slightly larger values of the test statistic, and hence smaller p-values, are obtained if $S = \pm 2$ in the presence of ties (because then $\text{var}(S) < 200$).

However, there are a number of situations for which it is possible to obtain $S = -1, 0,$ or 1, hence $Z_{sK} = 0$, and so, for a two-sided test, $p = 1$! These are:

---
[1] The formula is more complicated if there are multiple data per season (Gilbert 1987, p. 226 [118]; IDT 1998 [160]).
[2] See Problem 11.1.

1. There are no ties and an equal number (60) of terms in the S series have opposite signs.

2. There is an even number of ties and an equal number (< 60) of terms in the S series have opposite signs.

3. There is an odd number of ties and the remaining terms in the S series differ by 1 in their numbers of opposite signs.

11.1.3 Two anomalies

The seasonal Kendall trend test and the Sen slope estimator are independent calculations, and, in an extensive trend analysis, occasional anomalies can arise between them, as explained below.

A zero slope can be statistically significant! This arises when there are a disproportionate number of ties in the data, as can arise when a variable such as dissolved reactive phosphorus is reported from the laboratory on a very discrete scale at low concentrations. For example, analysis of 5 years of monthly dissolved reactive phosphorus data for a pristine site (on the Hurunui River, New Zealand, Smith *et al.* 1996 [288]) gave 53 positive slopes, 27 negative slopes, and 40 ties (12 groups of 2, 4 groups of 3, and 1 group of 4), so $S = 26$. These ties cause the median slope to be zero (it is straddled by ties), reducing var(S) from 200 to 152. So the test statistic is $|Z_{sK}| = \frac{25}{\sqrt{152}} = 2.028$. This is statistically significant at the 5% level (being greater than $Z_{1-\alpha/2} = 1.9600$). The problem here is that the slope estimator is not very appropriate in the presence of many ties. It may be better to take the median of all non-tied values, or perform some other slope estimation procedure. If there are a large number of ties resulting from replacement of "less than" data, it may not be meaningful to seek for trends at all, especially if the detection limit has changed with time.

Absolute certainty that there is no trend! As noted in Section 11.1.2, it is possible to obtain $p = 1$, implying absolute certainty that there is no trend. This arises whenever the test statistic is computed to be zero [from Equation (11.1)]. A simple way to resolve the problem is to report the test's p-value as CEILING($p*$), where $p*$ is the largest p-value that can be attained when the test statistic is nonzero and CEILING raises $p*$ to the next (higher) percentage point. Because $p*$ is always less than 1 this procedure works well. For example, as shown in Section 11.1.1, for trend analysis of monthly sampling for 5 years we have $p* = 0.9436$ and so the result should be reported as $p < 0.95$. For ten years of monthly sampling we have $p* = 0.9794$ and so the result should be reported as $p < 0.98$.

Five further considerations for trends are offered.

- The more data one has over a given time period, the more likely it is that a trend will be detected within that period.[3] This is the "power depends on sample size" issue that we have addressed many times now. For example, Stansfield (2001 [296]) reports that fewer trends were detected from a monthly dataset that was reduced to quarterly sampling.

- Failing to find a statistically significant result from a trend tests doesn't imply that there was no trend at all. We can always expect there to be some upward and downward changes.

- Detection limits may change with time, confounding the comparability of trend tests' results. Stansfield reports that trends detected using a low detection limit are often not detected when a higher limit is used.

- If analytical methods change, apparent step trends may be observed. If so, it is always wise to refer back to records of those methods, or to the laboratory analysts directly, to see if that is the case. While it is desirable that laboratories do not change methods in the middle of a trend monitoring exercise, there will of course be occasions when the pressure for a new and better method becomes irresistible. For that reason alone, it is also wise to include the analysts in the development of the trend monitoring program, so they may "buy-in" to its implementation. If those analysts have compelling reasons for changing their methodology, it is desirable to seek to implement an interim transitional arrangement whereby the old and the new methods are used in parallel for a time, to seek to establish a correlation between the two ("overlap studies", Newell & Morrison 1993 [240]). This correlation can be used to adjust the old method's results to those of the new methods, and vice versa.

- Trends in different directions can be evident for the same data, if one performs the analysis on the raw data and then also on the flow-adjusted data [see Helsel & Hirsch (2002 [147]) for a detailed discussion of this technique.] For example, compare Figures 2.4 and 11.1. Both can be instructive for water management— the former indicating trends in catchment response, the latter indicating trends in catchment loading.

11.1.4 Multi-site trends; a Bayesian approach

Let's consider a regional water quality trend analysis in which trend slope calculations have been carried out at many sites. How may we interpret the results in terms of regional patterns? For example, the groundwater nitrate trend test results in Table 11.1 (using the Sen slopes and linear regressions, "LR") have been reported for

---

[3]We can ignore the effects of serial correlation if our interest in trends is confined to the period of record, as explained in Section 2.11.1.

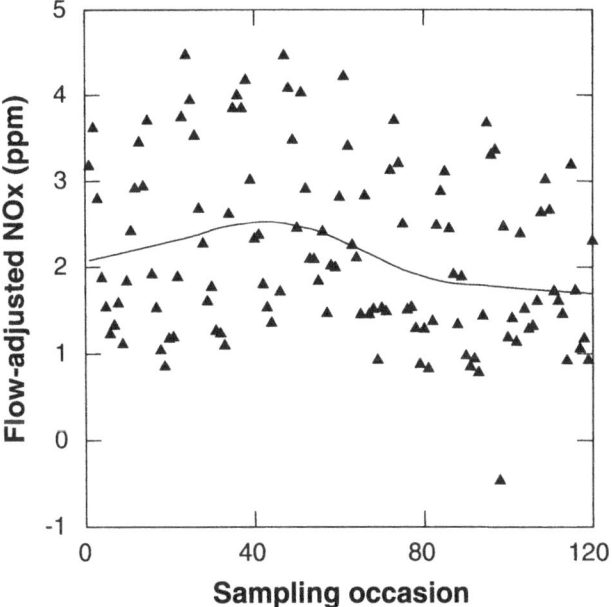

Fig. 11.1 Lowess fit for flow-adjusted Ngakaroa Stream data, prepared using Systat (SPSS 1998 [294]).

the Canterbury region in the South Island of New Zealand, for 5-year and 10-year periods.[4]

Table 11.1 Groundwater trends in Canterbury

Slope	5 years data		10 years data	
	Sen	LR	Sen	LR
<0 (downward)	41	52	51	64
>0 (upward)	74	76	53	64
= 0 (none)	14	1	24	0

A regional politician may ask the question: what is the probability that we have an upward trend? In doing so, we are forced to consider using Bayesian methods. This is because the question concerns the probability of a parameter (the change in mean groundwater nitrate levels over time), rather than the probability of data assuming that

---
[4]These results are regardless of whether "statistical significance" was attained in the frequentist methods used to obtain them.

there is no trend (the p-value), as would be the case using frequentist methods. The "penalty" in doing so is that we have to furnish a prior belief probability statement about the likelihood of a trend before seeing the data in the table. Let's take an impartial view, that an upward trend is held to be equally as likely as a downward trend, so we use a uniform distribution (between zero and one) for the proportion of upward trends. This prior distribution is also the beta function prior, $Be(1, 1)$. This immediately enables us to use the results established in Section 8.4.2. That is, the "Confidence of Compliance" [Equation (8.10)] tells us the probability that the proportion of the occurrence of a certain event is no more than X, given that we have measured e such events in n observations. So to answer the politician's question we could define those events to be upward trends at any individual site. Then, setting $X = \frac{1}{2}$ (which would be its value were there to be no regional trend), the probability that the true proportion of upward trends is no more than $\frac{1}{2}$ is

$$P_{\text{down}} = I_{X=\frac{1}{2}}(1 + e, 1 + n - e), \tag{11.2}$$

where I_X is the incomplete beta function. We could evaluate this function directly, with appropriate software. A simpler approach notes that we can expect that $n > e \gg 1$ and so $I_X(1+e, 1+n-e) \approx I_X(1+e, n-e)$. We can then use the identity between incomplete beta functions and the cumulative binomial distribution given by Equation (6.13), to obtain

$$P_{\text{down}} \approx 1 - B(e; n, \frac{1}{2}), \tag{11.3}$$

where B is the cumulative binomial distribution, and so the probability that the true upward trends is more than $\frac{1}{2}$ is

$$P_{\text{up}} \approx B(e; n, \frac{1}{2}). \tag{11.4}$$

Simplifying the statement "The probability that the true proportion of upward trends is more than $\frac{1}{2}$" to "The probability of a regional upward trend," we see that this probability for the first five years of linear regression slope data (ignoring the "no trend" results) is $B(76; 128, \frac{1}{2})$. This may be calculated to be 0.9866.[5] For the Sen Slope data the probability is $B(74; 115, \frac{1}{2})$. This may be calculated to be 0.9993%. So the politician can be told that in those five years there was a very high probability (>98%) that there truly was a rise in regional groundwater levels in those five years.

The picture is very different for the 10 years of data, however. For the Sen slope data the probability of an upward trend is $B(53; 104, \frac{1}{2}) = 0.6156$. Whatever trends occurred in the first 5 years have been largely reversed in the last 5 years, as may often be expected—simply reinforcing the point that trend analysis is scale-dependent.

[5] For example, this can be calculated by entering "=BINOMDIST(76, 128, 0.5, TRUE)" in an Excell cell.

11.2 IMPACTS AND NOMINAL DATA

Some water quality data, and related data, are nominal. They may be dichotomous (e.g., "clear", "turbid"), or polychotomous [e.g., octants of wind direction (N, NE, E, SE, S, SW, W, NW)]. These data can be more useful than may first appear.

For example, the city of Gisborne, North Island, New Zealand, operates a 6000-feet submarine outfall for the discharge of its treated sewage into a large Bay. Two substantial rivers also discharge into the bay (the Waipaoa and Turanganui rivers). Questions have arisen about the impact of the diluted wastewater plume on the shoreline and its relative importance compared with the river inflows—each river drains extensive agricultural landscapes.

The Gisborne authorities have also operated a substantial coastal water quality monitoring programme for microbiological fecal indicators (fecal coliforms and enterococci) over some years, both at shoreline and offshore sites. Records were kept of other variables that could be associated with the degree of contamination at these sites: tidal state (on a 1–4 scale), tidal range (in meters), average wind direction over the four hours preceding the sampling, and river flows. This enabled a statistical model to be constructed to predict the fecal indicator concentrations at various sites, using all these continuous and polychotomous data as independent variables. That model is an **analysis of covariance** (if all the predictors were continuous a regression model would be used; if all were discrete category labels, an analysis of variance model would be called for).

Inspection of the data reveled, not surprisingly, that the microbiological data and the river flow data were right-skewed. This skew was largely reduced by replacing these data by their logarithms. Collinearity of the independent variables (the tendency of pairs of them to rise or fall together) was low.

A working hypothesis is that contamination levels may be dominated by the river flows, except in conditions of low river flow (generally summer). Accordingly, models were run for year-round conditions first, and the pattern of p-values for each of the model coefficients were examined. Then models were then rerun for low river flow conditions only, and p-value patterns examined again. That is, p-values were used as a relative measure of importance of the various independent variables, not as an absolute measure, nor even to compare between the two models [because the low-flow model has a much smaller sample size (see the box on page 88)]. For the year-round model the dominant p values were associated with the river flows. But for low-flow conditions, p-values for wind speed and direction coefficients at the monitoring sites between the outfall and the shore became very significant. Furthermore, they were particularly significant for onshore winds.

This rather simple example well illustrates the wisdom in collecting nominal data as part of a monitoring program. They can be used to great effect in statistical models to help separate randomness from pattern.

11.3 CONCORDANCE ASSESSMENT

This is an example of where one-sided tests should be used, but seldom are.

11.3.1 Cohen's kappa for dichotomous variables

Some water quality variables are dichotomous. Assays showing the presence or absence of microorganisms are a common example. Even for methods that are not dichotomous, there may be an interest in seeing how well they perform in detecting the presence of *any* particles—for example, drinking-water standards often require the complete absence of pathogen indicators. Typically, this assessment is carried out by comparing a new method with a "gold standard" method, such as new multi-well coliform assay procedures versus tried-and-true multiple tube fermentation methods.

Many statistical procedures have been advanced for making such ("old" versus "new") assessments. For example, McNemar's test (Zar 1996 [349]). One test that seems particularly suited to the task uses Cohen's kappa statistic (Cohen 1960 [49]; Bishop *et al.* 1975 [25]; Fleiss 1981 [95]).[6] This statistic has two attractive features:

1. It is "chance-corrected" (one may expect a certain amount of agreement by chance alone).

2. Straightforward one-sided tests may be used to assess the degree of importance that the observations imply.

The chance-correction is constructed as follows. Consider any index that assumes the value 1 when there is complete agreement. Let p_0 denote the observed value of that index and let p_e denote its value expected on the basis of chance alone. Then the observed excess beyond chance is $p_0 - p_e$, whereas the maximum possible excess is $1 - p_e$. The ratio of these differences defines kappa (κ), that is,

$$\hat{\kappa} = \frac{p_0 - p_e}{1 - p_e}, \qquad (11.5)$$

where the circumflex over κ makes it clear that it is the observed value of κ, our estimate of its true value, based on the data at hand. We can summarize presence/absence data for two dichotomous variables ("Alternative" and "Gold standard") in the following table:
where the total number of data is $n = n_{pp} + n_{pa} + n_{ap} + n_{aa}$. These data are turned into frequencies by dividing by n, as follows.
where $a = \frac{n_{pp}}{n}$, $b = \frac{n_{pa}}{n}$, $c = \frac{n_{ap}}{n}$, and $d = \frac{n_{aa}}{n}$. (See Problem 11.2 for various accuracy measures associated with these variables.) The marginal frequencies in the

---

[6] A number of authors have questioned its utility, e.g., Thompson & Walter (1988 [315]), Kraemer & Bloch (1988 [174]), Spitznagel & Helzer (1985 [293]), especially because of its dependance on the "true prevalence rate." From such literature, it seems wise to aim to obtain a set of examples in which true negatives do not overwhelm true positives, and vice versa, in which case kappa performs as well as any other statistic—and better than many.

CONCORDANCE ASSESSMENT

Table 11.2 Presence/absence table

Alternative	Gold standard	
	Presence	Absence
Presence	n_{pp}	n_{pa}
Absence	n_{ap}	n_{aa}

Table 11.3 Frequency table for presence/absence data

Alternative	Gold standard		Total
	Presence	Absence	
Presence	a	b	p_1
Absence	c	d	q_1
Total	p_2	q_2	1

Table are defined as row and column sums (e.g., $p_1 = a + b$). We can use these marginal frequencies to construct a chance-expected table, Table 11.4.[7]

Table 11.4 Chance-correction table for presence/absence data

Alternative	Gold standard		Total
	Presence	Absence	
Presence	$p_1 p_2$	$p_1 q_2$	p_1
Absence	$q_1 p_2$	$q_1 q_2$	q_1
Total	p_2	q_2	1

Now the frequency table (Table 11.3) shows that the overall proportion of chance-expected agreement is

$$p_0 = a + d, \qquad (11.6)$$

and the chance-expected table (Table 11.4) shows that the overall proportion of chance-expected agreement is

---
[7] Its derivation is discussed in Problem 11.3.

$$p_e = p_1 p_2 + q_1 q_2, \tag{11.7}$$

and so Cohen's kappa is given by inserting Equations (11.6) and (11.7) into Equation (11.5),

$$\hat{\kappa} = \frac{a + d - p_1 p_2 + q_1 q_2}{1 - p_1 p_2 + q_1 q_2}, \tag{11.8}$$

which can be simplified to (see Problem 11.4)

$$\hat{\kappa} = \frac{2(ad - bc)}{p_1 q_2 + p_2 q_1}. \tag{11.9}$$

A value $\hat{\kappa} = 0$ indicates no agreement beyond that due to chance alone; $\hat{\kappa} = 1$ indicates perfect agreement. Negative values can arise, and denote discordance.

Equivalence criteria Landis & Koch (1977 [177]) have suggested the agreement descriptors for kappa shown in Table 11.5. Note that this table is in terms of the true value of κ, not its estimate $\hat{\kappa}$, so some account will need to be made for statistical sampling error. That is, should a permissive or a precautionary stance be taken? Because one needs firm grounds for concluding that a new method compares well with a "gold standard" method, a precautionary stance is desirable. So, for example, if the new method is claimed to be in "substantial agreement" with the standard, we should perform a one-sided hypothesis test **H**: $\kappa < 0.60$.[8] If **H** were to be rejected we would have strong grounds for asserting the agreement (concordance) is at least "substantial". Of course, in order to reject the hypothesis, $\hat{\kappa}$ would have to be some way above 0.60 for this to occur, and that is as it should be in a precautionary approach.

Table 11.5 Suggested agreement descriptors.

Kappa statistic	Strength of agreement
<0.00	Poor
0.00–0.20	Slight
0.21–0.40	Fair
0.41–0.60	Moderate
0.61–0.80	Substantial
0.81–1.00	Almost perfect

Source: Landis & Koch (1977 [177]).

---

[8]We discussed these tests in Chapter 5.1. As to the merits of testing a point-null hypothesis, see Problem 11.5.

To perform this test we can use a simple confidence interval approach.[9] The decision rule is simply

$$\text{Reject } \mathbf{H} \text{ if } \hat{\kappa} - 1.6449 S_{e,\hat{\kappa}} > \kappa^*, \qquad (11.10)$$

where κ^* is a criterion value (e.g., 0.6 in the example above) and $S_{e,\hat{\kappa}}$ is the standard error of the estimate, for which a formula is given in this chapter's Appendix.[10]

Example A new easily performed presence/absence microbiological assay procedure has been tested in 120 samples against a more resource-intensive standard method for drinking-water surveillance: 24 pairs of observations are "present/present," 8 are "present/absent," 5 are "absent/present," and 83 are "absent/absent." Then from Equation (11.6), $p_0 = \frac{24+83}{120} = 0.89167$, and, from Table 11.3, $p_1 = \frac{24+8}{120} = 0.26667$, $p_2 = \frac{24+5}{120} = 0.24167$, $q_1 = \frac{5+83}{120} = 0.73333$, $q_2 = \frac{8+83}{120} = 0.75833$. Then, from Equation (11.7), $p_e = 0.26667 \times 0.24167 + 0.73333 \times 0.75833 = 0.62056$. Finally, from Equation (11.5) Cohen's kappa is $\hat{\kappa} = \frac{0.89167 - 0.62056}{1 - 0.62056} = 0.71449$.[11] Applying the standard error formula [Equation (A.1), in this chapter's Appendix] we obtain $S_{e,\hat{\kappa}} = 0.07382$. Inserting this into into Equation (11.10) we obtain a lower one-sided 95% confidence limit of $0.71449 - 1.6449 \times 0.07382 = 0.59306$. That is, because κ is less than 0.6, we do not have sufficient evidence to confidently infer "Substantial Agreement."

This example shows that, under a precautionary approach, even though the observed kappa value is above the Landis & Koch boundary for "Substantial" agreement, it isn't far enough above to convey a convincing message of that degree of agreement. Had more paired observations been made it could be that the boundary would have been passed. But it is also possible that further sampling would result in a reduction in the observed value of kappa. In any event, a strategy of sampling until "statistical significance" is attained is "peeking", and, as such, cannot return such a result (as explained in Section 4.3.5).

A Bayesian interpretation Recall that the Bayesian approach enables us to make direct probability statements about a hypothesis. Noting that kappa is a continuous variable, we may make use of the result shown in the box on page 101, in Section 4.9. That is, if we assume a unimodal prior distribution that is symmetrical about κ^*, then p becomes the probability of $\mathbf{H}$ being true. This seems an entirely plausible assumption, so we could interpret the result of the example as follows: there is a greater than 5% probability that the tested hypothesis is true and so, under a precautionary approach, it is not accepted that the new method is equivalent to the gold standard method. A further advantage of this approach is that appears to avoid all the problems associated with "peeking", were we to embark on further sampling.[12]

---

[9]Note that this is a "one-sample" case; that is, our hypothesis concerns only one (κ) variable. So there is a one-to-one correspondence between a one-sided test and a one-sided confidence interval, as discussed in Section 5.6.

[10]The value 1.6449 is the 95%ile of the unit normal distribution.

[11]Alternatively, using Equation (11.9), $\hat{\kappa} = \frac{2\left(\frac{24}{120} \times \frac{83}{120} - \frac{8}{120} \times \frac{5}{120}\right)}{0.26667 \times 0.75833 + 0.24167 \times 0.73333} = 0.71449$.

[12]It is not entirely clear that this is so; this is an area subject under discussion (e.g., Hodges 1996 [152]).

11.3.2 Continuous variables—Lin's concordance correlation coefficient

Various measures have been used to assess the degree of agreement between continuous variables [Pearson correlation coefficient (r), paired t-tests, least-squares analysis for slope and intercept, coefficient of variation, intraclass correlation coefficient]. The concordance correlation coefficient (r_c) was proposed by Lin (1989 [184]), with the claim that it avoids all the shortcomings associated with the usual procedures. It also appears to be superior to the previously proposed limits-of-agreement procedure (Altman & Bland 1983 [4]; Bland & Altman 1986 [27]), as discussed by Steichen & Cox (2002 [297]). The full equations are listed in this chapter's Appendix. The concordance correlation coefficient can range from -1 to $+1$ and is robust on as few as 10 pairs of data (Lin 1989 [184]). It would appear that this measure could be examined for equivalence criteria in similar manner to Cohen's kappa. It also appears that the Landis & Koch Strength of Agreement criteria need modification before application to this statistic, in that they are too generous.[13]

11.4 DETECTION LIMITS

Reports of chemical analyses for compounds at low levels and of potential public health concern can contain "less than" results. That is, while the laboratory may measure a positive signal, it is sufficiently close to that obtained from a series of reference "blank samples" for the analyst to be uncertain of the result. So it is commonly reported as being less than the "detection limit." There are a number of ways in which this limit can be defined. The following is a comparison of two of the options.[14] It is included in this text not as a recommendation for how detection limits should be defined, but as an example clarifying the need for careful thinking when it comes to considering Types I and II error risks. See Helsel (2005 [146]) for a full discourse on all the issues involved.

11.4.1 Type A censoring

The detection limit (or "limit of detection"), denoted by L_A, is defined as some multiple (typically between 2 and 4) of the standard deviation of a series of blanks (S). Furthermore, the true standard deviation (σ) is assumed to be known (even though it has to be measured), so that S is replaced by σ and all calculations use the unit normal distribution, rather than more complex t and noncentral t-distributions. As a result the calculations are approximate, rather than exact.

The most common multiplier is 3 (Miller & Miller 1988 [227], Eurachem 1998 [88]). In that case, $L = 3S$. The origins of this multiplier are given by Helsel (2005 [146]). In Type A censoring all (blank-corrected) data measured below the detection

---
[13]This, also, is an area under current (2005) consideration.
[14]The two options are described a "Type A" and "Type B," following usage of these terms in the UK (Ellis 1989 [84]).

limit are reported as less than that limit. For example, if $S = 1.2$ units the detection limit is $= 3.6$ units and laboratory results of 4.5, 3.4, 3.0, 2.5, 8.9 would be reported as 4.5, <3.6, <3.6, <3.6 and 8.9.

This approach guards against the Type I error risk, of "detecting" something that wasn't there. That is, we test the inferiority hypothesis **H**: $x = 0$, where x is the compound's true concentration (negative true conentrations are impossible), and perform a one-sided inferiority test (see Table 5.2) against the alternative hypothesis **K**: $x > 0$.[15] If **H** is rejected then **K** is accepted and "detection" can be claimed. That is, it is very unlikely that there was none of the compound in the sample if a result at least as large as L_A was obtained. But this approach ignores the Type II error, of failing to detect a compound that was actually present, and its dependency on sample size.

11.4.2 Type B censoring

Compared with Type A censoring this approach is slightly more complex, and may produce less censored data.[16] It is an example of "informative censoring" (or "insider censoring") and as such can introduce a bias into all subsequent analyses, unless special care is taken. For that reason some (e.g., Helsel 2005 [146]) do not favor its use.

It proceeds by defining *two* limits: the "criterion of detection" (C) and also the detection limit (L_B). (C may also be called the "limit of decision," Miller & Miller 1988 [227].) The first limit has the property that if an observed blank-corrected concentration is greater than C then it is reported at face value (so it takes the role of L_A in Type A censoring). Once again, this is because there is only a small chance of getting an erroneous value at least as large as C if the compound in fact is absent (i.e., if $x = 0$). On the other hand, if an observed result is just higher than C, there's an even chance that the true value is less than C. So the confidence implied by reporting values just above C is concerned with the possibility of the compound being *absent*, not in it merely being less than C. The maximum probability of committing the Type I error (measuring a value greater than C when the compound was absent), is obtained from the properties of a 5% level test of **H**. The resulting formula is $C = 2.33S$ (note that the multiplier is 2.33, not 1.6449—the 95%ile of the unit normal distribution—this issue is addressed in Problem 11.7). However, if the laboratory result is less than C it is not reported as less than C but as less than the (higher value of) the detection limit (L_B). Here's the reason. If the true concentration should in fact be C there is a not-small probability of obtaining a value less than C and so failing to "detect" the compound (that probability is about 0.5).[17] Therefore if the true concentration is in

---

[15] Note that **H** posits an equality, not the inequality $x \leq 0$, because the true value of x cannot be zero. In contrast a measured concentration *can* be less than zero (see Problem 11.6).

[16] It will do so only if the criterion of detection (C—see the following text) is les than the limit L in Type A censoring.

[17] For example, take a case where $C = 1$ and the laboratory measured two blank-corrected values of 1.1 and 0.9, reporting them as "1.1" and "<1.0." Would you really be sure that the second true value is less than 1? Remember, C is calculated after first assuming that the true concentration is zero.

fact close to C and we report all data less than C as "<C," the Type II error risk will be large—not small. To maintain a small Type I error risk the limit of detection is defined as follows: if the true concentration were in fact at least L_B there should be a low probability of getting a result less than C. Setting "low" as 5%, the multiplier for this limit is 4.65, that is, $L_B = 4.65S$. Again, see Problem 11.7 for its derivation.

This type of censoring, while not as simple as the more common Type A, can produce more information. For example, again using $S = 1.2$ units, we have $C \approx 2.8$ units and $L_B \approx 5.6$ units and so the data 4.5, 3.4, 3.0, 2.5, and 8.9 would be reported as 4.5, 3.4, 3.0, <5.6, and 8.9. This record includes two uncensored data (3.4 and 3.0) below the limit of detection (but above the criterion of detection). This gives more information than the standard practice, which, as noted in the previous section, would report these two results as <3.6.

Two penalties are paid in Type B censoring.

- The limit reported in the Type B censored result (<5.6) is higher than for Type A (<3.6).

- Statistical methods applied to the results of Type B censoring are more complex (Ellis 1989, [84]). Indeed some writers discourage Type B censoring because of this complexity (Helsel 2005, [146]). However, where there are environmental "breakthrough" concerns, funders of laboratory analyses may be very pleased to see less censored data (as is the case for Type B censoring using 5% error risks versus Type A censoring using $L_A = 3S$).

11.4.3 Handling "less than" data

Three cases should be considered. These refer to data generated by Type A censoring (as noted, procedures for Type B censoring are more difficult).

Few "less than" data. When a dataset contains only a few data (say <10%) below L, a rough analysis of those data can proceed by replacing those data by $\frac{1}{2}L$, although this is less satisfactory than the next item.

Moderate amount of "less than" data. Use a statistical distribution fitting method (Helsel & Hirsch 2002 [147]), Helsel (2005 [146]), as depicted on Figure 11.2.

1. Fit a plausible statistical distribution to the data above L (e.g., using a probability plot).

2. Extrapolate that distribution below L to "fill-in" values below L.

3. Add up the concentrations.

Mostly, or all, "less than" data. If there are many "less thans" in a dataset neither of the above procedures can be used. For example, take a set of results for ten individual chemicals: <0.1, <0.1, <0.1, <0.1, <0.1, <0.1, <0.1, <0.1, 0.8, <0.1. What then is the total? Replacing each "<0.1" by 0.1 is implausible (could all nine "less thans" really be "knocking at the door"?), and we can't sensibly fit a distribution to just one datum. Even replacement by 0.05 seems implausible. Taking data at face value we could say that the range of total concentration is 0.8–1.7, where the former figure is obtained by replacing all the censored data by zeroes and the latter figure by replacing those data by the detection limits. Beyond that little statistical help is available (but see Helsel 2005 [146]), and one must rely on plausibility arguments. One should also note that it is much better practice to analyze the compounds with a method that has a lower limit of detection, reducing the number of measurements if budgets are limited.

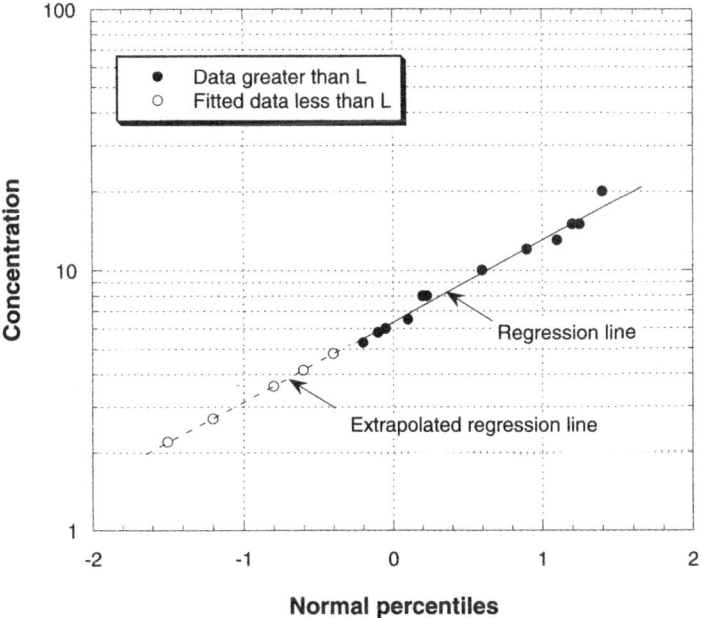

Fig. 11.2 Fitting a lognormal distribution to $>L$ data (where $L = 5$), and extrapolating back to obtain values of $<L$ data. Fill-in values (open circles) are selected randomly from the left tail of a lognormal distribution (figure derived from Helsel & Hirsch 2002 [147]).

Problems

11.1 Derive the p-value for the seasonal Kendall trend test when the test statistic is $|Z_{sK}| = \frac{1}{200} = 0.0707$.

11.2 Using the entries in Table 11.3, define the following terms: prevalence, overall proportion of agreement, sensitivity, specificity, false positive rate, false negative rate, overall proportion of disagreement, odds ratio.

11.3 Explain how Table 11.4 accounts for chance-correction.

11.4 Show that Equations (11.8) and (11.9) are equivalent.

11.5 What information would be gained by performing a point-null hypothesis test on kappa? Would an equivalence test be useful?

11.6 Laboratory results below the limit of detection can be negative. Can you explain this? What should be done about it?

11.7 Why isn't the multiplier for the criterion of detection in Type B censoring, using 5% significance level, equal to 1.6449 (being the 95%ile of the unit normal distribution)? Derive the multipliers for this criterion and for the detection limit in Type B censoring at the 5% level (2.33 and 4.65).

Appendix

Standard error for Cohen's kappa From Fleiss (1981, p. 221 [95]) the standard error of $\hat{\kappa}$ may be written as

$$S_{e,\hat{\kappa}} = \frac{\sqrt{A+B-C}}{(1-p_e)\sqrt{n}}, \qquad (A.1)$$

where n is the number of paired data, p_e is the chance-expected agreement proportion [Equation (11.7)], and

$$A = a\left[1 - (p_1 + p_2)(1-\hat{\kappa})\right]^2 + d\left[1 - (q_1+q_2)(1-\hat{\kappa})\right]^2, \qquad (A.2)$$
$$B = (1-\hat{\kappa})^2 \left[b(p_2+q_1)^2 + c(p_1+q_2)^2\right], \qquad (A.3)$$
$$C = \left[\hat{\kappa} - p_e(1-\hat{\kappa})\right]^2, \qquad (A.4)$$

where a, b, c, d and the marginal frequencies p_1, p_2, q_1 and q_2 are defined in Table 11.3. Bishop *et al.* (1975, p. 396 [25]) give a different (but equivalent) form,

$$S_{e,\hat{\kappa}} = \frac{\sqrt{E+F+G}}{(1-p_e)\sqrt{n}}, \qquad (A.5)$$

where

$$E = p_0(1-p_0), \quad F = 2(1-p_0)\left[\frac{2p_0 p_e - \theta_3}{1 - p_e}\right], \qquad (A.6)$$
$$G = (1-p_0)^2 \left[\frac{\theta_4 - 4p_e^2}{(1-p_e)^2}\right], \qquad (A.7)$$

with

$$\theta_3 = a(p_1+p_2) + d(q_1+q_2), \qquad (A.8)$$
$$\theta_4 = a(p_1+p_2)^2 + b(p_2+q_1)^2 + c(p_1+q_2)^2 + d(q_1+q_2)^2. \qquad (A.9)$$

Concordance correlation coefficient This coefficient (ρ_c) is stated in terms of squared vertical deviations from the perfect concordance line, i.e., a line with slope 1 proceeding from the origin. It is estimated by

$$r_c = \frac{2S_{XY}}{S_X^2 + S_Y^2 + (\bar{X} - \bar{Y})^2} \quad (A.10)$$

where the sample means, variances and covariance are: $\bar{X} = \frac{1}{n}\sum X_i, \bar{Y} = \frac{1}{n}\sum Y_i$, $S_X^2 = \frac{1}{n}\sum(X_i - \bar{X})^2, S_Y^2 = \frac{1}{n}\sum(Y_i - \bar{Y})^2, S_{XY} = \frac{1}{n}\sum(X_i - \bar{X})(Y_i - \bar{Y})$, and the summations are from $i = 1$ to n.

A standard error can be calculated for this estimate. In doing so it has been shown that the "Fisher transformation" is desirable to better meet the normal approximations to be invoked when confidence intervals are calculated or hypothesis tests are performed (Lin 1989 [184]). This transformation is

$$\hat{Z} = \tanh^{-1}(r_c) = \frac{1}{2}\ln\left(\frac{1+r_c}{1-r_c}\right), \quad (A.11)$$

and we obtain the sample standard error of estimate (of $\hat{Z}$) as

$$S_{e,\hat{Z}} = \sqrt{\frac{\frac{(1-r^2)r_c^2}{(1-r_c^2)r^2} + \frac{2r_c^3(1-r_c)u^2}{r(1-r_c^2)^2} - \frac{r_c^4 u^4}{2r^2(1-r_c^2)^2}}{n-2}} \quad (A.12)$$

where r is Pearson's correlation coefficient defined as $r = \frac{S_{XY}}{S_X S_Y}$ [cf. Equation (2.21)] and u is the "location shift relative to the scale" parameter, defined by

$$u = \left(\frac{n-1}{n}\right)\frac{\bar{X} - \bar{Y}}{\sqrt{S_X S_Y}} \quad (A.13)$$

The approximate lower one-sided confidence 95% confidence interval for Z is $L_Z = \hat{Z} - 1.6449 S_{e,\hat{Z}}$. Therefore, inverting the Z transformation, the 95% lower confidence limit for r_c is

$$L_{r_c} = \tanh(L_Z) \quad (A.14)$$

Note that the standard error equation in Lin (1989 [184]) has slightly different coefficients in the denominator under the square root. Equation (A.12) contains the corrections given by Lin (2000 [185]).

12
Answers to exercises

Chapter 2

Problem 2.1 $\Pr(A$ and $B)$ is the probability that both events occur, but they may do so simultaneously. So the "space" over which $\Pr(A$ and $B)$ is defined (the collection of all their possible outcomes) is the area occupied jointly by A and B. In contrast, $\Pr(A|B)$ is the probability that A occurs *given that we know that B has already occurred*. It occupies the same space, but is normalized by $\Pr(B)$. As an example, the probability of obtaining effluent concentrations of total phosphorus (TP) and suspended solids (SS) both in excess of their consent limits in the same compliance assessment period will be different from the probability that TP will be found to breach its limit, given that the SS is already known to have been in breach (typically, SS results are obtained earlier than TP results, because of laboratory methodological issues).

Problem 2.2 $\Pr(A$ and $B)$ is always smaller than $\Pr(A|B)$. This follows from Equation (2.8), where $\Pr(A|B)$ is obtained by dividing $\Pr(A$ and $B)$ by a number less than one [i.e., $\Pr(A)$]. So the probability of an effluent breaching its TP limit given that the SS is known to have been in breach is always higher than the probability of a TP breach occurring in absence of any prior knowledge about the SS concentrations.

Problem 2.3 Mutually exclusive events contain no elements in common. Independent events do, but the occurrence of one of them does not affect the occurrence probability of the others. Note that because the intersection of mutually exclusive events is empty, they cannot be independent (if they were, Equation (2.9) would require that one (or both) of $\Pr(A)$ and $\Pr(B)$ be zero. [See Bolstad (2004 [28]) for a fuller discussion.]

Problem 2.4 Denote the possible climate shift event by C. Also, denote by F the event that both B and A occur (strong westerly winds and at least moderate rainfall, during urbanization). We want to calculate $\Pr(F)$. To do so we use the data given in the problem to obtain the conditional probability $\Pr(F|C) = 0.15 \times 0.6 = 0.09$ (from the example in Section 2.1 we already have $\Pr(F|\text{not-}C) = 0.05$). From the rule of total probability (Equation 2.12), $\Pr(F) = \Pr(C)\Pr(F|C) + \Pr(\text{not-}C)\Pr(F|\text{not-}C) = 0.7 \times 0.09 + 0.3 \times 0.05 = 0.078$. As one might expect, this is just the weighted sum of the two deposition probabilities (with and without climate change).

Problem 2.5 Taking an impartial stance can be argued to mean that the prior probabilities should be taken as $\Pr(H) = \Pr(K) = \frac{1}{2}$. In that case Bayes' rule [Equation (2.14)] gives us $\Pr(H|E) = \frac{0.5 \times 0.9}{0.5 \times 0.9 + 0.5 \times 0.1} = 0.9$. So the posterior probability probability is the same as the likelihood obtained from the data presented to the tribunal after assigning prior probabilities, and the data have "spoken for themselves."[1]

Problem 2.6 Denote the required probability by Q. From the distributive rule for three events [Equation (2.7)] we have $Q = \Pr[(A \text{ and } B) \text{ or } (A \text{ and } C)] = \Pr[A \text{ and } (B \text{ or } C)]$. Now because A is sampled from a different population from B and C, the probability of its occurrence is independent of their probabilities. So we use the independence probability rule [Equation (2.9)] to obtain $Q = \Pr(A)\Pr(B \text{ or } C)$. Then from the probability rule for non-mutually exclusive events [Equation (2.4)] we have $Q = \Pr(A)[\Pr(B) + \Pr(C) - \Pr(A \text{ and } B)]$. But if B occurs then C cannot simultaneously occur (because they are sampled from the same population). Therefore $\Pr(A \text{ and } B) = 0$ and so $Q = \Pr(A)[\Pr(B) + \Pr(C)]$, as required.

Problem 2.7 If data are widely dispersed, as in Figures 1.1(a) and 1.1(b), precision of measurement is hardly affected by the number of them. In contrast, the precision of the estimate (e.g., of the mean) increases as the number of data increases. This is manifest, for example, in the width of confidence and credible intervals being narrower for a set of data than for a subset.

Problem 2.8 The abscissa for these distributions cover all possible events. The probability of *any* one of these events occurring is 1, since *something* must happen (and note that this includes a nil event, corresponding to zero on the abscissa).

Problem 2.9 From Equation (2.15) a random variable one standard deviation (σ) *above* the mean corresponds to a normal deviate of $Z = \frac{(\mu+\sigma)-\mu}{\sigma} = 1$. Using Equation (2.18) and Table A.1.1 we have the the required area as $G(1) - G(-1) =$

---
[1] This is an unusual case where truly impartial prior probabilities can be set, leaving subsequent data to speak for themselves.

$2G(1) - 1 = 2 \times 0.8413 - 1 = 0.6826$. For an area two standard deviations either side of the mean we calculate $G(Z = 2) - G(Z = -2) = 2 \times 0.9772 - 1 = 0.9544$. These calculations hold for any normal distribution, because any value of μ or σ can be used to calculate the normal deviate (this a major advantage in using this dimensionless term).

Problem 2.10 In Section 2.2.5 we defined the geometric mean of data as the nth root of those data, that is, $\bar{X}_g = \sqrt[n]{X_1 X_2 \ldots X_n}$. Then, from fundamental properties of logarithms and exponents, $\log(\bar{X}_g) = \frac{1}{n} \sum_{i=1}^{n} \log(X_i)$, being the mean of the logarithms. Anti-logging this expression we obtain the sample geometric mean as the antilog of the mean of the logarithms. For the MPN data the mean of the (base 10) logarithms is 0.834, and so the geometric mean is antilog(0.834) = $10^{0.834} = 6.8$ MPN / 100 mL. If any datum is zero then $\bar{X}_g = \sqrt[n]{X_1 X_2 \ldots X_n}$ is zero also. But in this case taking logarithms does not work, because $\log(0)$ is undefined.

Problem 2.11 If any datum is recorded as zero (as is sometimes done for "less than" data), the calculation $GM = \sqrt[n]{X_1 X_2 \ldots X_n}$ will return zero—hardly a good measure of central tendency.

Problem 2.12 The sample geometric mean overestimates the geometric mean of a lognormal distribution. See footnote 20 on page 31.

Problem 2.13 No, the correct result for the geometric mean for all the data is 160.55. This is obtained by grouping all the data and exponentiating the average of all 22 logarithms. The error arises because the logarithm of the averages does not equal the average of the logarithms. In contrast, *if each group has the same number of data*, the average of the two site averages does equal the average of the between-site averages [i.e., $\frac{1}{2}(387.27 + 923.36) = 655.32$, as required]. The moral of this story is: the safe option when computing summary statistics is to first group all the data, rather than taking shortcuts by averaging some of them first. Note too that the between-site sample geometric mean is about one quarter of the between-site arithmetic mean; this is typical for right-skewed data.

Problem 2.14 Neither equation should be used, because this is an errors-in-variables problem (both X and Y can be expected to be measured with similar precision). Accordingly, using "Ricker's Geometric Mean Regression" (see footnote 37 on page 47), we have $a = 0.86947, b = 1.0170, c = 0.94901$, and $d = -0.60288$. The better equation is therefore $Y = u + gX$, where $g = \sqrt{\frac{0.86947}{0.94901}} = 0.95718$ and $u = \bar{Y} - g\bar{X} = 2.81805 - 0.95718 \times 2.07146 = 0.83529$. This equation bisects the two original equations, as might be expected given that each of those equations ignores the imprecision of its abscissa. Note that all three equations pass through the point $(\bar{X}, \bar{Y})$.

Problem 2.15 Each coefficient b_i in a logistic regression can be interpreted as the logarithm of an odds ratio relating y to a particular variable x after adjustment for all the other x variables. Consider a model with three independent variables. If x_2 is binary (zero-one), then the relative odds ratio $\frac{\exp(b_1 x_1 + b_2 1 + b_3 x_3 + a)}{\exp(b_1 x_1 + 0 + b_3 x_3 + a)} = \exp(b_2)$. So the coefficient is the (natural) logarithm of the relative odds ratio. If x_2 is a continuous variable, a similar derivation shows that b_2 is the natural logarithm of the odds ratio relating the odds for y resulting from a one unit increase in x_2.

Chapter 3

Problem 3.1 If we consider a prior distribution for only positive values of chlorophyll concentrations, then the "vague" or "diffuse" prior postulates that before obtaining data, it is held that any positive value is equally probable. But our experience shows that 500 g m^{-3} cannot be exceeded. There are infinitely more numbers > 500 than between zero and 500. Therefore a vague prior postulates that "large" positive values are more likely than are small values. This cannot be claimed to be uninformative (in contrast to the estuarine sediment deposition example, where a vague prior really is uninformative—see the answer to Problem 2.5).

Problem 3.2 Equation (3.2) is obtained by applying Equation (3.1) to the logarithms of the data, and then antilog the resulting equation.

Problem 3.3 These data are right skewed ($g_1 = 2.8200$), suggesting that they have been drawn from a distribution that is approximately lognormal. So we may use Equation (3.2) to compute a 95% confidence interval. The sample geometric mean is $\bar{X}_g = 35.4610$, the standard deviation of the natural logarithms of the data is $S_Y = 1.1810$ and, from Table A.2.2, $t_{0.975,9} = 2.2622$. Therefore the upper and lower 95% confidence limits are 15.235 and 82.537. In contrast, the (inappropriate) 95% confidence interval around the mean [using Equation (3.1)] is 76.500 ± 92.416. The lower limit is less than zero, a strong signal that something is amiss.

Problem 3.4 Nonparametric methods use the ranks of data. The central tendency rank statistic corresponding to the mean is the median.

Problem 3.5 These equations are simply formed from Equation (3.10) by replacing the $\frac{\alpha}{2}$ subscript on the t value by α. So the equations are $(\bar{X} - \bar{Y}) + t_{1-\alpha,f}\left(S_p\sqrt{2/n}\right)$ and $(\bar{X} - \bar{Y}) - t_{1-\alpha,f}\left(S_p\sqrt{2/n}\right)$ for upper and lower one-sided limits, respectively.

Problem 3.6 For two independent samples of size n we can combine Equations (3.8) and (3.9) to give $S_{e,\bar{x}-\bar{y}} = \sqrt{\frac{1}{n}(S_x^2 + S_y^2)}$. Now $S_x^2 + S_y^2 = \frac{\sum(X_i - \bar{X})^2 + \sum(Y_i - \bar{Y})^2}{n-1} =$

$\frac{\sum X_i^2 - 2\bar{X}\sum X_i + n\bar{X}^2 + \sum Y_i^2 - 2\bar{Y}\sum Y_i + n\bar{Y}^2}{n-1}$, where the summation ($\Sigma$) is from $i = 1$ to n. Using the definition of the mean (Section 2.2.5, in which $\sum X_i = n\bar{X}$) this quantity can be simplified to $\frac{\sum X_i^2 - n\bar{X}^2 + \sum Y_i^2 - n\bar{Y}^2}{n-1}$. Substituting this result into the above equation for $S_{e,\bar{x}-\bar{y}}$ gives the desired equation.

Problem 3.7 The standard deviation of paired data is $S_d = \sqrt{\frac{\sum[(X_i - Y_i) - (\bar{X} - \bar{Y})]^2}{n-1}} = \sqrt{\frac{\sum[X_i^2 - 2X_iY_i + Y_i^2 - 2\bar{X}X_i + 2\bar{X}Y_i + 2\bar{Y}X_i - 2\bar{Y}Y_i] + n\bar{X}^2 - 2n\bar{X}\bar{Y} + n\bar{Y}^2}{n-1}}$. Again using the definition of the mean (Section 2.2.5, in which $\sum X_i = n\bar{X}$) and collecting terms we obtain $S_d = \sqrt{\frac{\sum X_i^2 + \sum Y_i^2 - n(\bar{X}^2 + \bar{Y}^2) - 2(\sum X_iY_i - n\bar{X}\bar{Y})}{n-1}}$. Substituting this expression into the definition of the standard error [Equation (3.12)] we obtain the required result.

Problem 3.8 No. If $n\bar{X}\bar{Y} > \sum X_iY_i$ the equation derived in Problem 3.7 will result in $S_{e,\overline{x-y}} > S_{e,\bar{x}-\bar{y}}$. Note that $\sum X_iY_i - n\bar{X}\bar{Y} = \sum(X_i - \bar{X})(Y_i - \bar{Y}) = (n-1)\text{Cov}(X,Y)$, where $\text{Cov}(X,Y)$ is the sample covariance, which can be positive or negative.

Problem 3.9 The interval for the mean difference becomes wider than the interval for the difference in means. The new values are $\bar{X} = 22.600$, $S_x = 4.949$, $S_d = 5.025$ and so the 95% confidence interval limits on the difference in means are 7.900 ± 3.338, whereas those for the mean difference are 7.900 ± 3.701.

Problem 3.10 The waist is at the mean value of X, that is, $\bar{X}$. This is seen by substituting $X = \bar{X}$ into Equations (3.16) and (3.17); the last term under the square root sign goes to zero, and all other terms are independent of X.

Problem 3.11 For an infinite sample size our estimate of the standard deviation *is* the population value; $S = \sigma$. For finite sample size, with unknown σ, there is uncertainty in its value. Further, for percentiles above the median, the slope of the unit normal curve is negative, meaning that the distribution of k is right skewed, as shown in Figure A.1 (on page 248).[2] The Figure displays the distribution of k for an upper one-sided tolerance interval on a normal 95%ile. The total shaded area for each curve is 0.05. This skew pulls the median k value to the right, and hence it is larger than the unit normal value. Converse arguments apply to percentiles below the median.

Problem 3.12 Interpolation for k values could be made by fitting a curve through pairs of n, k data on Table (A.4.2). For example, for the 95% upper one-sided confidence limit on a normal 95%ile, $k \approx 1.7630 + 8.5885n^{-0.8676}$ for $n \leq 10 \leq 200$, and $k \approx 1.6704 + 0.2538e^{-0.0016194n}$ for $200 < n \leq 10000$. This could be implemented using embedded IF statements in Excel. Similarly, for the 95% upper

---

[2]Reprinted from *Water Research* **37**(15), McBride, Confidence of Compliance: parametric versus non-parametric approaches, 3666–3671, 2003, with permission from Elsevier (reference [207]).

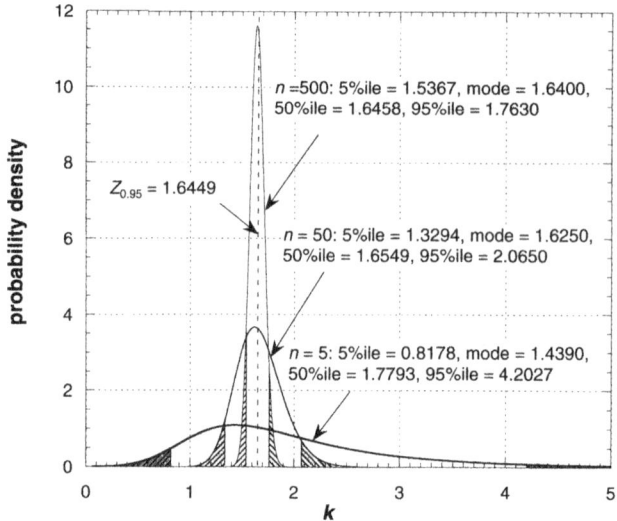

Fig. A.1 Asymmetry of one-sided confidence (or tolerance) intervals about a normal 95%ile.

one-sided confidence limit on a normal median, $k \approx 0.0737 + 3.1133n^{-0.7698}$ for $n \leq 10 \leq 200$, and $k \approx 0.0165 + 0.1563\mathrm{e}^{-0.0016557n}$ for $200 < n \leq 10000$.

Chapter 4

Problem 4.1 The language of the example strongly implies that a point-null hypothesis was tested. In that case its hypothesis ("no association") should not be "accepted," only "not rejected."[3] Indeed, the hypothesis is implausible (that there is no association between swimmers' health and fecal contamination of the water).

Problem 4.2 The p-value is the chance of getting data that would cast at least as much doubt on the point-null hypothesis (of unbiasedness), were that hypothesis to be true. So obtaining 5 "overs" in $n = 20$ trials, obtaining 0, 1, 2, 3 or 4 "overs" would cast even more doubt on that hypothesis. But so would obtaining 15, 16, 17, 18, 19 or 20 "overs". Now the p-value is the probability of getting results *at least as extreme* as have been obtained. So we need the probability, "under $\mathbf{H_0}$," of obtaining these few or these many "overs." This is $B(5; 20, \frac{1}{2}) + B(15; 20, \frac{1}{2})$. We can simplify this using the relationship derived in Problem 6.7, i.e., $B(k; n, \frac{1}{2}) = B(n-k; n, \frac{1}{2})$, and so $B(5; 20, \frac{1}{2}) + B(15; 20, \frac{1}{2}) = 2B(5; 20, \frac{1}{2})$, as required (this is confirmed by

---

[3]For further discussion see reference McBride (1993 [204]).

ANSWERS TO EXERCISES   249

inspection of Figure 4.3, where the shaded bars are symmetrical about the $k = 10$ bar).

Problem 4.3   Once again, the p-value is the chance of getting data that would cast at least as much doubt on the point-null hypothesis (of unbiasedness), were that hypothesis to be true. We need to consider two negative binomial probabilities: first, where 20 *or more* trials are required in order to obtain 5 "overs"; second, where 20 *or more* trials are required in order to obtain 5 "unders." Both these events cast doubt on the hypothesis and so both must be included in the p-value. The form in which they enter the p-value is given by Equation (2.5) for non-mutually exclusive events, noting that the probability of last probability in that equation (of one or the other event occurring) is zero, because we require them both to happen. Accordingly, the required probability is simply the addition of the two identical probabilities, that is, $2\mathcal{B}^*(20; 5, \frac{1}{2})$.

Problem 4.4   The probabilities for the vertical bars $[2\mathcal{B}^*(10; 5, \frac{1}{2})]$ on Figure 4.4 add to 1, and so $\mathcal{B}^*(10; 5, \frac{1}{2}) = \frac{1}{2}$. Now from Equation (6.18), setting $\theta = \frac{1}{2}$, we obtain $B^*(2k-1; k, \frac{1}{2}) = 1 - \mathcal{B}^*(2k; k, \frac{1}{2}) = \frac{1}{2}$, as required.

Problem 4.5   This is a simple extension of Problem 4.2.

Problem 4.6   Define D as the event that less than 6 or more than 14 "overs" occur in the first 20 trials. Let D_c be its complement, that between 6 and 14 "overs" occur in those trials. Define E as the event that less than 14 or more than 26 overs occur in 40 observations, that is, after a further 20 trials. Now p is the probability of hitting one of the the shaded parts of Figure 4.3 after 20 observations or, failing that, hitting the shaded part of Figure 4.5 after a further 20 observations. Therefore $p = \Pr(D) + \Pr(E \text{ and } D_c)$. From Problem 4.2 we have $\Pr(D) = 2B(5; 20, \frac{1}{2})$. Also we can calculate $\Pr(E \text{ and } D_c)$ by selecting in turn each nonshaded bar in Figure 4.3 (for $k = 6, 7, \ldots, 14$) and for each of them calculating the probability that there would be sufficiently few or many "overs" in the second batch of 20 trials to result in either fewer than 14 or more than 26 "overs" in the entire batch of 40 trials. That is

$$\begin{aligned}
\Pr(E \text{ and } D_c) =\ & \Pr[(6|20 \text{ and } \leq 7|20)] \\
& + \Pr[(7|20 \text{ and } \leq 6|20) \text{ or } (7|20 \text{ and } 20|20)] \\
& + \Pr[(8|20 \text{ and } \leq 5|20) \text{ or } (8|20 \text{ and } \geq 19|20)] \\
& + \Pr[(9|20 \text{ and } \leq 4|20) \text{ or } (9|20 \text{ and } \geq 18|20)] \\
& + \Pr[(10|20 \text{ and } \leq 3|20) \text{ or } (10|20 \text{ and } \geq 17|20)] \\
& + \Pr[(11|20 \text{ and } \leq 2|20) \text{ or } (11|20 \text{ and } \geq 16|20)] \\
& + \Pr[(12|20 \text{ and } \leq 1|20) \text{ or } (12|20 \text{ and } \geq 15|20)] \\
& + \Pr[(13|20 \text{ and } 0|20) \text{ or } (13|20 \text{ and } \geq 14|20)] \\
& + \Pr[(14|20 \text{ and } \geq 13|20)]
\end{aligned}$$

Now the two sequential stages of the "peeking" design are independent, and so [using Equation (2.9) in Chapter 2] the first probability in this equation is given by $\Pr(6|20 \text{ and } \leq 7|20) = \Pr(6|20)\Pr(\leq 7|20)$. Similarly the last probability is $\Pr(14|20)\Pr(\geq 13|20)$. For the (more complex) intervening probabilities we note that they are all of a form $\Pr[(S \text{ and } T) \text{ or } (S \text{ and } V)]$—for example, with $S =$ "7|20", $T =$ "≤6|20" and $V =$ "20|20". Now T and V describe events occurring in samples from the same population. So we know from Problem 2.6 that these probabilities are given by $\Pr(S)[\Pr(T) + \Pr(V)]$. Substituting these results into the equation $p = \Pr(D) + \Pr(E \text{ and } D_c)$ we obtain the required result ($p = 0.0799$).

Problem 4.7 The false positive rate is α and the false negative rate is β. See also Problem 11.2 for further definition of these (and other related) terms.

Problem 4.8 The formula results from simple repeated application of Bayes' rule [Equation (4.2)], where $\Pr(\mathbf{H}) = P_0$, $\Pr(E|\mathbf{H}) = 1 - \beta$, $\Pr(\mathbf{K}) = 1 - P_0$, and $\Pr(E|\mathbf{K}) = \alpha$. For $n = 3$, the posterior hazard probability is $\frac{0.9^3 \times 0.001}{0.9^3 \times 0.001 + 0.05^3 \times 0.999} = 0.854$—a dramatic imporvement on the prior hazard probability of 0.001.

Chapter 5

Problem 5.1 It is likely that the p-value would have decreased, because in a point-null test power depends on sample size (Section 5.2.2). The study would then have concluded that there *is* a statistically significant association between swimmers health and the presence of animal fecal residues in the water they swim in.

Problem 5.2 The mean difference and its standard deviation are $\bar{X} = 4.429$ and $S_d = 2.225$. The standard error of the difference is therefore $S_{e,\overline{x-y}} = 2.225/\sqrt{7} = 0.841$. The paired data constitute a one-sample test, so that the test statistic is $T = \frac{\bar{X}-0}{S_{e,\overline{x-y}}} = 4.429/0.841 = 5.395$. The degrees of freedom are $f = n - 1 = 6$, so we will reject the inferiority hypothesis if $|T| > t_{1-\alpha,f} = t_{0.05,6} = 1.943$. The inferiority hypothesis is again rejected, even more strongly than when treating the data as independent (in which case $|T| = 4.071$).

Problem 5.3 If the test rejects its hypothesis, then $|T| > t_{1-\alpha/2,f}$, and so either $T < -t_{1-\alpha/2,f}$ or $T > t_{1-\alpha/2,f}$. Substituting $T = \frac{\bar{X}-\mu_0}{S_e}$ and multiplying the result by S_e, we obtain either $\bar{X} - \mu_0 < -t_{1-\alpha/2,f}(S_e)$ or $\bar{X} - \mu_0 > t_{1-\alpha/2,f}(S_e)$. With $\mu_0 = 0$, and noting that for a single sample $S_e = \frac{S}{\sqrt{n}}$, we can simply rearrange these equations to give either $\bar{X} + t_{1-\alpha/2,f}(\frac{S}{\sqrt{n}}) < 0$ or $\bar{X} - t_{1-\alpha/2,f}(\frac{S}{\sqrt{n}}) > 0$. From Equation (3.1) the first of these two equations requires that the *upper* α-level two-sided confidence limit be less than zero, while the second requires that the *lower* limit of that interval be greater than zero. In other words, the confidence interval cannot contain zero.

Problem 5.4 This can be addressed using the method used to answer Problem 5.3.

Problem 5.5. The limiting form is a "spike in the ceiling." That is, a straight line across the top of the graph (at detection probability of 100%), except for a line at zero effect size extending down from the ceiling to a height α.

Problem 5.6 Two top hats, one of them upside down! The sides of the hats are the edges of the equivalence interval.

Problem 5.7 The standard errors in the *Two independent samples* section of Table 5.10 should be written as $S_{e,\overline{x-y}} = \frac{\sigma}{\sqrt{n}}$ (for variance known) and $S_{e,\overline{x-y}} = \frac{S_d}{\sqrt{n}}$ (for variance unknown), where S_d is the standard deviation of the differences. Also, the degrees of freedom becomes $f = n - 1$.

Problem 5.8 "Power" has no sensible meaning for an *individual* result—this is why Section 4.5 discourages the pursuit of the power of a performed test (retrospective power). Averaged over a large group of tests, a paired sample test is always more powerful. But that's not to say that there couldn't be an occasion on which a stronger result is obtained by not treating the data as paired.

Problem 5.9 The duo (n_i, d_e) means that both tests would reject their hypotheses: (i) the precautionary test of the inequivalence hypothesis would infer "equivalence," while (ii) the permissive test of the equivalence hypothesis finds "inequivalence." This cannot happen. In the first case, referring to Table 5.9, we would need to obtain $T_l \geq t_{1-\alpha,f}$ and $T_r \leq -t_{1-\alpha,f}$ (variance unknown), or $|T| \leq c_i$ (variance known). In the second case we would need $T_l < -t_{1-\alpha,f}$ or $T_r > t_{1-\alpha,f}$ (variance unknown), or $|T| > c_e$ (variance known). These conditions cannot be simultaneously satisfied (noting, from Table 5.9, that c_e is always greater than c_i).

Problem 5.10 The first two cases are covered by the answer to Problem 5.9. For the third case, note from Figure 5.13 that for the first two members of the trio to be "d" the detection region is everywhere beyond the equivalence interval. But, if the desired power is greater than 50%, the width of the nondetection region for the Power Approach never extends beyond that interval (as explained in Section 5.3.5). So the trio (d_i, d_e, n_p) is impossible.

Chapter 6

Problem 6.1 $F(x; \mu_y, \sigma_y, \tau) = \int_\tau^x \frac{1}{\sqrt{2\pi}\sigma_y(\xi-\tau)} e^{-\frac{1}{2}v(\xi)^2} d\xi$ is the CDF of the lognormal distribution, where $v(\xi) = \frac{\ln(\xi-\tau)-\mu_y}{\sigma_y}$ and ξ is a dummy variable of integration. The lower integration limit is τ, not zero, because the lognormal distribution [Equation (6.3)] is not defined for $x < \tau$. Now $\frac{dv}{d\xi} = \frac{1}{\sigma_y}\left(\frac{1}{\xi-\tau}\right)$ and so $d\xi = \sigma_y(\xi-\tau)\,dv$. Substituting this into the CDF equation, noting that $\ln((x-\tau) \to 0) \to -\infty$, we obtain $F(x; \mu_y, \sigma_y, \tau) = G(v(x))$, where G is the CDF of the unit normal distribution, as required.

Problem 6.2 For the normal distribution the coefficient of variation is 1.616; for the lognormal it is 1.073. The lognormal's value is lower because it is right-skewed. Conversely, for a given coefficient of variation the lognormal has a higher ratio of the 90%ile to the median than does the normal percentile. This has implications for setting microbiological standards (or for any other right-skewed water quality variables) using medians and upper percentiles: The ratio needs to be applied to the appropriate distribution.

Here's how these results are calculated. For the normal distribution the median and the mean (μ) are identical—see Section 6.2.1. (In this Problem, and in Problem 6.3, we'll take $\mu > 0$, noting that percentile ratios have no practical meaning if $\mu \leq 0$.) And from Equation (2.16) and Table A.1.2 the 90%ile of the normal distribution is $\mu + 1.2816\sigma$, where σ is the distribution's standard deviation. So the ratio (r) of the 90%ile to the median (50%ile) is $r = \frac{\mu+1.2816\sigma}{\mu} = 1 + 1.2816\eta$ where $\eta = \frac{\sigma}{\mu}$ is the distribution's coefficient of variation (CV). Now the problem's data gives us $r = \frac{43}{14} = 3.071$ and so we can solve for the CV to get $\eta = \frac{r-1}{1.2816} = \frac{2.071}{1.2816} = 1.616$. For the lognormal distribution we work with the logarithms of the fecal coliform concentration (y) being normally distributed. So the population's 90%ile is $\exp(\mu_y + 1.2816\sigma_y)$ giving the ratio as $r = \frac{\exp(\mu_y+1.2816\sigma_y)}{\exp(\mu_y)} = \exp(1.2816\sigma_y)$. We now need to get σ_y in terms of η. To do that we know that $\eta = \sqrt{\exp(\sigma_y^2) - 1}$ (see Section 6.2.2, with $\tau = 0$). Manipulating this we get $\sigma_y = \sqrt{\ln(1+\eta^2)}$ and so $r = \exp[1.2816\sqrt{\ln(1+\eta^2)}\,]$. Finally, solving for η we have $\eta = \sqrt{\exp\left(\frac{\ln(r)}{1.2816}\right)^2 - 1}$, and so, inserting $r = 3.071$, we find that $\eta = 1.073$.

Problem 6.3 The ratio of the 99%ile to the 95%ile for the unit normal and two-parameter lognormal distributions, both with coefficient of variation = 0.5, are 1.187 and 1.380, respectively. If that coefficient is 1.0 the ratios are 1.258 and 1.764.

Here's how these results are calculated. For the normal distribution the ratio of an upper and a lower percentile is $r = \frac{\mu+Z_{upper}\sigma}{\mu+Z_{lower}\sigma}$, where Z_{upper} and Z_{lower} are the upper and lower percentiles of the unit normal distribution (e.g., for the ratio of the 99%ile to a 95%ile we have $Z_{upper} = 2.3263$ and $Z_{lower} = 1.6449$). This is easily manipulated to give $r = \frac{1+Z_{upper}\eta}{1+Z_{lower}\eta}$, and so $\eta = \frac{r-1}{Z_{upper}-rZ_{lower}}$. Note that it is possible for the percentiles of a normal distribution to have different signs (the upper percentile being positive and the lower being negative), in which case such ratios have little meaning. In contrast this cannot happen for the (strictly positive) two-parameter lognormal distribution. In this case the ratio of the percentiles is $r = \frac{\exp(\mu_y + Z_{\text{upper}})}{\exp(\mu_y + Z_{\text{lower}})} = \exp(Z_{\text{upper}}\sigma_y - Z_{\text{lower}}\sigma_y)$, where μ_y and σ_y are the mean and standard deviation of logarithms of the water quality variable. As in Problem 6.2 we have $\eta = \sqrt{\exp(\sigma_y^2) - 1}$, and so $r = \exp[(Z_{\text{upper}} - Z_{\text{lower}})\sqrt{\ln(1 + \eta^2)}\,]$. Finally, manipulating this we have $\eta = \sqrt{\exp\left(\frac{\ln(r)}{Z_{\text{upper}} - Z_{\text{lower}}}\right)^2 - 1}$.

Problem 6.4 $F(x;\alpha,\beta) = \int_0^x \frac{1}{\beta^\alpha \Gamma(\alpha)} \xi^{\alpha-1} e^{-\xi/\beta}\, d\xi$ is the CDF of the gamma distribution, where ξ is a dummy variable of [see Equation (6.4)]. Substituting another dummy variable $v = \xi/\beta$ we obtain $d\xi = \beta\, dv$. Substituting this into the CDF, $F(x;\alpha,\beta) = \int_0^{x/\beta} \frac{1}{\beta^\alpha \Gamma(\alpha)} \beta^{\alpha-1} e^{-v} \beta\, dv = P(\alpha, x/\beta)$, as required.

Problem 6.5 This is a simple application of the fundamental properties of logarithms and exponents, motivated by the desire to avoid computer roundoff errors.

Problem 6.6 Integrating the beta probability mass [Equation (6.5)] up to a value $x(<1)$ we have $F(x;\alpha,\beta) = \int_0^x \frac{1}{B(\alpha,\beta)} \xi^{\alpha-1}(1-\xi)^{\beta-1}\, d\xi$. This is the required incomplete beta function ratio.

Problem 6.7 A symmetric distribution [e.g., Figures 2.11(g) and 2.11(j)], has its skewness coefficient equal to zero. The binomial distribution's skewness coefficient (Section 6.2.5) is $\frac{1-2\theta}{\sqrt{n\theta(1-\theta)}}$ which is zero only for $\theta = \frac{1}{2}$. From the cumulative binomial probability relationship given at the end of that section [Equation (6.12)] we have $\mathcal{B}(k;n,\theta) = 1 - B(k-1;n,\theta)$ [call this Equation (Q)]. Substituting $n-k$ for k this becomes $\mathcal{B}(n-k;n,\theta) = 1 - B(n-k-1;n,\theta)$. But another of those relationships [Equation (6.10)] gives us $B(k;n,\theta) = 1 - B(n-k-1;n,1-\theta)$, and so $B(n-k-1;n,\theta) = 1 - B(k;n,1-\theta)$. Substituting this into Equation (Q) we obtain $\mathcal{B}(n-k;n,\theta) = B(k;n,1-\theta)$. The right-hand side of this equation equals $B(k;n,\theta)$ only when $\theta = \frac{1}{2}$—that is, when the distribution is symmetric. This establishes the required result.

Problem 6.8 For the normal approximation, note that $n\theta$ and $n(1-\theta)$ are both greater than 5, so the approximation is justified. We obtain $\mu = n\theta = 100$, $\sigma = \sqrt{n\theta(1-\theta)} = 9.48683$, and so $Z = \frac{90-100}{9.4863} = -1.05409$. Inserting these values into Equation (6.1), we obtain $b(90; 1000, 0.1) \approx f(-1.05409; 100, 9.48683) = 0.02413$. The exact value is $b(90; 1000, 0.1) = 0.02483$. For the Poisson approximation we have a large n (1000) and small θ (0.01), so this approximation is justified. Here, $\mu = n\theta = 10$ and inserting this value into the Poisson pmf—Equation (6.14)—

we have $b(3; 1000, 0.01) \approx p(3; 10) = \frac{10^3 e^{-10}}{3!} = 0.00757$. The exact value is $b(3; 100, 0.01) = 0.00739$.[4]

Problem 6.9 The algorithm in Section 6.2.5 is *recursive*, in which a new value of the (binomial) probability is calculated from the old value. The counter for "old" and "new" is k. This recursion of course needs a starting value, and this is obtained by setting $k = 0$ in the binomial pmf [Equation (6.6)], so that $b(0; n, \theta) = \bar{\theta}^n$ where $\bar{\theta} = 1 - \theta$. This establishes step 1 of the algorithm. The recursion is multiplicative, and is also obtained from the definition of the binomial pmf: $b(k + 1; n, \theta) = \binom{n}{k+1} \theta^{k+1} (\bar{\theta})^{n-(k+1)} = \frac{n!}{(k+1)!(n-k-1)!} \theta^{k+1} (\bar{\theta})^{n-k-1}$. So we have $b(k+1; n, \theta) = \frac{n!(n-k)}{(k+1)k!(n-k)!} \theta \theta^k \frac{(\bar{\theta})^{n-k}}{\bar{\theta}} = \frac{\theta(n-k)}{(k+1)\bar{\theta}} b(k; n, \theta)$. To avoid rounding errors we accumulate logarithms and then antilog as required—turning multiplications and divisions of potentially large numbers into additions and subtractions of much smaller numbers. This gives the kernel of the algorithm.

The (slight) advantage of this algorithm is that it makes no approximations and uses only standard functions, whereas the direct algorithm requires access to a method for calculating logarithms of the gamma function.

Problem 6.10 Noting that for integer n its factorial is related to the gamma function by $\Gamma(n + 1) = n!$, we have $\binom{n}{k} = \frac{\Gamma(n+1)}{\Gamma(k+1)\Gamma(n-k+1)}$. The required formula simply follows by taking logarithms and then antilogging. This requires access to an approximation to the logarithm of gamma functions; occasionally, texts have a table of these for an integer argument (e.g., Zar (1996 [349])). One could also calculate $B(n; k; \theta)$ by calculating $b(i; n, \theta)$ for $i = 0, 1, \ldots, n$ and summing all the these terms [see Equation (6.7)], so the answer to the second part of this problem is "yes."

Problem 6.11 From the Poisson pmf [Equation (6.14)] we have $p(0; \mu) = e^{-\mu}$ and the recursion $p(k + 1, \mu) = \frac{\mu \mu^k e^{-\mu}}{(k+1)k!} = \frac{\mu}{k+1} p(k; \mu)$. So the algorithm is:

1. Set $i = 0$, $\ln(p) = -\mu$, $p = \exp[\ln(p)]$, $P = p$. If $k = 0$, then stop, else go to Step 2.

2. Set these variables, in the order given: $\xi = \frac{\mu}{i+1}$, $\ln(p) = \ln(\xi) + \ln(p)$, $p = \exp[\ln(p)]$, $P = P + p$, $i = i + 1$.

3. Repeat Step 2 until $i = k$.

Problem 6.12 If you have done all the previous exercises, showing that $b^*(n; k, \theta) = \frac{k}{n} b(k; n, \theta)$ should now be trivial.

Problem 6.13 From Section 6.2.7 the skewness coefficient of the geometric distribution is that of the negative binomial distribution with $k = 1$, and so is $g_1 = $

---

[4] These results demonstrate the approximate nature of these results; it would be much better to use the accurate algorithms or software described herein.

ANSWERS TO EXERCISES   255

$J\left(\frac{2-\theta}{\sqrt{\theta}}\right)$, where $J = \frac{1}{\sqrt{\mu}}$. Differentiating with respect to θ we obtain $\frac{\partial g_1}{\partial \theta} = -J\left(\frac{\theta+2}{2\theta^{3/2}}\right)$, which, noting that $0 < \theta < 1$, is a strictly decreasing function of θ. Therefore the skew of the geometric distribution always increases as θ decreases. This is also true for the negative binomial, because its coefficient of skew can also be expressed as a function of μ and θ only.

Problem 6.14   From Equations (6.17) and (6.18) we have $B^*(n; k, \theta) = 1 - B(k-1; n, \theta)$. Furthermore, Equation (6.13) allows us to express $B(k-1; n, \theta)$ as the incomplete beta function ratio $I_{1-\theta}(n-k+1, k)$. Substituting $n = 2k-1$ we obtain the result $B^*(2k-1; k, n) = 1 - I_{1-\theta}(k, k)$. Now the incomplete beta function ratio has the property that $I_\theta(a, b) = 1 - I_{1-\theta}(b, a)$, and so we have $B^*(2k-1; k, n) = 1 - I_{1-\theta}(k, k) = I_\theta(k, k)$. For $\theta = \frac{1}{2}$ this reduces to $I_{1/2}(k, k) = \frac{1}{2}$, as required. Note that had we used the alternative form of the negative binomial distribution discussed at the end of Section 2.6.3 we would state this special case as $B_s^*(k-1; k, \frac{1}{2}) = \frac{1}{2}$.

Problem 6.15   Define $s = n - k$ and substitute $n = s + k$ into the pmf [Equation (6.15)], noting the property of the binomial coefficient that $\binom{a}{b} = \binom{a}{b-a}$. Also note from Section 2.6.3 that the mean is related to n and θ by $\mu = \frac{k(1-\theta)}{\theta}$ from which we obtain $\theta = \frac{k}{\mu+k}$ and $1 - \theta = \frac{\mu}{\mu+k}$.

Problem 6.16   From the negative binomial pmf (Problem 6.15) we have $b_s^*(0; k, \theta) = \left(\frac{k}{\mu+k}\right)^k$, and the recursion $b_s^*(s+1; k, \theta) = \left(\frac{s+k}{s+1}\right)\left(\frac{\mu}{\mu+k}\right) b_s^*(s; k, \theta)$. The algorithm is therefore

1. Set $i = 0$, $\ln(b_s^*) = k \ln\left(\frac{k}{\mu+k}\right)$, $b_s^* = \exp[\ln(b_s^*)]$, $B_s^* = b_s^*$. If $s = 0$, then stop, else go to Step 2.

2. Set these variables, in the order given: $\xi = \left(\frac{i+k}{i+1}\right)\left(\frac{\mu}{\mu+k}\right)$, $\ln(b_s^*) = \ln(\xi) + \ln(b_s^*)$, $b_s^* = \exp[\ln(b_s^*)]$, $B_s^* = B_s^* + b_s^*$, $i = i + 1$.

3. Repeat Step 2 until $i = s$.

Problem 6.17   Hypergeometric distribution calculations are not as straightforward as the previous cases—dealing with finite populations is often a difficult task (Casella & Berger 2002, page 87 [45]). Nevertheless we can proceed by noting that the starting value for a recursive algorithm is given by $h_0 = h(0; n, N, m) = \frac{(N-m)!(N-n)!}{N!(N-m-n)!}$. So $\ln(h_0) = \ln[\Gamma(N-m+1)] + \ln[\Gamma(N-n+1)] - \ln[\Gamma(N+1)] - \ln[\Gamma(N-m-n+1)]$. Then, from the hypergeometric pmf [Equation (6.19)] we use the recursion $h(k+1; n, N, m) = \frac{(n-k)(m-k)}{(k+1)(N-m-n+k+1)} h(k; n, N, m)$. The recursive algorithm is then

1. Set $i = 0$, $\ln(h) = \ln(h_0)$, $h = \exp[\ln(h)]$. If $n = 0$, then stop, else go to Step 2.

2. Set these variables, in the order given: $\xi = \frac{(n-k)(m-k)}{(k+1)(N-m-n+k+1)}$, $\ln(h) = \ln(\xi) + \ln(h)$, $h = \exp[\ln(h)]$, $i = i+1$.

3. Repeat Step 2 until $i = n$.

Problem 6.18 From Sections 6.2.3 and 6.2.11 in Chapter 6 we know that the chi-square ordinates are given by $p(x) = \frac{1}{2^{\frac{f}{2}}\Gamma(\frac{f}{2})} x^{\frac{f-2}{2}} e^{-\frac{x}{2}}$. Inserting $f = 6$ and the first HDR abscissa value in Figure 3.5 we obtain the ordinate as $p(0.607) = \frac{1}{2^3 2!}(0.607)^2 e^{-0.607/2} = 0.0170$. Likewise, using the second HDR abscissa value (12.802) we obtain the ordinate as $p(12.802) = \frac{1}{2^3 2!}(12.802)^2 e^{-12.802/2} = 0.0170$, a numerical demonstration of the theorem given in footnote 18 of Chapter 3.

Problem 6.19 The simplest way to construct an algorithm to find the (HDR) of an asymmetric unimodal distribution is an iterative procedure that makes successive guesses at the value of the ordinate that corresponds to the HDR, calculating the total tail area it cuts off, comparing that with the desired tail area (α) and adjusting the ordinate up or down to come ever-closer to a total tail area = α.

Problem 6.20 When testing the equivalence hypothesis $\mathbf{H_e}$ with known variance we need to calculate the value of the critical value of the test statistic, c_e (see Table 5.9). If the equivalence interval width shrinks to zero we have $\mu_L \to \mu_U$ (one sample case) or $\delta_L \to \delta_U$ (two-sample case), with terms defined as in Table 5.10. In either case, the equivalence hypothesis ($\mathbf{H_e}$) becomes the point-null hypothesis ($\mathbf{H_0}$). Also, as defined in that Table, the noncentrality parameter γ_U is zero and so, using the definition of the critical c_e value in Table 5.9, we find that it is the solution of $G(c_e) - G(-c_e) = 1 - \alpha$. That solution is $c_e = Z_{1-\alpha/2}$ [using Equation (2.17)]. Finally, note that the test statistic (T) for testing the equivalence hypothesis when the variance is known (in Table 5.10) is identical to the point-null test statistic (T). So the decision rule is: Reject $\mathbf{H_e}$ if $|T| > Z_{1-\alpha/2}$. The decision rule for $\mathbf{H_e}$ is then identical to that for the point-null hypothesis $\mathbf{H_0}$ given Table 5.7.[5]

When the variance is unknown this coalescing of decision rules does *not* occur.[6] First, note that the equivalence tests' statistics T_l and T_r (in Table 5.10) both become identical to the point-null test statistic (T) when the interval becomes infinitely narrow. From Table 5.9 the decision rule for the test of the equivalence hypothesis becomes: Reject $\mathbf{H_e}$ if $T < -t_{1-\alpha,f}$ or if $T > t_{1-\alpha,f}$, that is, if $|T| > t_{1-\alpha,f}$. However, for the point-null hypothesis test the rule is: Reject $\mathbf{H_0}$ if $|T| > t_{1-\alpha/2,f}$ (see Table 5.7). This result shows that in the limit where the equivalence interval shrinks to zero the true level of significance test of this test of the equivalence hypothesis is 2α (but it quickly descends to α as the interval is widened, so that the two tests' decision rules are increasingly alike).

---

[5] We might have expected this result, given that both tests (with variance known) are in the UMPU family of tests (see footnote 10 in Chapter 5 and footnote 15 in Chapter 6).
[6] We might have expected this result also, given that there are no UMPU tests of the inequivalence hypothesis with variance unknown (see see footnote 13 in Chapter 5).

Problem 6.21 When testing the inequivalence hypothesis H_i with known variance we need to calculate the value of the critical value of the test statistic, c_i (see Table 5.9). For the reasons given in the answer to Problem 6.20, c_i becomes the solution of $G(c_i) - G(-c_i) = \alpha$ as the equivalence interval width shrinks to zero. That solution is $c_i = Z_{(1+\alpha)/2}$. If we forbid the Type I error (by setting $\alpha = 0$ and so $c_i = 0$), the tested hypothesis ($\mu \neq \mu_0$, or $\delta \neq \delta_0$) could never be rejected, which is as expected (because it is always true).[7]

When the variance is unknown, the test of the inequivalence hypothesis for an infinitely narrow equivalence interval requires us to satisfy both $T \geq t_{1-\alpha,f}$ and $T \leq -t_{1-\alpha,f}$ to enable H_i to be rejected (see Table 5.9). This can only be satisfied when $t_{1-\alpha,f} = 0$, in which case $\alpha = \frac{1}{2}$. In the limit where the equivalence interval shrinks to zero, the true level of significance test of this test of the inequivalence interval is 0—but it quickly ascends to α as the interval is widened, so that the two tests' decision rules are increasingly alike.

Problem 6.22 The development of the detection probability equation for the inferiority hypothesis test is similar to that given for the superiority hypothesis, in Section 6.4.1. In that case we would commit the Type II error with probability $\beta = \Pr(T \leq t_{1-\alpha,f} | \mu > \mu_0)$. Therefore, Power = $1 - \beta = 1 - \Pr(T \leq t_{1-\alpha,f} | \mu > \mu_0) = 1 - \Pr(t' \leq t_{1-\alpha,f} | \gamma)$, when $\gamma > 0$. To simplify this expression we note that the t'-distribution has the property that $\Pr(t' \leq t_{1-\alpha,f} | \gamma) = 1 - \Pr(t' \leq -t_{1-\alpha,f} | -\gamma)$ (Owen 1968 [251]), so Power = $\Pr(t' \leq -t_{1-\alpha,f} | -\gamma)$. Finally, noting that $\Pr(x \leq y) = Pr(x < y)$ for continuous variables, we write the power formula as Power = $\Pr(t' < -t_{1-\alpha,f} | -\gamma)$.

Problem 6.23 The top of the region occurs at $d = 0$. The required result follows immediately by substituting $d = 0$ into Equation (6.4.8).

Problem 6.24 The teepee with chimney becomes a vase, as in Figure A.2 (on page 258). In fact, if the vase is plotted for larger standard errors its mouth can be wider than its base—i.e., wider than the equivalence interval! (In the present case, for a 5% level test with a $\pm 20\%$ equivalence interval, this happens once $S_{e,\bar{x}-\bar{y}} > 318.32$.) Furthermore the neck of the vase is at its narrowest when the standard error is exactly equal to the half-width of the interval (i.e., when $S_{e,\bar{x}-\bar{y}} = \delta_U = 20$). It is easily shown (from the quantities on Table 5.10) that at this neck the partial noncentrality parameter is unity (i.e., $\gamma_U = 1$, because $\delta_U = S_{e,\bar{x}-\bar{y}}$), and that $\gamma_U = \delta_U / S_{e,\bar{x}-\bar{y}} > 1$ below the neck. The widening of the detection region above the neck—to be even larger than the equivalence interval width—is of course inconsistent with the expected properties of testing the equivalence interval. From this behavior it appears that this test, with variance known, is only valid if the partial noncentrality parameter is greater

---

[7]This need not happen in a Bayesian formulation of such problems (Berger & Delampady 1987 [17]), because the prior probability can assign a spike of probability (a pmf) at the point where the null hypothesis is true—but this can give rise to "Lindley's paradox", in which remarkably different results can be obtained from frequentist and Bayesian approaches (Oakes 1986 [245]).

than 1; that is, the standard error is no greater than the half-width of the equivalence interval. If $\gamma_U < 1$ the interval is too narrow for the test to work—there is too much variability in too few data to secure the test's desired performance while forcing its size to be maintained at α.

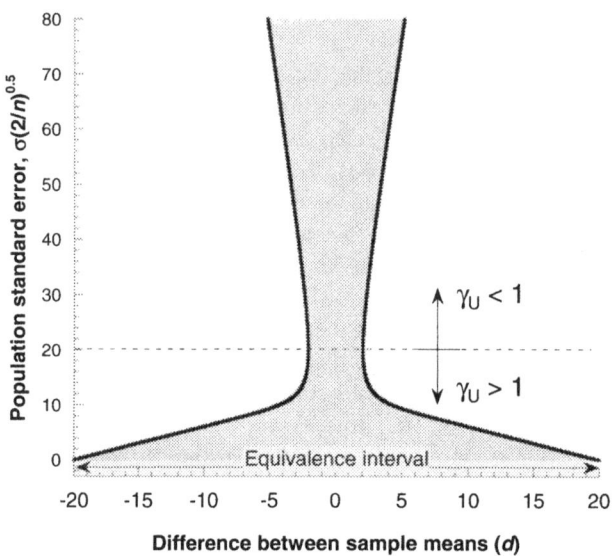

Fig. A.2   The vase.

Chapter 7

Problem 7.1   There is always the potential for dispute, and debate, when it is claimed that wastewater discharges of a certain type, scale and location cause only "minor" effects. Often this is because the cumulative effect of "minor" discharges ceases to be minor. This debate can be conducted, and valiant attempts made to resolve the issues, when the water quality management plan proposing their authorization is being developed—so adequate time needs to be made for this debate to occur. Statistical issues are seldom to the fore in that debate.

Problem 7.2   Generally stormwater discharges are authorized by a management plan (or even by statute), there being so many of them that case-by-case compliance assessments are impractical. General consents may also be declared, subject to specific design and operational criteria.

Problem 7.3. The more filters a plant has the higher the supplier's risk, and this is unfair. Here's why. Under the proposed rule, all filters at a plant are each required to deliver water with turbidity less than 0.3 NTU for 95% of each day. If all but one of the filters are well within this requirement but one is in breach of it, the supply may have a better quality than is the case when all filters are within it but a number of them are marginal for compliance. That is, a complying plant may produce an inferior product to a noncomplying plant, and this problem is exacerbated the more filters a plant operates—the supplier's risk increases with the size of the plant. Fairness arguments indicate that there should be some compensation for all the well-performing filters. Techniques akin to multi-determinand compliance (Ellis 1985 [82]) can help here. Let's assume three categories of filter performance on any one day: bad, borderline, and good. Let's also assume that it is desirable to have compliance assessed daily. Then each day count up the total number of minutes that filter outputs are above 0.3 NTU, and multiply that time by the proportion of bad + borderline filters. If that product is greater than 72 minutes (5% of a day) then the plant is non-complying. Let's call that product the "plant transgression time."

For example, define a filter as "bad" if it exceeds 0.3 NTU for >5% of a day; define a filter as "marginal" if it exceeds 0.2 NTU (but not 0.3 NTU) for >5% of the day; define a filter as "good" if it does not produce 0.2 NTU for more than 5% of the day. Then consider a 10-filter plant, with one "bad" filter, exceeding 0.3 NTU for 188 minutes. Another two are "borderline" and the other seven are "good." Then the plant transgression time is $188 \times 0.3 = 56.4$ minutes, so the plant complies. If there are two bad filters, with the second exceeding 0.3 NTU for 88 minutes, and another is borderline, then the plant transgression time is $(188 + 88) \times 0.3 = 82.8$ and the plant is noncomplying. This is a fairer rule, and simple to implement. Its fine detail could be assessed by running straightforward Monte Carlo simulations for various mixtures of good, marginal, and bad filters.

Placing the standard on the combined filter output only is unwise, because one badly performing filter may be delivering an inordinate number of oocysts while its contribution to the turbidity of the combined filters' outputs may not cause a breach of the standard.

Problem 7.4 Ideally one should be able to place only effluent standards in an effluent consent. Such a consent is then readily enforceable because the consent holder should have full control and responsibility over what is discharged. The holder does not have full control over what happens in the receiving water, because its quality is also influenced by other sources of contamination and also by natural phenomena. However, sometimes the science necessary to predict the effect of an effluent is quite uncertain, and so a receiving water standard is also used as a "back-stop." In any event some ongoing monitoring of conditions in the receiving water is called for, to attempt to check that environmental effects of the effluent, and other wastewater discharges, do not compromise the receiving water standards.

Problem 7.5 An effluent consent that contains only receiving water standards is unenforceable. If receiving water standards are breached and it *seems* to be the result of a particular discharge whose consent contains only receiving water standards, the consent holder has the option of claiming that it was other discharges that were the cause, or even that is was the result of natural processes. That is, while some may claim that environmental effects are correlated or associated with the effluent, establishing causation is a much more difficult task.

Problem 7.6 Fundamentally, this is because a maximum can seldom be estimated—whatever estimate is made will eventually be exceeded if enough samples are taken. For this reason, Barnett & O'Hagan 1997, p. 23 [12] say "...it is not unusual to see a standard specify that 'the level shall not exceed x' .... This is completely nonsensical." For example, take an effluent BOD_5 concentration. Occasional high values can be obtained by reason of short-term wastewater treatment plant malfunction caused by an upstream spill of toxic compounds into a sewer. The regulatory agency is then unlikely to be successful in having any penalty imposed, especially in the common case that no deleterious effect has been reported. So, nothing is done. Further discussion of these issues is given by Ellis (1986 [83]).

Problem 7.7 This is a many-faceted question, as also are the questions raised in the following two similar problems, having to do with both statistical realities and also with public perception. (Some of the issues have already been alluded to in Section 2.13.4.) The statistical realities are that many effluent water quality data exhibit occasional high values that may need to be accommodated in a compliance rule, especially if they are the result of either measurement error (e.g., fresh bird droppings contaminating a sample, or laboratory contamination), accident (e.g., power supply failure) or heterogeneity (e.g., streaming in a UV disinfection process). There are difficulties in using maximum standards to accommodate these features—see Problem 7.6. On the other hand, failing to require compliance with a maximum limit can be read by some members of the public as a "license to pollute."

In the case of effluent standards there are strong grounds for stipulating a limit on a central tendency statistic (e.g., a median) *and* on a high percentile (e.g., a 95%ile), but not on a maximum. Such standards indicate important information to funders, designers, and operators of wastewater treatment plants—the median indicating what is expected in the normal course of events, with the 95%ile indicating what can be tolerated by way of unusual circumstances. When such limits are being met it is not feasible to operate a plant so as to greatly exceed the 95%ile limit for the remaining 5% of the time—unless by deliberate action, and that should be penalized. If a maximum is set and it is "near" the 95%ile limit (e.g., double that limit), for many contaminants it will eventually be exceeded.

If this does not assuage public concerns, a requirement could be built into a consent that any exceedances of a standard's limit should be "explained to the satisfaction of the regulatory agency."

Problem 7.8 Even here the case for a maximum standard is not strong. However note that, in contrast to the case of effluent standards, a precautionary proof of safety approach should be taken (see Section 7.1). A consequence of this is that sampling for compliance with 95%iles can require that no exceedances should occur in 38 to 76 samples, and only one in 77 to 108 samples,... (see Tables 8.1 and 8.2). This is very close in appearance to a maximum standard, yet it does allow an occasional exceedance in a large number of samples. Importantly, it also provides a rational basis for designing a compliance sampling program.

Problem 7.9 The case for maximum standards in receiving waters is stronger, especially because enforcement action is not necessarily an automatic consequence of the exceedance. For example, exceedance of an upper limit for health risk indicators in recreational waters can be taken as a signal that increased surveillance, sanitary surveys and public notification of potential health hazard should be implemented (MoH/MfE 2003 [223]).

Problem 7.10 "Running" compliance rules penalize a consent holder more than once for a breach. For example, consider a rule stating that no more than 3 exceedances of a limit should occur in a rolling 12 week period of twice-weekly sampling. That is, compliance is assessed from the data from week 1 to week 12, then for week 2 to week 13, and so on. If 4 exceedances occurred in a block, that block will appear in more than one (eight!) of these rolling periods. On the other hand using adjacent periods means assessing compliance for the periods week 1 to 12, week 13 to 24, and so on. If the block of 4 exceedances is split between the end of one of these periods and the beginning of the next, breach will not be inferred at all.

Problem 7.11 Take an 80%ile SS standard, that is, 80% of the time the suspended solids concentration should be below a value X. A precautionary proof of safety approach could allow only a "small" (e.g., 5%) chance of being wrong if compliance compliance is inferred. Therefore the *upper* one-sided 95% confidence limit on SS should be compared with X; if that limit exceeds X a breach would be inferred. Conversely, a permissive proof of safety approach would allow only a 5% chance of being wrong if breach is inferred. In this case the *lower* one-sided 95% confidence limit on SS should be compared with X. A face value stance would compare the estimated 80%ile with X directly.

Problem 7.12 In supplying drinking water it is the consumer's risk, not the producer's risk, that must always be minimized. If we were to keep the producer's risk small the quality of the drinking water could often deteriorate unacceptably *before* being detected, with consequent serious impacts on human health. This means that we need a compliance rule that is *always* precautionary. But we have seen in Section 5.2.4 that point-null hypothesis tests don't satisfy this property: At low sample sizes they are permissive but swing to being ultra-precautionary at large sample sizes. More generally, a point-null test examines the wrong hypothesis. What we have here is essentially a one-sided problem and compliance rules should be based on testing the *superiority hypothesis* (a proof of safety approach). That is, in devising the sampling requirements for a compliance rule it should be assumed that the supply is in breach of its standard (i.e., in fact contaminant concentrations are higher than (superior to) the standard's requirement). Power calculations would then show that the supplier must collect many samples in order to be able to routinely reject the hypothesis and so confidently conclude that the water supply is safe. This approach is much more satisfactory than the alternative—of testing the inferiority hypothesis (a proof of hazard approach)—in which case there is a disincentive to collect so many samples.

Problem 7.13 The question "How many samples should be taken to assess compliance with a water quality standard?" is solely statistical only once the definition of the compliance assessment period and the burden of proof have been agreed. What is needed then is the number of samples necessary to meet that burden of proof within the agreed compliance assessment period. But it is important to note that setting the standard is not a solely statistical issue—it is not even a solely scientific issue.

Chapter 8

Problem 8.1 Being a long-term value, the historical 95%ile covers many compliance assessment periods (e.g., six months, as in the stated problem). This long-term result is an average of these many results, some of which will be greater than the long-term value and others below it. For a very long historical period, and if samples were taken from a stationary symmetrical distribution, the proportion of overs and unders would be very close to $\frac{1}{2}$. However, the length of historical record will be short, not stationary (there will be some trend), and the distribution being sampled asymmetrical, to at least some degree. Therefore "about half" of the future 95%iles will be over the long-term value. Splitting the historical record into seasons and requiring compliance with seasonal 95%iles may actually *worsen* the problem, by creating the need for more compliance assessments. Some allowance for sampling error must be made in such cases.

Problem 8.2 From Equation (8.1) the Hazen rank of the 95%ile when $n = 10$ is 10. That is, the estimated 95%ile is the maximum of the data. This represents an uncomfortable amalgam of statistics. If possible, more observations should be made in order to arrive at a more satisfactory estimate.

Problem 8.3 The logarithms of the data can be assumed to have come from a somewhat normal population. As presented on page 69, we can estimate the 95%ile of normal distribution with unknown variance from the formula $\bar{Y} + kS$, where $\bar{Y}$ and S are the mean and standard deviation of the logarithms of the observations, and k is read from Table A.4.2. To get a face value estimate of that percentile, choose the 50% level of confidence in that table. Then we find that $k = 1.7016$ for $n = 10$ observations, and $k = 1.6498$ for $n = 100$ observations. Because we must take antilogs to get the desired result, we get a more dramatic reduction in the estimated 95%ile than the differences between these two k values might otherwise suggest. (The mean and standard deviation of the logarithms at each site are different, so that, formally, a Monte Carlo analysis could be performed to demonstrate this point from the data shown.)

Problem 8.4 In a proof of hazard approach, the benefit of the doubt is given to the "producer"—for example, a the operator of a wastewater treatment plant effluent discharge. If the proportion of exceedances among samples is ($\frac{e_h}{n}$) we would have the "face value" stance. To move the burden of proof further away from the producer, we must raise that ratio. (A formal proof of this result is difficult.[8])

Problem 8.5 This is the converse of the answer to Problem 8.4.

Problem 8.6 Substituting Equation (6.13) into Equation (8.4), we obtain $p_h = 1 - B(e_h; n, X) = I_X(e_h + 1, n - e_h)$. From Equation (8.10) we obtain the same incomplete beta function ratio by taking $\alpha_{\text{prior}} = 1$ and $\beta_{\text{prior}} = 0$, that is, $\text{Be}(1, 0)$.

Chapter 9

Problem 9.1 Consider a beach with occasional presence of adenoviruses in its water. Clinical trial data indicate that this pathogen is very infective: the probability of infection when inhaling just one of them is about 34% (see Figure 9.3). Also, let's perform one million QMRA calculations for one person, as a "representative" of the community at a beach on a given exposure day. Therefore, on average, on about one third of the days that the individual inhales an adenovirus that individual will become infected; on the other two-thirds of those days the individual is not infected at all. Now if that person is "representative" of the group of people at the beach, then inhalation of an adenovirus by that person implies that about one third of those people will be predicted to be infected. If the individual does not inhale any adenoviruses, nobody will be infected.[9] This sudden jump in the risk profile from zero to one-third is not realistic, because on any given day each person at the beach will inhale a different amount of water. Just because one of their number has inhaled an unusually large amount of water and got unlucky does not mean that so many will suffer the same fate. But if the "representative" is replaced by 1000 people exposed on 1000 days, the differential exposures of those individuals on any one day is explicitly accounted for, and so the risk profile rises smoothly from zero. This problem is much less severe for less infective pathogens. For example, data for a strain of *Campylobacter jejuni* indicates a median infective dose of about 900 (see Figure 9.3), and in this region (and for much lesser doses) the dose-response curve is quite flat. So an individual

---

[8]That is, if $\sum_{e_h+1}^{n} \binom{n}{e} X^e (1-X)^{n-e} < \frac{1}{2}$ where e, e_h, and n are all non-negative integers, with $e_h < n$ and $0 \leq X \leq 1$, then $\frac{e_h}{n} > X$, always.
[9]Strictly, this invokes the "ergodic hypothesis"—that time averages and ensemble averages are interchangeable.

ingesting three times the volume of another faces about the same infection risk. These features are easily demonstrated in numerical experiments.

Problem 9.2 The simplest distribution to construct is triangular; it is bounded above and below and need only three points to describe it: minimum, mode and maximum. Therefore if we use the given data (minimum = 10, median = 50, maximum = 100) as those three points, the median has in fact been reinterpreted as the mode. They would only be identical if the maximum was 90.

Problem 9.3 Using Equation (9.12) with the values used on Figure 9.2, we have $\tilde{f} = -0.903 + 75.072 \left(\frac{1.2046}{3.2046}\right)^{1/2.2046} = 47.262$ grams.

Problem 9.4 Using the nomenclature of Section 9.3.1, the probability is $Pe^{-\lambda}$. This is obtained by setting $e = 0$ and $\mu = \lambda$ in Equation (6.14), the probability mass function for the Poisson distribution, and multiplying by the proportion of contaminated samples.

Problem 9.5. The sum is Equation (9.6), as required, that is, $P(1 - e^{-\lambda})$ (because, from the exponential series, $\sum_{k=1,2,\ldots}(\lambda^k/k!) = e^\lambda - 1$).

Problem 9.6 Referring to Figure 9.4, the area of the left shaded triangle is 0.5, from which Equation (9.9) follows immediately. Similarly, Equation (9.11) follows by noting that the area of the right shaded triangle (q), but it can't be computed until X_P is known. This is obtained by noting that the unshaded area $= s = (X_P - X_{50})h_2 + \frac{1}{2}(X_P - X_{50})(h_1 - h_2) = \frac{1}{2}(X_P - X_{50})(h_1 + h_2)$. After substituting $h_2 = \frac{2q}{X_{100} - X_P}$, noting that $s + q = \frac{1}{2}$ and expanding terms, we obtain the quadratic $X_P^2 - \left(X_{50} + X_{100} + \frac{1}{h_1}\right)X_P + \left[X_{50}X_{100} + 2\left(\frac{sX_{100} + qX_{50}}{h_1}\right)\right] = 0$. Equation (9.10) then follows as the solution of this quadratic, necessarily subtracting the discriminant form the first term.

Problem 9.7 We need to expand Equation (9.16) by grouping terms into recognizable chunks: $P_{\text{inf}}(d) = e^{-dr}\left[\sum_{k=1}^\infty \frac{(dr)^k}{k!}\right]\left[\sum_{i=k}^\infty \frac{[d(1-r)]^{i-k}}{(i-k)!} e^{-d(1-r)}\right]$. The first bracket is the exponential series for e^{rd}, less one. The second bracket is the infinite sum of a Poisson series (Section 6.2.6)—if we set $m = i - k$ and $z = d(1-r)$ the term becomes $\sum_{m=0}^\infty \frac{z^m}{m!}e^{-z}$. So that bracket is unity. Therefore we have $P_{\text{inf}}(d) = e^{-dr}(e^{dr} - 1)(1) = 1 - e^{-dr}$, an elegant result.

Problem 9.8 Substituting Equation (6.5) (with x replaced by r) into Equation (9.19) we obtain $P_{\text{inf}}(d) = 1 - \int_0^1 e^{-dr}\frac{r^{\alpha-1}(1-r)^{\beta-1}}{B(\alpha,\beta)} dr$. Kummer's confluent hypergeometric function is defined as $_1F_1(a,b,z) = \frac{\Gamma(b)}{\Gamma(a)\Gamma(b-a)}\int_0^1 x^{a-1}(1-x)^{b-a-1}e^{zx}\,dx$. We get the desired result if we substitute $a = \alpha, b = \alpha + \beta, x = r$ and $z = -d$ [noting that $B(\alpha,\beta) = \frac{\Gamma(\alpha)\Gamma(\beta)}{\Gamma(\alpha+\beta)}$].

ANSWERS TO EXERCISES   265

Problem 9.9 To derive Equation (9.23) we assume that $\bar{r}$ follows the beta distribution. Therefore $P_{\text{inf}}^0(i) = \int_0^1 [1-(1-\bar{r})^i] \left[\frac{\bar{r}^{\alpha-1}(1-\bar{r})^{\beta-1}}{\text{B}(\alpha,\beta)}\right] \, d\bar{r} = 1 - \int_0^1 \frac{\bar{r}^{\alpha-1}(1-\bar{r})^{\beta+i-1}}{\text{B}(\alpha,\beta)} \, d\bar{r}$
$= 1 - \frac{\text{B}(\alpha,\beta+i)}{\text{B}(\alpha,\beta)}$.

Problem 9.10 Gauss's hypergeometric function is a little more complex than the Kummer function, that is, $_2F_1(a,b;c;z) = \frac{\Gamma(c)}{\Gamma(b)\Gamma(c-b)} \int_0^1 x^{b-1}(1-x)^{c-b-1}(1-xz)^{-a} \, dx$. It has two useful properties: $_2F_1(a,b;c;z) = {_2F_1}(b,a;c;z)$ and $_2F_1(a,b;c;z) = (1-z)^{-b} {_2F_1}(b,c-a;c;\frac{z}{z-1})$ [Equation (15.3.5) of Abramowitz & Stegun 1972 [1]]. The required result is obtained by making the substitutions $a = \alpha, b = k, c = \alpha + \beta$ and $z = -\frac{d}{k}$.

Chapter 10

Problem 10.1 The total volume examined in Table 10.3 is 333 mL, so if the mode of the distribution of $\Pr(i,j,k|n)$ [Equation (10.10)] is at $n = 1$, the MPN per 100 mL is $\frac{n \times 100}{333} \approx 0.3$. Three patterns can give rise to this result: 1-0-0, 0-1-0, 0-0-1. Nine patterns can give rise to a mode at $n = 4$, in which case the MPN is 1.2 per 100 mL. In seven of these cases there are four positive tubes (e.g., 1-3-0, 1-1-2), in which case each positive tube contains just one microorganism. However, it is much more likely that one tube will contain two microorganisms, so that there are only three positive tubes and the most likely patterns are 2-1-0 and 2-0-1.

Problem 10.2 Equation (10.2) can be written as $\Pr(r|R,n) = \frac{1}{R^n} \frac{R!}{(R-r)!} S_n^{(r)}$, where $S_n^{(r)} = \frac{1}{r!} \sum_{i=0}^r \binom{r}{i}(-1)^i (r-i)^n$ is "Stirling's number of the second kind" [its upper summation limit can also be taken as $r - 1$, because $(r-i)^n = 0$ when $i = r$]. Equation (10.6) then follows by noting that this Stirling number has a recurrence relation $S_n^{(r)} = r S_{n-1}^{(r)} + S_{n-1}^{(r-1)}$ (e.g., see Abramowitz & Stegun 1972 [1]).

Problem 10.3 The recurrence relation Equation (10.6) is strictly additive, containing no negative terms, so should not be confounded by computational subtractive cancelation errors. These errors could arise with Equation (10.7), because its $(-1)^i$ term causes successive terms to oscillate. However, the direct equation is much faster, because it is summed over r only, whereas the recursive equation is to be summed over both r and n. This may be important in extensive calculations for large MPN setups. Experience shows that the direct algorithm is accurate unless the calculated probability is tiny ($<10^{-6}$). So a good strategy is to first calculate probabilities using the direct algorithm, and only revert to the recursive form if the (maximum) calculated probability is less than $<10^{-6}$.

Problem 10.4 For a single series with R tubes the occurrence probability of getting r positive tubes containing a total of n microorganisms is given directly by Equation (10.2). With two sets of tubes (I and J) the equation is a simplified form of Equation (10.10), that is, $\Pr(i,j|n) = \sum_{n_I=0}^{n} \Pr(i|I,n_I)\Pr(j|J,n_J)\Pr(n_I,n_J|n)$. For four sets of tubes I, J, K and L a more complex form of Equation (10.2) turns up, that is, $\Pr(i,j,k,l|n) = \sum_{n_I=0}^{n}\sum_{n_J=0}^{n-n_I}\sum_{n_K=0}^{n-(n_I+n_J)} \Pr(i|I,n_I)\Pr(j|J,n_J)\Pr(k|K,n_K)$
$\times \Pr(l|L,n_L)\Pr(n_I,n_J,n_K,n_L|n).$[10]

Chapter 11

Problem 11.1 Because this is a two-sided test, the p-value is the probability that the normal deviate is greater than 0.0707, or is less than -0.0707. An approximate value of the first probability is obtained by simple linear interpolation on the cumulative unit normal probability table (Table A.1.1). That gives $\Pr(Z \leq 0.0707) \approx 0.5279 + \frac{0.0707-0.0700}{0.0800-0.0700} \times (0.5319 - 0.5279) = 0.5282$. Therefore $\Pr(Z > 0.0707) = 0.4718$ and so $p = 2 \times 0.4718 = 0.9436$. This agrees with the exact result—for example, obtained by entering =2*(1-NORMSDIST(0.0707)) in Excel.

Problem 11.2 Using the entries in Table 11.3, prevalence $= a+c$, overall proportion of agreement $p_0 = a+d$, sensitivity $= \frac{a}{a+c}$ (see Section 4.4), specificity $= \frac{d}{b+d}$, false-positive rate $= \frac{b}{b+d}$, false-negative rate $= \frac{c}{a+c}$, overall proportion of disagreement $= b+c$, odds ratio $= \frac{ad}{cb}$.

Problem 11.3 Consider the case where the number of "present" results obtained by the standard method is independent of the number obtained by the proposed alternative method (and so, similarly, for "absent" results). In that case row totals in the frequency table (Table 11.3) are independent of the column totals. Now, even if that were the case one can still obtain chance-agreement in individual trials (that is, "present"/"present" or "absent"/"absent"). The probabilities of this happening, to be inserted into the north-west and south-east cells of the chance-expected table (Table 11.4), follow from the definition of statistical independence, i.e., Equation (2.9). Therefore the probabilities in those cells—the frequencies of chance-agreement in a long run of trials—are the product of the appropriate row (p_1 and p_2) and column (q_1 and q_2) totals, that is, $p_1 p_2$ and $q_1 q_2$. A similar argument applies for chance-*dis*agreement, in which case the probabilities to be inserted into the northeast and southwest cells of the table are $p_1 q_2$ and $q_1 p_2$. Note that this distribution of chance-expected probabilities

---

[10] Some care must be exercised when computing the probabilities in these equations, because for very large n (typically $n > 1500$) their factorial ratios can exceed a computer's memory capacity, yet their quotient is in fact small. This problem is simply avoided by exponentiating the logarithms of such terms, turning multiplication and division of very large numbers into addition and subtraction of much smaller values, as discussed in Section 6.1.

still preserves the row and column totals of Table 11.3, for example, in the first row of Table 11.4, $p_1 p_2 + p_1 q_2 = p_1(p_2 + q_2) = p_1$.

Problem 11.4 The denominator of Equation (11.8) can be written as $1 - p_1 p_2 - q_1 q_2 = 1 - p_1(1-q_2) - q_1(1-p_2) = p_1 q_2 + q_1 p_2$, as required (noting that $p_1 + q_1 = 1$). Simplifying the numerator is a little more tedious. It can be written as $a + d - (a+b)(a+c) - (c+d)(b+d) = (a+d) - (a^2 - d^2) - 2d^2 - 2bc - (a+d)(c+b)$. Noting from Table 11.3 that $c + d = 1 - (a+d)$ we can state the denominator as $(a+d)\{1 - (a-d) - [1-(a+d)]\} - 2d^2 - 2bc = (a+d)(2d) - 2d^2 - 2bc = 2(ad-bc)$, as required.

Problem 11.5 Very little would be learned if a point-null test were performed on kappa. Testing **H**: $\kappa = 0$ would in many cases, even with few samples, result in a "statistically significant" result, because $\hat{\kappa}$ is often considerably above zero. But all that implies is that there are good grounds for saying that κ is not near zero, which can be expected for a method that seeks to mimic the performance of another. An equivalence test may indicate a likely range of κ, but our interest is surely in whether it exceeds a stated criterion, and that is a "one-sided" problem.

Problem 11.6 A laboratory result comes from an *individual* instrument reading subtracted by an *average* signal from tests on a set of blanks. By definition some of those blanks will have been greater than the average and so some low-level blank-corrected concentrations can turn out to be negative. Data users would not see these if the laboratory censors its low-level results, because they would merely be reported as less than the detection limit. Occasionally a laboratory may not censor such data—for example, if the client wants early warning of a breakthrough of organic groundwater contaminants, and all blank-corrected results are made available. Explanation of the negative result is then crucial—because the immediate reaction of many is to say that "there must be a mistake."

Problem 11.7 The essential reason is that the data are blank-corrected, and so the variance of the concentrations must also account for the variance of the blank samples (Wilson 1961 [343]; Kirchmer 1983 [173]). To calculate the criterion of detection we note that it seems plausible that the variance of the blanks should be about the same as the variance of samples when only low concentrations of a trace compound are present. Because those variances are additive [see Equation (3.8)], we have the standard deviation of the blank-corrected concentrations as $S_{\text{corrected}} = \sqrt{S^2 + S^2} = \sqrt{2}S$, where S is the standard deviation of the blanks. Therefore the distribution of observed concentration values (X) when the true value (x) is zero is obtained by noting that the normal deviate $Z = \frac{X-0}{S_{\text{corrected}}} = \frac{X}{\sqrt{2}S}$ follows the unit normal distribution. For a 5% one-sided inferiority test we know that $Z_{0.95} = 1.6449$ (see page 29). Simple algebra then shows that the criterion of detection is defined as $X = C$, where $C = \sqrt{2}(1.6449)S \approx 2.33S$.

To calculate the detection limit factor we need to note that to determine it we are dealing with a two-sample one-sided inferiority hypothesis test procedure at the 5%

level (so it would only reject the hypothesis of zero concentration 5% of the time if in fact the concentration was zero) and with 95% power to detect a value L. From the detection probability (D) given in the northeast cell of Table 5.3 we therefore have $D = 0.95 = G(-Z_{0.95} + \gamma)$, where γ is the noncentrality parameter and G is the cumulative unit normal distribution function. From Table 5.4 for variance known with two independent samples each of size 1 (i.e., one observed concentration subtracted by the mean of the blanks) and setting $\delta = L$ and $\delta_0 = 0$, we have $\gamma = L/\sqrt{2}S$. We can then invert the detection equation, noting that $D^{-1}(0.95) = Z_{0.05}$, to give $L = 2\sqrt{2}Z_{0.05}S \approx 4.65S$. Note that this means that that if the true concentration were at least L, then the probability of getting a result less than C is at most 0.05, as depicted in Figure A.3 on page 268 (where each of the shaded areas $= 0.05$).

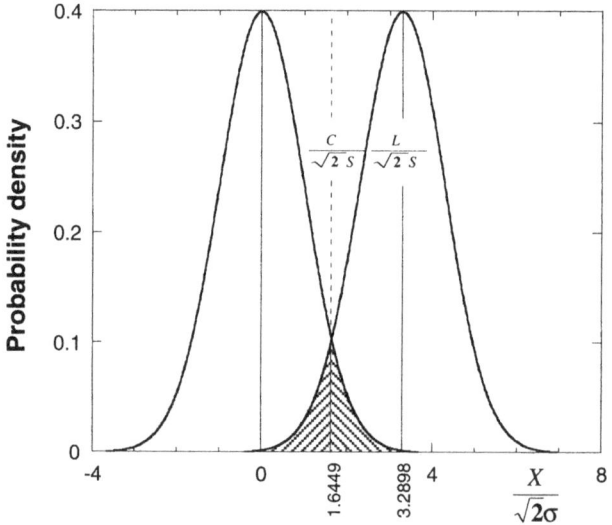

Fig. A.3 Detection distributions.

Note too that we can appeal to these results, as shown on the figure, to show that it is very unlikely that we would get an observed value below C if in fact the true value were greater L. But this does not immediately permit us to claim that the true value is then most probably less than L. That is, $\Pr(X < C | x \geq L) = 0.05$ does not guarantee that $\Pr(x \geq L | X < C) = 0.05$.[11] The latter should be recognized as a Bayesian probability, because it has to do with the probability of a state of nature, given our data. So to make the desired claim we must use Bayes' rule (Equation 2.13). That is, define event A as "$X < C$" and event B as "$x \geq L$". Then from Equation (2.13) we have $\Pr(A|B)\Pr(B) = \Pr(B|A)\Pr(A)$. To obtain $\Pr(A|B) = \Pr(B|A)$ we must adopt as prior belief that $\Pr(B) = \Pr(A)$, i.e., it is as likely that we will get

---

[11]The '|' symbol means "given that."

an observed datum below C as it is that the true concentration is above L. That seems plausible. Accordingly we can say that $\text{Prob}(x \geq L | X < C) = 0.05$ and so, using the probability rule for mutually exclusive events [Equation (2.3)], $\Pr(x < L | X < C) = 0.95$: if the observed datum is below C, we can be confident that the true value is below L.

References

1. Abramowitz, M.; Stegun, I.A. (1972). *Handbook of Mathematical Functions.* 9th printing. Dover, New York.

2. Aldenberg, T.; Jaworska, J.S. (2000). Uncertainty of the hazardous concentration and fraction affected for normal species sensitivity distributions. *Ecotoxicology and Environmental Safety* **46**: 1–18.

3. Aldenberg, T.; Slob, W. (1993). Confidence limits for hazardous concentrations based on logistically distributed NOEC toxicity data. *Ecotoxicology and Environmental Safety* **25**: 48–63.

4. Altman, D.G.; Bland, J.M. (1983). Measurement in medicine: the analysis of method comparison studies. *Statistician* **32**: 307–317.

5. Anderson, S.; Hauck, W.W. (1983). A new procedure for testing equivalence in comparative bioavailability and other clinical trials. *Communications in Statistics-Theory and Methods* **12**(23): 2663–2692.

6. Anderson, S.; Hauck, W.W. (1985). Letter to the Editor. *Biometrics* **41**: 561–563.

7. APHA (1998). Standard methods for the examination of water and wastewater. 20th ed. American Public Health Association, American Water Works Association, Water Environment Federation, Washington D.C.

8. APHA (2001). Compendium of methods for the microbiological examination of foods. 4th ed. American Public Health Association, Washington D.C.

9. Baker, M.; Russell, N.; Roseveare, C.; O'Hallahan, J.; Palmer, S.; Bichan, A. (1998). Outbreak of cryptosporidiosis linked to Hutt Valley swimming pool. *The New Zealand Public Health Report* **5**(6): 41–45.

10. Barnard, G.A. (1958). Thomas Bayes's essay towards solving a problem in the doctrine of chances. *Biometrika* **45**(3&4): 293–315.

11. Barnett, V. (1999). *Comparative Statistical Inference*. 3rd ed. Wiley, New York.

12. Barnett, V.; O'Hagan, A. (1997). *Setting Environmental Standards: The Statistical Approach to Handling Uncertainty and Variation*. Chapman and Hall, London.

13. Beliaeff, B.; Mary, J.-Y. (1993). The "most probable number" estimate and its confidence limits. *Water Research* **27**(5): 799–805.

14. Benjamini, Y.; Hochberg, Y. (1995). Controlling the false discovery rate: a practical and powerful approach to multiple testing. *Journal of the Royal Statistical Society* **B**(57): 289–300 (see http://www.math.tau.ac.il/ roee/index.htm).

15. Berger, J.O. (1986). Are P-values reasonable measures of accuracy? Francis, I.S.; Manly, B.F.J.; Lam, F.C. (eds.), Proceedings of the Pacific Statistical Congress: 21–27.

16. Berger, J.O.; Berry, D.A. (1988). Statistical analysis and the illusion of objectivity. *American Scientist* **76**: 159–165.

17. Berger, J.O.; Delampady, M. (1987). Testing precise hypotheses (with discussion). *Statistical Science* **2**(3): 317–352.

18. Berger, R.L.; Hsu, J.C. (1996). Bioequivalence trials, intersection-union tests and equivalence confidence sets. *Statistical Science* **11**(4): 283–319 (with discussion).

19. Berger, R.L; Sellke, T. (1987). Testing a point null hypothesis: the irreconcilability of p-values and evidence (with discussion). *Journal of the American Statistical Association* **82**: 112–122.

20. Berkson, J. (1938). Some difficulties of interpretation encountered in the application of the chi-square test. *Journal of the American Statistical Association* **33**: 526–542.

21. Berkson, J. (1942). Tests of significance considered as evidence. *Journal of the American Statistical Association* **37**: 325–335.

22. Berry, D.A. (1996). *Statistics: a Bayesian Perspective*. Duxbury, CA.

23. Berthouex, P.M.; Brown, L.C. (1980). *Statistics for Environmental Engineers*. Lewis Publishers, Boca Raton, FL.

24. Best, D.J. (1990). Optimal determination of most probable numbers. *International Journal of Food Microbiology* **11**: 159-166.

25. Bishop, Y.M.M.; Fienberg, S.E.; Holland, P.W. (1975). *Discrete Multivariate Analysis.* The MIT Press, Cambridge, MA.

26. Black, R.E.; Levine, M.M.; Clements, M.L.; Highes, T.P.; Blaser, M.J. (1988). Experimental *Campylobacter jejuni* infection in humans. *The Journal of Infectious Diseases* **157**: 472-479.

27. Bland, J.M.; Altman, D.G. (1986). Statistical methods for assessing agreement between two methods of clinical measurement. *The Lancet,* **February 8**: 307–310.

28. Bolstad, W.M. (2004). *Introduction to Bayesian Statistics.* Wiley, New York.

29. Bondy, W.H. (1969). A test of an experimental hypothesis of negligible difference between means. *The American Statistician* **23**: 28–30.

30. Bower, B. (1997). Null science. Psychology's statistical status quo draws fire. *Science News* **151**: 356–357.

31. Bowker, A.H.; Lieberman, G.J. (1972). *Engineering Statistics*, 2nd ed. Prentice-Hall, Englewood Cliffs, NJ.

32. Box, G.E.P. (1979). Some problems of statistics and everyday life. *Journal of the American Statistical Association* **74**(365): 1–4.

33. Box, G.E.P.; Tiao, G. (1973). *Bayesian Inference in Statistical Analysis.* Addison-Wesley, Reading, MA (re-issued in 1992 by Wiley, New York).

34. Bross, I.D. (1985). Why proof of safety is much more difficult than proof of hazard. *Biometrics* **41**: 785–793.

35. Buhl-Mortensen, L. (1996). Type-II statistical errors in environmental science and the precautionary principle. *Marine Pollution Bulletin* **32**(7): 528–531.

36. Burkhardt, W.; Calci, K.R. 2000. Selective accumulation may account for shellfish-associated viral illness. *Applied and Environmental Microbiology*, **66**(4): 1375–1378.

37. Burnham, K.P.; Anderson, D.R. (2002). *Model Selection and Multimodel Inference: A Practical Information-Theoretic Approach.* 2nd ed., Springer, New York.

38. Cabelli, V.J. (1989). Swimming-associated illness and recreational water quality criteria. *Water Science & Technology* **21**(2): 13–21.

39. Calderon, R.L.; Mood, E.W.; Dufour, A.P. (1991). Health effects of swimmers and nonpoint sources of contaminated water. *International Journal of Environmental Health Research* **1**: 21–31.

40. Carlin, B.P.; Louis, T.A. (2000). *Bayes and Empirical Bayes Methods for Data Analysis*. Chapman and Hall/CRC, Boca Raton, FL.

41. Carver, R.P. (1978). The case against statistical significance testing. *Harvard Education Review* **48**: 378–399.

42. Carver, R.P. (1993). The case against statistical significance testing, revisited. *Journal of Experimental Education* **61**(4): 287–292.

43. Cascio, W.F.; Zedeck S. (1983). Open a new window in rational research planning: adjust alpha to maximise statistical power. *Personnel Psychology* **36**: 517–526.

44. Casella, G.; Berger, R.L. (1987). Reconciling Bayesian and frequentist evidence in the one-sided testing problem. *Journal of the American Statistical Association* **82**: 106–111 (with discussion following the subsequent paper, by Berger and Sellke).

45. Casella, G.; Berger, R.L. (2002). *Statistical Inference*. 2nd ed. Duxbury, Pacific Grove, CA.

46. Chow, S.-C.; Liu, J.P. (1992). *Design and Analysis of Bioavailability and Bioequivalence Studies*. Marcel Dekker, New York.

47. Chow, S.L. (1996). *Statistical Significance: Rationale, Validity and Utility*. Sage Publications, London.

48. Cochran, W.G. (1950). Estimation of bacterial densities by means of the "Most Probable Number". *Biometrics* **6**: 105–116.

49. Cohen, J. (1960). A coefficient of agreement for nominal scales. *Educational and Psychological Measurement* **20**(1): 37–46.

50. Cohen, J. (1988). *Statistical Power Analysis for the Behavioral Sciences*, 2nd ed. Lawrence Erlbaum, Hillsdale, NJ.

51. Cohen, J. (1992). A power primer. *Psychological Bulletin* **112**: 155–159.

52. Cohen, J. (1994). The earth is round ($p < .05$). *American Psychologist* **49**(12): 997–1003 (reproduced in Harlow *et al.* 1997 [136]).

53. Cole, P.C.; Blair, R.C. (1999). Overlapping confidence intervals. *Journal of the American Academy of Dermatology* **41**(6): 1051–1052.

54. Cole, R.G.; McBride. (2004). Assessing impacts of dredge spoil disposal using equivalence tests: implications of a precautionary (prrrof of safety) approach. *Marine Ecology Progress Series* **279**: 63–72.

55. Cole, R.G.; McBride.; G.B.: Healy, T.R. (2001). Equivalence tests and sedimentary data: dredge spoil disposal at Pine Harbour Marina, Auckland. *Journal of Coastal Research Special Issue* **34**: 611–622.

56. Conover, W.J. (1980). *Practical Nonparametric Statistics*. 2nd ed. Wiley, New York.

57. Cooper, B.E. (1985a). Algorithm AS 3. The integral of Student's t-distribution. In P. Griffiths and I.D. Hill (eds.), *Applied Statistics Algorithms*. Ellis Horwood, Chichester, UK, pp. 38–39 (published for the Royal Statistical Society).

58. Cooper, B.E. (1985b). Algorithm AS 5. The integral of the non-central t-distribution. In P. Griffiths and I.D. Hill (eds.), *Applied Statistics Algorithms*. Ellis Horwood, Chichester, UK, pp. 40–42 (published for the Royal Statistical Society).

59. Cornfield, J. (1966). Sequential trials, sequential analysis, and the likelihood principle. *The American Statistician* **29**(2): 18–23.

60. Cotter, A.J.R. (1985). Water quality surveys: a statistical method based on determinisim, quantiles and the binomial distribution. *Water Research* **19**(9): 1179–1189.

61. Cox, D.R. (1987) Comment on Berger, J.O.; Delampady, M. Testing precise hypotheses. *Statistical Science* **2**(3): 335–336.

62. Crabtree, K.D.; Gerba, C.P.; Rose, J.B.; Haas, C.N. (1997). Waterborne adenovirus: a risk assessment. *Water Science & Technology* **35**(11–12): 1–6.

63. Craun, G.F.; Hubbs, S.A.; Frost, F.; Calderon, R.L.; Via, S.H. (1998). Waterborne outbreaks of cryptosporidiosis. *Journal of the American Water Works Association* **90**(9): 81–91.

64. David, F.N.; Barton, D.E. (1962). *Combinatorial Choice*. Griffin, London. 356 p.

65. Davis, P.J.; Hersh, R. (1981). *The Mathematical Experience*. Birkhäuser, Boston. 440 p.

66. Dayton, P.K. (1998). Reversal of the burden of proof in fisheries management. *Science* **279**: 821–822.

67. DeGroot, M.H. (1973). Doing what comes naturally: interpreting a tail area as a posterior probability or as a likelihood ratio. *Journal of the American Statistical Association* **68**: 966–969.

68. de Man, J.C. (1975). The probability of most probable numbers. *European Journal of Applied Microbiology* **1**: 67–78.

69. de Man, J.C. (1977). MPN tables for more than one test. *European Journal of Applied Microbiology* **4**: 307–316.

70. de Man, J.C. (1983). MPN tables, corrected. *European Journal of Applied Microbiology* **17**: 301–305.

71. Dennis, B. (1996). Discussion: should ecologists become Bayesian? *Ecological Applications* **6**: 1095–1103.

72. Diletti, E.; Hauschke, D.; Steinijans, V.W. (1991). Sample size determination for bioequivalence assessment by means of confidence intervals. *International Journal of Clinical Pharmacology, Therapy and Toxicology* **29**: 1–8 (repeated in vol. **30**, Supplement No. **1**: S51–S58).

73. Dixon, P.M. (1998). Assessing effect and no effect with equivalence tests. In M.C. Newman and C.L. Strojan (eds.), *Risk Assessment: Logic and Measurement*. Ann Arbor Press, Chelsea, MI, pp. 275–301.

74. Dixon, P.M.; Garrett, K.A. (1994). Statistical issues for field experimenters. In R.J. Kendall and T.E. Lacher, Jr. (eds.), *Wildlife and Population Modelling: Integrated Studies of Agroecosystems*. Lewis, Boca Raton, FL, pp. 439–450.

75. Dupont, W.D. (1983). Sequential stopping rules and sequentially adjusted P values: does one require the other? *Controlled Clinical Trials* **4**: 3–10.

76. Edgington, E.S. (1987). *Randomization Tests*. 2nd ed. Marcel Dekker, New York.

77. Edwards, A.W.F. (1972). *Likelihood*. Cambridge University Press, London.

78. Edwards, A.W.F. (2000). Book review: Statistical Evidence, by R. Royall. *Statistics in Medicine* **19**: 879–885.

79. Edwards, W.; Lindman, H.; Savage, L.J. (1963). Bayesian statistical inference for psychological research. *Psychological Review* **70**: 193–242.

80. Eisenhart, C.; Wilson, P.W. (1943). Statistical methods and control in bacteriology. *Bacteriological Reviews* **7**: 57–137.

81. Elliott, J.C. (1983). Some methods for the statistical analysis of samples of benthic invertebrates. Freshwater Biological Association Scientific Publication No. 25, The Ferry House, Ambleside, Cumbria, UK.

82. Ellis, J.C. (1985). Determination of pollutants in effluents. Part A. Assessment of alternative definitions of effluent compliance. Report **TR 230**, Water Research Centre, Medmenham, England, January.

83. Ellis, J.C. (1986). Determination of pollutants in effluents. Part B. Alternative forms of effluent consents: some statistical considerations. Report **TR 235**, Water Research Centre, Medmenham, England, April.

84. Ellis, J.C. (1989). Handbook on the design and interpretation of monitoring programmes. Report **NS 29**, Water Research Centre, Medmenham, England (first published April 1990).

85. Ellison, A.M. (2004). Bayesian inference in ecology. *Ecology Letters* **7**: 509–520.

86. El-Sharaawi, A.H. (1985). Some goodness-of-fit methods for the Poisson plus added zeroes distribution. *Journal of Applied Microbiology* **49**(5): 1304–1306.

87. Erickson, W.P.; McDonald, L.L. (1995). Tests for bioequivalence of control media and test media in studies of toxicity. *Environmental Toxicology and Chemistry* **14**(7): 1247–1256.

88. Eurachem (1998). The fitness for purpose of analytical methods. A laboratory guide to method validation and related topics. Eurachem, Teddington, Middlesex, UK.

89. Fairweather, P.G. (1991). Statistical power and design requirements for environmental monitoring. *Australian Journal of Marine and Freshwater Research* **42**: 555–567.

90. Ferguson, T.S. (1967). *Mathematical Statistics: A Decision Theoretic Approach*. Academic Press, New York.

91. Fisher, R.A. (1925). *Statistical Methods for Research Workers*. Oliver and Boyd, Edinburgh.

92. Fisher, R.A. (1930). Inverse probability. *Proceedings of the Cambridge Philosophical Society* **26**: 528–535.

93. Fisher, R.A. (1935–1971). *The Design of Experiments*. Editions 1–8. Hafner, New York.

94. Fisher, R.A. (1959). *Statistical Methods and Scientific Inference*. 2nd ed. Oliver and Boyd, Edinburgh.

95. Fleiss, J.L. (1981). *Statistical Methods for Rates and Proportions*. 2nd ed. Wiley, New York.

96. Fleiss, J.L. (1986). Significance tests have a role in epidemiologic research: reactions to A.M. Walker (different views). *American Journal of Public Health* **76**: 559–560.

97. Fleiss, J.L. (1987). Letter to the editor: Some thoughts on two-tailed tests. *Controlled Clinical Trials* **8**: 394.

98. Forster, L.I. (2003). Measurement uncertainty in microbiology. *Journal of AOAC International* **86**(5): 1089–1094.

99. Fraser, G.; Cooke, K.R. (1991). Endemic giardiasis and municipal water supply. *American Journal of Public Health* **81**(6): 760–762.

100. Freedman, P.L.; Dilks, D.W.; Holmberg, H.P.; Moskus, P. E.; McBride, G.; Hickey, C.; Smith, D.G.; Striplin, P.L. 2003. Navigating the TMDL process:

method development for addressing narrative criteria. Water Environment Research Foundation Report **01-WSM-1**. Alexandria, VA.

101. Freund, J.E. (1992). *Mathematical Statistics*. 5th ed. Prentice-Hall, Upper Saddle River, NJ.

102. Frick, R.W. (1995). Accepting the null hypothesis. *Memory & Cognition* **23**(1): 132–138.

103. Furumoto, W.A.; Mickey, R. (1967). A mathematical model for the infectivity-dilution curve of tobacco mosaic virus: theoretical considerations. *Virology* **32**: 216–223.

104. Gale, P. (1998). Simulating *Cryptosporidium* exposures in drinking water during an outbreak. *Water Science & Technology* **38**(12): 7–13.

105. Gale, P. (2003). Developing risk assessments of waterborne microbial contaminations. In D. Mara and N. Horan (eds.), *The Handbook of Water and Wastewater Microbiology*. Academic Press, Amsterdam, pp. 263–280.

106. Gale, P.; van Dijk, P.A.H.; Stanfield, G. (1997). Drinking water treatment increases microorganism clustering; the implications for microbiological risk assessment. *Journal of Water Supply: Research and Technology – Aqua* **46**(3): 117–126.

107. García, L.V. (2003). Controlling the false discovery rate in ecological research. *TRENDS in Ecology and Evolution* **18**(11): 553–554.

108. Gardner, M.J.; Altman, D.G. (1989). *Statistics with Confidence-Confidence Intervals and Statistical Guidelines*. British Medical Journal, London.

109. Garrett, K.A. (1997). Use of statistical tests of equivalence (bioequivalence tests) in plant Pathology. *Phytopathology* **87**: 372–374.

110. Gelman, A.; Carlin, J.B.; Stern, H.S.; Rubin, D.B. (2000). *Bayesian Data Analysis*. Chapman and Hall/CRC, Boca Raton, FL.

111. Gerard, P.D.; Smith, D.R.; Weerakkody, G. 1998. Limits of retrospective power analysis. *Journal of Wildlife Management* **62**(2): 801–807.

112. Gerba, C.P.; Rose, J.B.; Haas, C.N.; Crabtree, K.D. 1996. Waterborne rotavirus: a risk assessment. *Water Research* **30**(12): 2929–2940.

113. Germano, J.D. (1999). Ecology, statistics, and the art of misdiagnosis: the need for a paradigm shift. *Environmental Reviews* **7**(4): 167–190.

114. Germano, J.D. (2001). Reflections of statistics, ecology, and risk assessment. In J.Y. Aller, S.A. Woodin and R.C. Aller (eds.), *Organisms-Sediment Interactions*. University of South Carolina Press, Columbia, SC.

115. Gerrodette, T.; Dayton, P.K.; Macinko, S.; Fogarty, M.J. (2002). Precautionary management of marine fisheries: moving beyond the burden of proof. *Bulletin of Marine Science* **70**(2): 657–668.

116. Gibbons, J.D.; Pratt, J.W. (1975). P-values: interpretation and methodology. *The American Statistician* **29**: 20–25.

117. Gibbons, R.D. (1994). Statistical methods for groundwater monitoring. Wiley, New York.

118. Gilbert, R.O. (1987). *Statistical Methods for Environmental Pollution Monitoring*. Van Nostrand Reinhold, New York, NY.

119. Gill, J. (2002). *Bayesian Methods. A Social and Behavioural Approach*. Chapman and Hall/CRC, Boca Raton, FL.

120. Goldstein, S.T.; Juranek, D.D.; Ravenbolt, O.; Hightower, A.W.; Martin, D.G.; Mesnik, J.L.; Griffiths, S.D.; Bryant, A.J.; Reich, R.R.; Herwaldt, M.D. (1996). Cryptosporidiosis: an outbreak associated with drinking water despite state-of-the-art water treatment. *Annals of Internal Medicine* **124**(5): 459–468.

121. Good, I.J. (1982). Standardized tail-area probabilities. *Journal of Statistical Computation and Simulation* **16**: 65–66.

122. Goodman, S.N. (1993). p values, hypothesis tests, and likelihood: implications for epidemiology of a neglected historical debate. *American Journal of Epidemiology* **137**(5): 485–501 (includes discussion).

123. Goodman, S.N. (1999a). Toward evidence-based medical statistics. 1: The P value fallacy. *Annals of Internal Medicine* **130**(12): 995–1004.

124. Goodman, S.N. (1999b). Toward evidence-based medical statistics. 2: The Bayes factor. *Annals of Internal Medicine* **130**(12): 1005–1013.

125. Goodman, S.N.; Royall, R. (1988). Evidence and scientific research. *American Journal of Public Health* **78**(12): 1568–1574.

126. Gray, J.S. (1990). Statistics and the precautionary principle. *Marine Pollution Bulletin* **21**(4): 174–176.

127. Green, R.H. (1979). *Sampling Design and Statistical Methods for Environmental Biologists*. Wiley, New York.

128. Greenwood, M., Jr.; Yule, G.U. (1917). On the statistical interpretation of some bacteriological methods employed in water analysis. *Journal of Hygiene* **16**: 36–54.

129. Guttman, I. (1970). *Statistical Tolerance Regions*. Griffin, London.

130. Haas, C.N. (1996). How to average microbial densities to characterize risk. *Water Research* **30**(4): 1036–1038.

131. Haas, C.N. (2002). Conditional dose-response relationships for microorganisms: development and application. *Risk Analysis* **22**(3): 455–463.

132. Haas, C.N.; Rose, J.B.; Gerba, C.P. (1999). *Quantitative Microbial Risk Assessment*. John Wiley, New York, 449 pp.

133. Hacking, I. (1965) *Logic of Scientific Inference*. Cambridge University Press, Cambridge.

134. Hagen, R.L. (1997). In praise of the null hypothesis statistical test. *American Psychologist* **52**(1): 15–24.

135. Hamilton, M.A.; Collings, B.J. (1991). Determining the appropriate sample size for nonparametric tests for location shift. *Technometrics* **33**(3): 327–337.

136. Harlow, L.L.; Muliak, S.A.; Steiger J.H. (eds.) (1997). *What If There Were No Significance Tests?* Lawrence Erlbaum, Mahwah, NJ.

137. Harris, R.J. (1997a). Reforming significance testing via three-valued logic. In L.L. Harlow, S.A. Muliak and J.H. Steiger (eds.), *What If There Were No Significance Tests?* Lawrence Erlbaum, Mahwah, NJ, pp. 145–174.

138. Harris, R.J. (1997b). Significance tests have their place. *Psychological Science* **8**(1): 8–11.

139. Harris, R.J. (2001). *A Primer of Multivariate Statistics*. 3rd ed. Lawrence Erlbaum, Mahwah, NJ.

140. Hartnett, E.; Paoli, G.; Fazil, A.; Lammerding, A.; Anderson, S.; Rosenquist, H.; Christensen, B.B.; Nauta, M. (2001). Joint FAO/WHO Activities on Risk Assessment of Microbiological Hazards in Foods: Preliminary Report – Hazard identification, hazard characterization and exposure assessment of *Campylobacter* spp. in broiler chickens. Report MRA01/05.

141. Hauck, W.W.; Anderson, S. (1984). A new statistical procedure for testing equivalence in two-group comparative bioavailability trials. *Journal of Pharmacokinetics and Biopharmaceutics* **12**(1): 83–91.

142. Hauck, W.W.; Anderson, S. (1996). Comment on Berger, R.L. and Hsu, J.C. Bioequivalence trials, intersection-union tests and equivalence confidence sets. *Statistical Science* **11**(4): 303.

143. Hauschke, D.; Steinijans, V.W.; Diletti, E. (1990). A distribution-free procedure for the analysis of bioequivalence studies. *Journal of Clinical Pharmacology, Therapy and Toxicology* **28**: 72–78.

144. Hauschke, D.; Hothorn, L. (1998). Safety assessment in toxicological studies: proof of safety versus proof of hazard. In S.C. Chow and J.P. Liu (eds.), *Design and Analysis of Animal Studies in Pharmaceutical Development*. Marcel Dekker, New York, pp. 197–225.

145. Helms, M., Vastrup, P., Gerner-Smidt, P. & Mølbak, K., (2003). Short and long term mortality associated with foodborne bacterial gastrointestinal infections: registry based study. *British Medical Journal* **326**: 357–361.

146. Helsel, D.R. (2005). *Nondetects And Data Analysis: Statistics for censored environmental data.* Wiley, New York. 288 pp.

147. Helsel, D.R.; Hirsch, R.M. (2002). *Statistical Methods in Water Resources.* USGS Techniques of Water Investigations Book 4, Chapter A3, 510 pp.[12]

148. Hickey, C.W.; Quinn, J.M.; Davies-Colley, R.J. (1987). Effluent characteristics of domestic sewage oxidation ponds and their potential impacts on rivers. *New Zealand Journal of Marine and Freshwater Research* **23**: 585–600.

149. Hilborn, R.; Mangel, M. (1997). *The Ecological Detective: Confronting Models with Data.* Princeton University Press, Princeton, NJ.

150. Hill, A.B. (1965). The environment and disease: association or causation? *Proceedings of the Royal Society of Medicine* **58**: 295–300.

151. Hodges, A. (1983). *Alan Turing: The Enigma.* Simon and Schuster, New York.

152. Hodges, J.S. (1996). The effect of partial-ordering utilities on Bayesian design of sequential experiments. In D.A. Berry, K.M. Chaloner, J.K. Geweke (eds.), *Bayesian Analysis in Statistics and Econometrics: Essays in Honor of Arnold Zellner.* Wiley, New York, pp. 515–525.

153. Hoenig, J.M.; Heisey, D.M. (2001). The abuse of power: the pervasive fallacy of power calculations for data analysis. *The American Statistician* **55**: 19–24.

154. Hoffman, P. (1998). *The Man Who Loved Only Numbers.* Fourth Estate, London.

155. Hollingdale, S. (9189). *Makers of Mathematics.* Penguin, London.

156. Hoxie, N.J.; Davis, J.P.; Vergeront, J.M.; Nashold, R.D.; Blair, K.A. (1997). Cryptosporidiosis-associated mortality following a massive waterborne outbreak in Milwaukee, Wisconsin. *American Journal of Public Health* **87**(12): 2032–2035.

157. Howson, C.; Urbach, P. (1991). Bayesian reasoning in science. *Nature* **350**: 371–374.

158. Hrudey, S.E.; Rizak, S. (2004). Discussion of: Rapid analytical techniques for drinking water security investigations. *Journal of the American Water Works Association* **96**(9):110–113.

159. Hurlbert, S.H. (1984). Pseudoreplication and the design of ecological experiments. *Ecological Monographs* **54**(2): 187–211.

---

[12] http://water.usgs.gov/pubs/twri/twri4a3/

160. IDT (1998). WQSTAT PLUS User's Guide. Intelligent Design Technologies Ltd., Longmont, CO.

161. Iman, R.L.; Conover, W.J. (1983). *A Modern Approach to Statistics*. Wiley, New York.

162. Inman, H.F. (1994). Karl Pearson and R. A. Fisher on statistical tests: a 1935 exchange from *Nature*. *The American Statistician* **48**(1): 2–11.

163. ISO (2001). Water quality – Guidance on Validation of Microbiological Methods. Technical Report **ISO/TR 13843**, International Organization for Standardization, Geneva, Switzerland.

164. Jeffreys, H. (1961). *Theory of Probability*, Oxford University Press and Calrendon Press, Oxford, UK.

165. Johl, M.; Kausch, S.; Eichhorn, D.; Burger, G. (1995). Zur Einschätzung des gesundheitlichen Risikos der Neuen Donau bei deren Nutzung als Badegewässer aus virologischer Sicht. *Forum-Städte-Hygiene* **46**: 284–289 (cited by Schernewski & Jülich 2001).

166. Johnson, D.H. (1995). Statistical sirens: the allure of nonparametrics. *Ecology* **76**: 1998–2000.

167. Johnson, D.H. (1999). The insignificance of statistical significance testing. *Journal of Wildlife Management* **63**(3): 763–772.

168. Johnson, J.B.; Omland, K.S. (2004). Model selection in ecology and evolution. *TRENDS in Ecology and Evolution* **19**(2): 101–108.

169. Johnson, N.L.; Kotz, S. (1970). *Continuous Univariate Distributions-2*. Houghton-Mifflin, Boston.

170. Jones, L.V.; Tukey, J.W. (2000). A sensible formulation of the significance test. *Psychological Methods* **5**(4): 411–414.

171. Jowett, I.G.; Richardson, J. (2003). Fish communities in New Zealand rivers and their relationship to environmental variables. *New Zealand Journal of Marine and Freshwater Research* **37**: 347–366.

172. Kateman, G.; Müskens, P.J.W.M. (1978). Sampling of internally correlated lots. The reproducibility of gross samples as a function of sample size, lot size, and number of samples [1]. Part II. Implications for practical sampling and analysis. *Analytica Chimica Acta* **103**: 11–20.

173. Kirchmer, C.J. (1983). Quality control in water analyses. *Environmental Science and Technology* **17**(4): 174A–181A.

174. Kraemer, H.C.; Bloch, D.A. (1988). Kappa coefficients in epidemiology: an appraisal of a reappraisal. *Journal of Clinical Epidemiology* **41**(10): 959–968.

175. Krantz, D.H. (1999). The null hypothesis testing controversy psychology. *Journal of the American Statistical Association* **44**(448): 1372–1381.

176. Krebs, C.J. (1989). *Ecological Methodology*. Harper and Row, New York.

177. Landis, J.R.; Koch, G.G. (1977). The measurement of observer agreement for categorical data. *Biometrics* **33**: 159–174.

178. Larned, S.T.; Scarsbrook, M.R.; Snelder, T,H,; Norton, N.L.; Biggs, B.J.F. (2004). Water quality in low-elevation streams and rivers of New Zealand: recent state and trends in contrasting land-cover classes. *New Zealand Journal of Marine and Freshwater Research* **38**: 347–366.

179. Lee, P.M. (1997). *Bayesian Statistics: An Introduction*. Arnold, London.

180. Legendre, P. (1993). Spatial autocorrelation: trouble or a new paradigm? *Ecology* **74**(6): 1659–1673.

181. Legendre, P.; Legendre, L. (1998). *Numerical Ecology*, 2nd ed. Developments in Environmental Modelling 20. Elsevier, Amsterdam.

182. Lehmann, E.L. (1959). *Testing Statistical Hypotheses*. Chapman and Hall, New York.

183. Lehmann, E.L. (1986). *Testing Statistical Hypotheses*. 2nd ed. Wiley, New York.

184. Lin, L.I.-K. (1989). A concordance correlation coefficient to evaluate reproducibility. *Biometrics* **45**: 255–268.

185. Lin, L.I.-K. (2000). A note on the concordance correlation coefficient. *Biometrics* **56**: 324–325.

186. Lindley, D.V. (1971). *Bayesian Statistics, A Review*. Society for Industrial and Applied Mathematics, Philadelphia.

187. Lindley, D.V. (2000). The philosophy of statistics. *The Statistician* **49**(3): 293–337.

188. Loftis, J.C. (1989). An evaluation of trend detection techniques for use in water quality monitoring programs, EPA-600-3-89-037. U.S. Environmental Protection Agency Office of Research and Development, Environmental Research Laboratory, Corvallis, Oregon.

189. Loftis, J.C.; Harris, J.; Montgomery, R.H. (1987). Detecting changes in ground water quality at regulated facilities. *Ground Water Monitoring Review* **7**(1): 72–76.

190. Loftis, J.C.; McBride, G.B.; Ellis, J.C. (1991). Considerations of scale in water quality monitoring and data analysis. *Water Resources Bulletin* **27**(2): 255–264.

191. Loftis, J.C.; MacDonald, L.H.; Streett, S.; Iyer, H.K.; Bunte, K. (2001). Detecting cumulative watershed effects: the statistical power of pairing. *Journal of Hydrology* **251**: 49–64.

192. Loftus, G.R. (1991). On the tyranny of hypothesis testing in the social sciences. *Contemporary Psychology* **36**: 102–105.

193. Loyer, M.W.; Hamilton, M.A. (1984). Interval estimation of the density of organisms using a serial-dilution experiment. *Biometrics* **40**: 907–916.

194. MacKenzie, W.R.; Hoxie, N.J.; Proctor, M.E.; Gradus, M.S.; Blair, K.A.; Peterson, D.E.; Kazmierczak, J.J.; Fox, K.R.; Addias, D.G.; Rose, J.B.; Davis, J.P. (1994). A massive outbreak in Milwaukee of Cryptosporidium infection transmitted through the public water supply. *New England Journal of Medicine* **331**(3): 161–167.

195. Magnusson, W.E. (2000). Commentary. Error bars: Are they the King's clothes? *Bulletin of the Ecological Society of America* **81**(2): 147–150.

196. Mandallaz, D.; Mau, J. (1981). Comparison of different methods for decision-making in bioequivalence assessment. *Biometrics* **37**: 213–222.

197. Manly, B.F.J. (1997). *Randomisation, Bootstrap and Monte Carlo Methods in Biology.* 2nd ed. Chapman and Hall, London.

198. Manly, B.F.J. (2001). *Statistics for Environmental Science and Management.* Chapman and Hall/CRC Press, Boca Raton, FL.

199. Manly, B.F.J. (2004). One-sided tests of bioequivalence with non-normal distributions and unequal variances. *Journal of Agricultural, Biological and Environmental Statistics* **9**(3): 270–283.

200. Manly, B.F.J.; Francis, R.I.C.C. (2002). Testing for mean and variance differences with samples from distributions that may be non-normal with unequal variances. *Journal of Statistical Computation and Simulation* **72**: 633–646.

201. Mapstone, B.D. (1995). Scalable decision rules for environmental impact studies: effect size, Type I and Type II errors. *Ecological Applications* **5**: 401–410.

202. Martin Andrés, A. (1990). On testing for bioequivalence. *Biometrical Journal* **32**(1): 125–126.

203. Mathsoft (1997). S-Plus4 Guide to Statistics. MathSoft, Inc., Seattle, Washington.

204. McBride, G.B. (1993). Discussion of "Health effects of swimmers and nonpoint sources of contaminated water," by Calderon et al. *International Journal of Environmental Health Research* **3**: 115–116.

205. McBride, G.B. (1999). Equivalence tests can enhance environmental science and management. *Australian & New Zealand Journal of Statistics* **41**(1): 19–29.

206. McBride, G.B. (2002). Statistical methods helping and hindering environmental science and management. *Journal of Agricultural, Biological and Environmental Statistics* **7**(3): 300–305.

207. McBride, G.B. (2003). Confidence of Compliance: parametric versus nonparametric approaches. *Water Research* **37**(15): 3666–3671.

208. McBride, G.B.; Loftis, J.C.; Adkins, N.C. (1993). What do significance tests really tell us about the environment? *Environmental Management* **17**(4): 423–432 (errata in **18**(2): 317).

209. McBride, G.B.; Ellis, J.C. (2001). Confidence of Compliance: a Bayesian approach for percentile standards. *Water Research* **35**(5): 1117–1124.

210. McBride, G.B.; Salmond, C.E.; Bandaranayake, D.R.; Turner, S.J.; Lewis, G.D.; Till, D.G. (1998). Health effects of marine bathing in New Zealand. *International Journal of Environmental Health Research* **8**: 173–189.

211. McBride, G.B.; Till, D.; Ryan, T.; Ball, A. Lewis, G.; Palmer, S.; Weinstein, P. (2002). Freshwater Microbiology Research Programme. Pathogen occurrence and human health risk assessment analysis. Ministry for the Environment Technical Publication. 93 pp.[13]

212. McBride, G.B.; McWhirter, J. L.; Dalgety, M.H. (2003). Accounting for uncertainty in Most Probable Number enumerations for microbiological assays. *Journal of AOAC International* **86**(5): 1084–1088.

213. McCrady, M.H. (1915). The numerical interpretation of fermentation-tube results. *The Journal of Infectious Diseases* **17**: 183–212.

214. McCrady, M.H. (1918). Tables for rapid interpretation of fermentation-tube results. *The Public Health Journal (Canada)* **9**(5): 201–220.

215. McCullagh, P.; Nelder, J.A. (1995). *Generalized Linear Models*. 2nd ed. Chapman and Hall, London.

216. McDonald, L.L.; Erickson, W.P. (1994). Testing for bioequivalence in field studies: has a disturbed site been adequately reclaimed? In D.J. Fletcher and B.F.J. Manly (eds.), *Statistics in Ecology and Environmental Monitoring*. University of Otago Press, Dunedin, New Zealand, pp. 183–197.

217. McDowell, R.M.; McElvaine, M.D. (1997). Long-term sequelae to foodborne disease. *Scientific and Technical Review, World Organisation for Animal Health* **16**(2): 337–341.

---

[13] www.mfe.govt.nz/publications/water/freshwater-microbiology-nov02

218. McGill, R.; Tukey, J.W.; Larsen, W.A. (1978). Variations of box plots. *The American Statistician* **32**(1): 12–16.

219. Medema, G.J.; Teunis, P.F.M.; Havelaar, A.H.; Haas, C.N. (1996). Assessment of dose-response relationship of *Campylobacter jejuni*. *International Journal of Food Microbiology* **30**: 101–111.

220. Menand, L. (2001). *The Metaphysical Club*. Flamingo (Harper-Collins Publishers), London.

221. Messner, M.J.; Chappell, C.L.; Okhuysen, P.O. (2001). Risk assessment for *Crypotosporidium*: a hierarchical Bayesian analysis of human dose response data. *Water Research* **35**(16): 3934–3940.

222. Metzler, C.M. (1974). Bioavailability: A problem in equivalence. *Biometrics* **30**: 309–317.

223. MfE/MoH (2003). Microbiological Water Quality Guidelines for Marine and Freshwater Recreational Areas. Ministry for the Environment and Ministry of Health, Wellington, New Zealand.[14]

224. Millard, S.P. (1987). Proof of safety versus proof of hazard. *Biometrics* **43**: 719–725.

225. Millard, S.P.; Lettenmaier, D.P. (1986). Optimal design of biological sampling programs using annalysis of variance. *Estuarine, Coastal and Shelf Science* **22**: 637–656.

226. Millard, S.P.; Neerchal, N.K. (2001). *Environmental Statistics with S-Plus*. CRC Press, Boca Raton, FL.

227. Miller, J.C.; Miller, J.N. (1988). *Statistics for Analytical Chemistry*, 2nd ed. Ellis Horwood, New York.

228. Miller, I.; Freund, J.E. (1985). *Probability and Statistics for Engineers*. 3rd ed. Prentice-Hall, Englewood Cliffs, NJ.

229. MoH (2000). Drinking-Water Standards for New Zealand 2000. Ministry of Health, Wellington, New Zealand.

230. Mood, A.M.; Graybill, F.A. (1963). *Introduction to the Theory of Statistics*. 2nd ed., McGraw-Hill, New York.

231. Morris, R.D.; Naumova, E.N.; Levin, R.; Munasinghe, R. (1996). Temporal variation in turbidity and diagnosed gastroenteritis in Milwaukee. *American Journal of Public Health* **86**(2): 237–239.

---

[14] http://www.mfe.govt.nz/publications/water/microbiological-quality-jun03/

232. Morrison, D.E.; Henkel, R.E. (eds.) (1970). *The Significance Test Controversy – A Reader*. Aldine, Chicago, IL.

233. Muliak, S.A.; Raju, N.S; Harshman, R.A. (1997). There is a time and a place for significance testing. In L.L. Harlow, S.A. Muliak, and J.H. Steiger (eds.), *What If There Were No Significance Tests?* Lawrence Erlbaum, Mahwah, NJ.

234. Mumby, P.J. (2002). Statistical power of non-parametric tests: a quaick guide for designing sampling strategies. *Marine Pollution Bulletin* **44**: 85–87.

235. Murdoch, N. (2001). Justified design evaluation for percentile standards. *Environmental Modelling & Software* **16**: 725–738.

236. Müskens, P.J.W.M.; Kateman, G. (1978). Sampling of internally correlated lots. The reproducibility of gross samples as a function of sample size, lot size, and number of samples. Part I. Theory. *Analytica Chimica Acta* **103**: 1–9.

237. Nachamkin, I.; Allos, B.M.; Ho, T. (1998). *Campylobacter* species and Guillain Barré Syndrome. *Clinical Microbiology Reviews* **11**(3): 555–567.

238. Nelder, J.A. (1999). Statistics for the millennium: from statistics to statistical science. *The Statistician* **48**(2): 257–269.

239. Nester, M.R. (1996). An applied statistician's creed. *Applied Statistics* **45**: 401–410.

240. Newell, A.D.; Morrison, M.L. (1993). Use of overlap studies to evaluate method changes in water chemistry protocols. *Water, Air, and Soild Pollution* **67**: 433–456.

241. Neyman, J.; Pearson, E.S. (1928). On the use and interpretation of certain test criteria for the purposes of statistical inference. *Biometrika* **20**: 175–240.

242. Neyman, J.; Pearson, E.S. (1933). On the problem of the most efficient tests of statistical hypotheses. *Philosophical Transactions of the Royal Society, Series A* **231**: 289–337.

243. Noether, G.E. (1987). Sample size determination for some common nonparametric tests. *Journal of The American Statistician Association* **82**: 645–647.

244. Nunnally, J.C. (1960). The place of statistics in psychology. *Educational and Psychological Measurement* **20**: 641–650.

245. Oakes, M. (1986). *Statistical Inference: A Commentary for the Social and Behavioural Sciences*. Wiley, New York.

246. Odeh, R.E.; Owen, D.B. (1980). *Tables for Normal Tolerance Limits, Sampling Plans and Screening*. Marcel Dekker, New York.

247. Okhuysen, P.C.; Chappell, C.L.; Crabb, J.H.; Sterling, C.R.; DuPont, H.L. (1999). Virulence of three distinct *Cryptosporidium parvum* isolates for healthy adults. *The Journal of Infectious Diseases* **180**: 1275–1281.

248. Owen, D.B. (1956). Tables for computing bivariate normal probabilities. *Annals of Mathematical Statistics* **35**: 762–772.

249. Owen, D.B. (1962). *Handbook of Statistical Tables*. Addison-Wesley, Reading, MA.

250. Owen, D.B. (1965). A special case of a bivariate non-central t-distribution. *Biometrika* **52**: 437–446.

251. Owen, D.B. (1968). A survey of properties and applications of the non-central t-distibution. *Technometrics* **10**: 445–477.

252. Palisade Corporation. (2000). @RISK. Advanced Risk Analysis for Spreadsheets. Version 4.0.5. Newfield, New York.

253. Parkhurst, D.F. (2001). Statistical significance tests: equivalence and reverse tests should reduce misinterpretation. *Bioscience* **51**(12): 1051–1057.

254. Parkhurst, D.F.; Stern, D.A. (1998). Determining average concentrations of *Cryptosporidium* and other pathogens in water. *Environmental Science and Technology* **32**: 3424–3429.

255. Patel, H.I.; Gupta, G.D. (1984). A problem of equivalence in clinical trials. *Biometrical Journal* **26**(5): 471–474.

256. Pearson, E.S.; Hartley, H.O. (eds.) (1972). *Biometrika Tables for Statisticians*, Vol. II. Cambridge University Press, London.

257. Pedini, M. (1976). Council Directive of 8 December 1975 concerning the quality of bathing water. *Official Journal of the European Communities* **L31**: 1–7.

258. Phillips, K.F. (1990). Power of the Two One-Sided Tests procedure in bioequivalence. *Journal of Pharmacokinetics and Biopharmaceutics* **18**(2): 137–144.

259. Ponce, S.L. (1982). The use of paired-basin technique in flow-related wildland water-quality studies. Report WSDG-TP-00004. USDA Forest Service, Fort Collins, CO.

260. Poole, C. (1988). Editorial: Feelings and frequencies: two kinds of probability in public health research. *American Journal of Public Health* **78**: 1531–1532.

261. Poole, C. (2001). Low P-values or narrow confidence intervals: which are more durable? *Epidemiology* **12**(3): 291–294.

262. Pratt, J.W. (1965). Bayesian interpretation of standard inference statements (with discussion). *Journal of the Royal Statistical Society, Series B* **27**: 169–203.

263. Pratt, J.W.; Raiffa, H.; Schlaifer, R. (1995). *Introduction to Statistical Decision Theory*. MIT Press, Cambridge, MA.

264. Press, S.J. (2003). *Subjective and Objective Bayesian Statistics: Principles, Models and Applications*. Wiley, New York.

265. Press, W.H.; Teukolsky, S.A.; Vetterling, W.T.; Flannery, B.P. (1992). *Numerical recipes in FORTRAN. The art of scientific computing*. 2nd ed. Cambridge University Press, Cambridge UK.

266. Quinn, J. M.; Davies-Colley, R.J.; Hickey, C.W.; Vickers, M.L.; Ryan, P.A. (1992). Effects of clay discharges in streams. 2. Benthic invertebrates. *Hydrobiologia* **248**: 235–247.

267. Reckhow, K.H. (1996). Improved estimation of ecological effects using an empirical Bayes method. *Water Resources Bulletin* **32**: 929–935.

268. Reckhow, K.H.; Chapra, S.C. (1983). *Engineering Approaches for Lake Management. Volume 1: Data Analysis and Empirical Modelling*. Butterworth, Boston.

269. Reckhow, K.H.; Kepford, K.; Hicks, W.W. (1993). Statistical methods for the analysis of lake water quality trends. EPA # 841-R-93-003, United States Environmental Protection Agency, Office of Water, Washington DC.

270. Robinson, A.P.; Froese, R.E. (2004). Model validation using equivalence tests. *Ecological Modelling* **176**: 349–358.

271. Rocke, D.M. (1984). On testing for bioequivalence. *Biometrics* **40**: 225–230.

272. Rodda, B.E.; Davis, R.L. (1980). Determining the probability of an important difference in bioavailability. *Clinical Pharmacology and Therapy* **28**: 252–257.

273. Rose, J.B.; Sobsey, M.D. (1993). Quantitative risk assessment for viral contamination of shellfish and coastal waters. *Journal of Food Protection* **56**(12): 1043–1050.

274. Roussanov, B.; Hawkins, D.M.; Tatini, S.R. (1996). Estimating bacterial density from tube dilution data by a Bayesian method. *Food Microbiology* **13**(5): 341–363.

275. Royall, R.M. (1997). *Statistical Evidence. A Likelihood Paradigm*. Chapman and Hall/CRC, Boca Raton, LA.

276. Rozeboom, W.W. (1960). The fallacy of the null-hypothesis significance test. *Psychological Bulletin* **57**(5): 416–428.

277. Russell, D.G., Parnell, W.R., Wilson, N.C. *et al.* (1999). NZ Food: NZ People. Key results of the 1997 National Nutrition Survey. Ministry of Health, Wellington, New Zealand.

278. Rutherford, J.C. (1994). *River Mixing*. Wiley, New York.

279. Sabatti, C.S.; Service, S.; Freimer, N. (2003). False discovery rates in linkage and association genome screens. *Genetics* **164**(2): 829–833.

280. Schernewski, G.; Jülich, W.-D. (2001). Risk assessment of virus infections in the Oder estuary (southern Baltic) on the basis of spatial transport and virus decay simulations. *International Journal of Hygiene and Environmental Health* **203**: 317–325.

281. Schervish, M.J. (1996). P values: what they are and what they are not. *The American Statistician* **50**(3): 203–206.

282. Schuirmann, D.J. (1981). On hypothesis testing to determine if the mean of a normal distribution is contained in a known interval (abstract). *Biometrics* **37**: 617.

283. Schuirmann, D.J. (1987). A comparison of the two one-sided tests procedure and the power approach for assessing the equivalence of average bioavailability. *Journal of Pharmacokinetics and Biopharmaceutics* **15**: 657–680.

284. Schuirmann, D.J. (1996). Comment on: Bioequivalence trials, intersection-union tests and equivalence confidence sets, by R.L. Berger and J.C. Hsu. *Statistical Science* **11**(4): 312–313.

285. Sellke, T.; Bayarri, M.J.; Berger, R.L (2001). Calibration of p values for testing precise null hypotheses. *The American Statistician* **55**(1): 62–71.

286. Selwyn, M.R.; Hall, N.R. (1984). On Bayesian methods for bioequivalence. *Biometrics* **40**: 1103–1108.

287. Shabman, L.; Smith, E.P. (2003). Implications of applying statistically based procedures for water quality assessment. *Journal of Water Resources Planning and Management* **129**: 330–336.

288. Smith, D.G.; McBride G.B.; Bryers G.G.; Wisse J.; Mink D.F.J. (1996). Trends in New Zealand's national river water quality network. *New Zealand Journal of Marine and Freshwater Research* **30**: 485–500.

289. Smith, E.P.; Ye, K.; Hughes, C.; Shabman L. (2001). Statistical assessment of violations of water quality standards under section 303(d) of the Clean Water Act. *Environmental Science & Technology* **35**: 606–612.

290. Smith, E.P.; Zahran, A.; Mahmoud, M.; Ye, K. (2003). Evaluation of water quality using acceptance sampling by variables. *Environmetrics* **14**: 373–386.

291. Snedecor, G.W.; Cochran, W.G. (1980). *Statistical Methods*. 7th ed. The Iowa State University Press, Ames, IA.

292. Sokal, R.R.; Rohlf, F.J. (1981). *Biometry*. 2nd ed. W.H. Freeman, New York.

293. Spitznagel, E.L.; Helzer, J.E. (1985). A proposed solution to the base rate problem in the kappa statistic. *Archives of General Psychiatry* **42**: 725–728.

294. SPSS (1998). Systat, version 9.01. SPSS Inc., Chicago, IL.

295. Spriet, A.; Beiler, D. (1979). When can 'non significantly different' treatments be considered as 'equivalent'? *British Journal of Clinical Pharmacology* **7**: 623–624.

296. Stansfield, B. (2001). Effects of sampling frequency and laboratory detection limits on the determination of time series water quality trends. *New Zealand Journal of Marine and Freshwater Research* **35**: 1071–1075.

297. Steichen, T.J.; Cox, N.J. (2002). A note on the concordance correlation coefficient. *Stata Journal* **2**(2): 183–189.

298. Steinijans, V.W.: Hauck, W.W.: Diletti, E.: Hauschke, D.; Anderson, S. (1992). Effect of changing the bioequivalence range from (0.80, 1.20) to (0.80, 1.25) on the power and sample size. *International Journal of Clinical Pharmacology, Therapy and Toxicology* **30**: 571–575.

299. Steinijans, V.W.; Hauschke, D. (1993). International harmonization of regulatory bioequivalence requirements. *Clinical Research and Regulatory Affairs* **10**(4): 203–220.

300. Stewart-Oaten, A. (1995). Rules and judgements in statistics: three examples. *Ecology* **76**(6): 2001–2009.

301. Stewart-Oaten, A. (1996). Goals in environmental monitoring. In R.J. Schmitt and C.W. Osenberg (eds.), *Detecting Ecological Impacts in Coastal Habitats*. Academic Press, San Diego, CA, pp. 16–27.

302. Stewart-Oaten, A.; Murdoch, W.W.; Parker, K.R. (1986). Environmental impact assessment: "pseudoreplication" in time? *Ecology* **67**(4): 929–940.

303. Stuart, A.; Ord, J.K. (1994). *Kendall's Advanced Theory of Statistics*, Vol. 1. Distribution Theory. 6th ed. Arnold, London.

304. Stuart, A.; Ord, J.K.; Arnold, S. (1999). *Kendall's Advanced Theory of Statistics*, Vol. 2A. Classical Inference and the Linear Model. 6th ed. Arnold, London.

305. Stunkard, C.L. (1994). Tests of proportional means for mesocosm studies. In R.L. Graney, J.H Kennedy and J.H Rogers, Jr. (eds.), *Aquatic Mesocosm Studies in Ecological Risk Assessment*, Lewis, Boca Raton, FL, pp. 71–83.

306. Taper, M.L.; Lele, S.R. (eds.) (2004). *The Nature of Scientific Evidence: Statistical, Philosophical, and Empirical Considerations*. University of Chicago Press, Chicago.

307. Teunis P.F.M.; Havelaar A.H. (2000). The beta Poisson dose-response model is not a single-hit model. *Risk Analysis* **20**(4): 513–520.

308. Teunis, P.F.M.; Havelaar, A.H.; Medema, G.H. (1995). A literature survey on the assessment of microbiological risk for drinking water. Report no. 734301006, RIVM, Bilthoven, The Netherlands

309. Teunis, P.F.M.; van der Heijden, O.G.; van der Giessen, J.W.B.; Havelaar, A.H. (1996). The dose-response relation in human volunteers for gastro-intestinal pathogens. Report no. 284550002, RIVM, Bilthoven, The Netherlands, May.

310. Teunis, P.F.M.; Chappell, C.L.; Ockhuysen, P.C. (2002a). Cryptosporidium dose response studies: variation between isolates. *Risk Analysis* **22**(1): 175–183.

311. Teunis, P.F.M.; Chappell, C.L.; Ockhuysen, P.C. (2002b). Cryptosporidium dose response studies: variation between hosts. *Risk Analysis* **22**(3): 475–485.

312. Thomas, H.A. (1942). Bacterial densities from fermentation tube tests. *Journal of the American Water Works Association* **34**: 572–576.

313. Thomas, L.; Krebs, C.J. (1997). A review of statistical power analysis software. *Bulletin of the Ecological Society of America* **78**(2): 126–139.

314. Thompson, W.D. (1987). Statistical criteria in the interpretation of epidemiologic data (Different Views). *American Journal of Public Health* **77**: 191–194.

315. Thompson, W.D.; Walter, S.D. (1988). A reappraisal of the kappa coefficient. *Journal of Clinical Epidemiology* **41**(10): 949–958 (response to discussion on pp. 969–970).

316. Till, D.G.; McBride, G.B. (2004). Potential public health risk of *Campylobacter* and other zoonotic waterborne infections in New Zealand. Chapter 12 in J.A. Cotruvo; A. Dufour; G. Rees; J. Bartram; R. Carr; D.O. Cliver; G.F. Craun; R. Fayer and V.P.J. Gannon (eds.), *Waterborne Zoonoses: Identification, Causes and Control*, IWA Publishing, London, for the World Health Organization, pp. 191–207.

317. Tillett, H.E. (1995). Comment on The 'most probable number' estimate and its confidence limits. *Water Research* **29**(4): 1213–1214.

318. Tillett, H.E.; Coleman, R. (1985). Estimated numbers of bacteria from non-homogeneous bodies of water. *Journal of Applied Bacteriology* **59**: 381–388.

319. Townend, J. (2002). *Practical Statistics for Environmental and Biological Scientists*. Wiley, New York.

320. Tukey, J.W. (1960). Conclusions vs decisions. *Technometrics* **2**(4): 423–433.

321. Tukey, J.W. (1962). The future of data analysis. *Annals of Mathematical Statistics* **33**: 1–67.

322. USDHHS (1993). National Shellfish Sanitation Programme Manual of Operations, Parts I and II. U.S. Department of Health and Human Services, Washington D.C.

323. USDoH (1964). Smoking and Health. Public Health Service Publication **1103**. Report of the Advisory Committee to the Surgeon General, U.S. Department of Health, Education and Welfare, Washington D.C.

324. USEPA (1986). Quality Criteria for Water 1986. Report **EPA 440/5-86-001**. United States Environmental Protection Agency, Office of Water, Regulations and Standards, Washington D.C.

325. USEPA (1999). 1999 Update of Ambient Water Quality Criteria for Ammonia. Report **EPA 822-R-99-014**. United States Environmental Protection Agency, Office of Water, Washington D.C.

326. USEPA (2000). Stressor Identification Guidance Document. Report **EPA/822/B-00/025**. United States Environmental Protection Agency, Office of Research and Development, Washington D.C.

327. van Belle, G. (2002). *Statistical Rules of Thumb*. Wiley, New York.

328. van Dijk, P.A.H.; Ellis, J.C. (1995). User Guide to Aardvark for Windows. Release 2.01, Water Research Centre, Medmenham, UK.

329. Vardeman, S.B. (1992). What about the other intervals. *The American Statistician* **46**(3): 193–197.

330. Velleman, P.F. (1997). DataDesk, version 6, Statistics Guide. Data Description Inc., Ithaca, New York.

331. von Mises, R. (1964). *Mathematical Theory of Probability and Statistics*. Academic Press, New York.

332. Wald, A.; Wolfowitz, J. (1946). Tolerance limits for the normal distribution. *Annals of Mathematical Statistics* **17**: 208–215.

333. Walker, A.M. (1986). Reporting the results of epidemiologic studies (Different Views). *American Journal of Public Health* **76**: 556–558.

334. Ward, R.C. (1996). Water Quality Monitoring: Where's the Beef? *Water Resources Bulletin* **32**(4): 673–680.

335. Ward, R.C.; Loftis, J.C. (1983). Incorporating the stochastic nature of water quality into management. *Journal of the Water Pollution Control Federation* **55**(4): 408–414.

336. Ward, R.C.; Loftis, J.C.; McBride, G.B. (1986). The "data-rich but information poor" syndrome in water quality monitoring. *Environmental Management* **10**(3): 291–297.

337. Ward, R.C.; Loftis, J.C., DeLong, H.P.; Bell, H.F. (1988). Groundwater quality: a data analysis protocol. *Journal of the Water Pollution Control Federation* **60**(11): 1938–1945.

338. Ward, R.C.; Loftis, J.C.; McBride, G. B. (1990). *Design of Water Quality Monitoring Systems*. Van Nostrand Reinhold, New York.

339. Warn, A.E.; Brew, J.S. (1980). Mass balance. *Water Research* **14**: 1427–1433.

340. Wellek, S. (2003). *Testing Statistical Hypotheses of Equivalence*. Chapman and Hall/CRC, Boca Raton, FL.

341. Westlake, W.J. (1981). Response to T.B.L. Kirkwood on bioequivalence testing–a need to rethink. *Biometrics* **37**: 589–594.

342. WHO (2003). Guidelines for safe recreational water environments. Volume 1: Coastal and fresh waters. World Health Organization, Geneva.

343. Wilson, A.L. (1961). The precision and limit of detection of analytical methods. *Analyst* **86**: 22–25.

344. Wolfram, S. (1999). *The Mathematica® Book*. Cambridge University Press, Cambridge, UK.

345. Woodward, R.L. (1957). How probable is the Most Probable Number? *Journal of the American Water Works Association* **49**: 1060–1068.

346. Wymer, L.J.; Dufour, A.P. (2002). A model for estimating the incidence of swimming-related gastrointestinal illness as a function of water quality indicators. *Environmetrics* **13**: 669–678.

347. Young, J.C.; Minder, Ch.E. (1985). Algorithm AS 76. An integral useful in calculating non-central t and bivariate normal probabilities. In P. Griffiths and I.D. Hill (eds.), *Applied Statistics Algorithms*. Ellis Horwood, Chichester, UK (published for the Royal Statistical Society), pp. 145–148.

348. Zar, J.H. (1984). *Biostatistical Analysis*. 2nd ed. Prentice-Hall, Englewood Cliffs, NJ.

349. Zar, J.H. (1996). *Biostatistical Analysis*. 3rd ed. Prentice-Hall, Upper Saddle River, NJ.

Author Index

Abramowitz, M., 140, 144, 156, 265, 303
Aldenberg, T., 154
Altman, D.G., 58, 134, 236
Anderson, D.R., 104
Anderson, S., 82, 130–131
Baker, M., 197
Barnard, G.A., 12
Barnett, V., 12, 173, 175, 260
Barton, D.E., 222
Beiler, D., 130
Beliaeff, B., 218–219, 236
Benjamini, Y., 97
Berger, J.O., 88, 87, 104, 130, 257
Berger, R.L., 14, xxiii, 71, 12, 102, 104, 124, 131, 148, 159, 163, 190, 255
Berkson, J., 103
Berry, D.A., 88, 101
Berthouex, P.M., xxi
Best, D.J., 215
Bishop, Y.M.M., 232, 240
Black, R.E., 202, 205
Blair, R.C., 133
Bland, J.M., 236
Bloch, D.A., 232
Bolstad, 04, 244
Bolstad, W.M., xxi, 101
Bondy, W.H., 130
Bower, B., 103
Bowker, A.H., 38–39, 154
Box, G.E.P., 71, 169

Brew, J.S., 100
Bross, I.D., 81, 122
Brown, L.C., xxi
Buhl-Mortensen, L., 128
Burkhardt, W., 200
Burnham, K.P., 104
Cabelli, V.J., 173
Calci, K.R., 200
Calderon, R.L., 105
Carlin, B.P., 101, 218
Carver, R.P., 103
Cascio, W.F., 104
Casella, G., 14, xxiii, 71, 12, 102, 148, 159, 190, 255
Chapra, S.C., 47, 59
Chow, S.-C., 82, 128, 131, 164
Chow, S.L., 104
Cochran, W.G., xxi, 38, 114, 215, 218
Cohen, J., 81, 95, 232
Cole, P.C., 133
Cole, R.G., 131
Coleman, R., 222
Collings, B.J., 95
Conover, W.J., xxi, 34, 41, 49, 70, 40, 60–61, 99, 143
Cooke, K.R., 135
Cooper, B.E., 149, 156
Cornfield, J., 91
Cotter, A.J.R., 6
Cox, D.R., 85

295

Cox, N.J., 236
Crabtree, K.D., 196, 202
Craun, G.F., 197
David, F.N., 222
Davis, P.J., 81
Davis, R.L., 134
Dayton, P.K., 128
de Man, J.C., 215, 217–219
DeGroot, M.H., 87, 102, 190
Delampady, M., 130, 257
Dennis, B., 12
Diletti, E., 131
Dixon, P.M., 81, 131
Dufour, A.P., 48
Dupont, W.D., 103
Edgington, E.S., 100
Edwards, A.W.F., 12, 87, 102–103, 107
Edwards, W., 102, 190
Eisenhart, C., 215, 218
Elliott, J.C., 147
Ellis, J.C., xxiv, 44, 46, 3, 5, 180, 183–185, 188, 192, 236, 238, 259–260
Ellison, A.M., 12
El-Sharaawi, A.H., 206
Erickson, W.P., 131
Fairweather, P.G., 128
Ferguson, T.S., xxiii, 82, 130
Fisher, R.A., 12, 102–103
Fleiss, J.L., 80, 104, 116, 232, 240
Forster, L.I., 211
Francis, R.I.C.C., 115
Fraser, G., 135
Freedman, P.L., 172
Freund, J.E., xxiii, 38, 108, 144, 157
Frick, R.W., 104
Froese, R.E., 105, 131
Furumoto, W.A., 210
Gale, P., 36, 201, 203, 211
García, L.V., 97
Gardner, M.J., 58, 134
Garrett, K.A., 81, 131
Gelman, A., xxiii
Gerard, P.D., 95
Gerba, C.P., 197
Germano, J.D., 103
Gerrodette, T., 128
Gibbons, J.D., 104
Gibbons, R.D., xxi, 87
Gilbert, R.O., xxi, 49, 3, 31, 38–40, 61, 83, 99, 141, 180, 225–226
Gill, J., 54
Goldstein, S.T., 197
Good, I.J., 104
Goodman, S.N., 103–104
Gray, J.S., 128
Graybill, F.A., 111, 159

Green, R.H., xxi, 65
Greenwood, M.Jr., 215, 218
Gupta, G.D., 131
Guttman, I., 57, 72
Haas, C.N., 196, 201–203, 205, 209–213
Hacking, I., 12
Hagen, R.L., 104
Hall, N.R., 130
Hamilton, M.A., 95, 218
Harlow, L.L., xxiii
Harris, R.J., 81, 104
Hartley, H.O., 164
Hartnett, E., 212
Hauck, W.W., 82, 130–131
Hauschke, D., 131
Havelaar, A.H., 147, 202–204, 211
Heisey, D.M., 95
Helms, M., 198
Helsel, D.R., xxi, 3, 99, 228, 236–239
Helzer, J.E., 232
Henkel, R.E., xxiii
Hersh, R., 81
Hickey, C.W., 182
Hilborn, R., 104, 147
Hill, A.B., 135
Hirsch, R.M., xxi, 3, 99, 228, 238–239
Hochberg, Y., 97
Hodges, A., 101
Hodges, J.S., 235
Hoenig, J.M., 95
Hoffman, P., 54
Hollingdale, S., 169
Hothorn, L., 131
Howson, C., 59
Hoxie, N.J., 197
Hrudey, S.E., 93
Hsu, J.C., 124, 131, 163
Hurlbert, S.H., 65
Iman, R.L., xxi, 34, 40, 60–61, 99
Inman, H.F., 103
Jülich, W.-D., 198
Jaworska, J.S., 154
Jeffreys, H., 85, 103
Johl, M., 198
Johnson, D.H., 99, 103
Johnson, J.B., 104
Johnson, N.L., 148, 150, 152–153, 157
Jones, L.V., 81
Jowett, I.G., 97
Kateman, G., 46
Kirchmer, C.J., 267
Koch, G.G., 234
Kotz, S., 148, 150, 152–153, 157
Kraemer, H.C., 232
Krantz, D.H., 81
Krebs, C.J., xxi, 95

Landis, J.R., 234
Larned, S.T., 183
Lee, P.M., xxiii, 85, 101, 134, 186–190
Legendre, L., 42
Legendre, P., 46, 42
Lehmann, E.L., xxiii, 78, 82
Lele, S.R., 13
Lettenmaier, D.P., 65
Lieberman, G.J., 38–39, 154
Lin, L.I.-K., 236, 241
Lindley, D.V., 13, 58
Liu, J.P., 82, 128, 131, 164
Loftis, J.C., 46, 6, 65, 225
Loftus, G.R., 103
Louis, T.A., 101, 218
Loyer, M.W., 218
Müskens, P.J.W.M., 46
MacKenzie, W.R., 197
Magnusson, W.E., 66
Mandallaz, D., 130
Mangel, M., 104, 147
Manly, B.F.J., xxi, 100, 115, 131
Mapstone, B.D., 104
Martin, Andrés, A., 131
Mary, J.-Y., 218–219
Mau, J., 130
McBride, G.B., xxii, 25, 31, 103, 122, 131, 163, 183–185, 188, 193, 197, 204, 206, 218, 221–222, 247–248
McCrady, M.H., 215, 217
McCullagh, P., 48
McDonald, L.L., 131
McDowell, R.M., 198
McElvaine, M.D., 198
McGill, R., 24
Medema, G.J., 202, 213
Menand, L., 12, 27
Messner, M.J, 205
Metzler, C.M., 130
Mickey, R., 210
Millard, S.P., xxi, 70, 65, 81
Miller, I., 108, 144
Miller, J.C., 236–237
Miller, J.N., 236–237
Minder, Ch.E., 156
Mood, A.M., 111, 159
Morris, R.D., 197
Morrison, D.E., xxiii
Morrison, M.L., 228
Muliak, S.A., 104
Mumby, P.J., 95
Murdoch, N., 52, 100
Nachamkin, I., 198
Neerchal, N.K., xxi, 70
Nelder, J.A., 48, 12, 104
Nester, M.R., 103

Newell, A.D., 228
Neyman, J., 103
Noether, G.E., 95
Nunnally, J.C., 103
Oakes, M., 257
Odeh, R.E., 72, 154, 156
O'Hagan, A., 173, 175, 260
Okhuysen, P.C., 205
Omland, K.S., 104
Ord, J.K., 148
Owen, D.B., 72, 125, 152, 154–156, 158, 163–165, 257
Parkhurst, D.F., 8, 81
Patel, H.I., 131
Pearson, E.S., 103, 164
Pedini, M., 172
Phillips, K.F., 131, 163
Ponce, S.L., 65
Poole, C., 72, 12, 85
Pratt, J.W., 87, 104, 190
Press, S.J., 13, 12, 101, 186
Press, W.H., 140, 142, 144, 189
Quinn, J.M., 112
Reckhow, K.H., 46–47, 59, 218
Richardson, J., 97
Rizak, S., 93
Robinson, A.P., 105, 131
Rocke, D.M., 131
Rodda, B.E., 134
Rohlf, F.J., xxi
Rose, J.B., 199
Roussanov, B., 218
Royall, R.M., 79–80, 103
Rozeboom, W.W., 103
Russell, D.G., 199–200
Rutherford, J.C., 100
Sabatti, C.S., 97
Schernewski, G., 198
Schervish, M.J., 87, 104
Schuirmann, D.J., 82, 122, 128, 130–131, 163–165
Sellke, T., 87, 102, 104, 130
Selwyn, M.R., 130
Shabman, L., 176
Slob, W., 154
Smith, D.G., 88, 225, 227
Smith, E.P., 172, 176, 180, 183
Snedecor, G.W., xxi, 38, 114
Sobsey, M.D., 199
Sokal, R.R., xxi
Spitznagel, E.L., 232
Spriet, A., 130
Stansfield, B., 228
Stegun, I.A., 140, 144, 156, 265, 303
Steichen, T.J., 236
Steinijans, V.W., 131

Stern, D.A., 8
Stewart-Oaten, A., xxii, 79, 65
Stuart, A., 58, 12, 111, 118, 122, 130, 146, 148
Stunkard, C.L., 131
Taper, M.L., 13
Teunis, P.F.M., 147, 201–205, 209–211
Thomas, H.A., 216–217
Thomas, L., 95
Thompson, W.D., 72, 232
Tiao, G., 71
Till, D.G., 197
Tillett, H.E., 215, 222
Townend, J., xxi
Tukey, J.W., 1, 79, 81
Urbach, P., 59
van Belle, G., 13, 80, 133
van Dijk, P.A.H., 44
Vardeman, S.B., 72
Velleman, P.F., 25, 58

von Mises, R., 12
Wald, A., 154
Walker, A.M., 72
Walter, S.D., 232
Ward, R.C., 8, xxi, 5–6, 9
Warn, A.E., 100
Wellek, S., xxiii, 82, 98, 130–131, 159, 164
Westlake, W.J., 130
Wilson, A.L., 267
Wilson, P.W., 215, 218
Wolfowitz, J., 154
Wolfram, S., 156
Woodward, R.L., 217–219
Wymer, L.J., 48
Young, J.C., 156
Yule, G.U., 215, 218
Zar, J.H., xxi, 41, 58, 72, 22, 39, 61, 64, 81, 95–96, 98, 114, 232, 254
Zedeck, S., 104

Topic Index

A

Abraham De Moivre, 222
Accuracy, 7
After-trial evaluation, 13, 95
Algorithms, 139
Aliasing, 51
Analysis of variance, 27, 38
Arithmetic mean, 1
Autocorrelation, 43
Autocorrelation function, 46
Axiomatic approach, 13

B

Bar graph, 17, 23
Bayes' rule, 14
Bayesian methods, 2, 9, 12, 101
   conjugate distributions, 102
   estuary example, 15, 101
   highest posterior density (HPD) region, 71
   MPN, 218
   likelihood, 101
   posterior distribution, 101
   prior distribution, 101
Before-trial betting, 12, 94
Benefit of doubt, 5
Beta distribution, 26, 32
   mathematical details, 142

Bias, 51
Binomial distribution, 26, 33, 89, 92
   mathematical details, 143
   underdispersed, 143
Boxplots, 24
Bradford Hill criteria
   association versus causation, 134
Burden of proof, 5
   face value, 5
   for standards, 171
   policy decision, 176
   proof of hazard, 5, 120–121
   proof of safety, 5, 120, 122

C

Causality, 3
Censored data
   less than
      analyzing less-than data, 238
      blank-corrected, 237
      many options, 236
      Type A and B censoring, 236
Central limit theorem, 60
Central tendency, 2
Chance experiment, 11
Chi-square distribution, 27, 32, 37
   mathematical details, 149
Circular scale, 16
Classical probability, 12

Coefficient of skewness, 22
Coefficient of variation, 22
Concordance, 42
  Cohen's kappa, 232
    Baysian interpretation, 235
    chance correction, 232
    equivalence criteria, 234
  continuous variables
    Lin's concordance correlation coefficient, 236
Conditional probabilities, 14
Confidence intervals, 13, 50, 59
  MPN, 218
Confidence limits
  normal percentiles
    one-sided, 154
    two-sided, 152
Confidence of Compliance, 33, 187
Confidence of Failure, 187
Confidence versus credible intervals
  MPN, 218
Confusion matrix, 77
Continuous data, 16
Continuous distributions, 27
Controversy
  Bayesian methods, 12
Correlation, 41
Correlogram, 44
Coverage, 70
Credible intervals, 58, 71, 62
  MPN, 218
Critical value, 28
Cryptosporidium, 36
Cumulative binomial distribution, 34
Cumulative distribution function, 19
Cumulative frequency, 17
Cumulative probability, 2
Cumulative unit normal distribution, 28

D

Data Analysis Protocol, 9
Decision rule, 76, 83
Decisions, 78
Degrees of freedom, 22, 37
Designed experiments, 3
Detection formulae
  equivalence hypothesis
    known variance, 159
    unknown variance, 162
  inequivalence hypothesis
    known variance, 161
    unknown variance, 162
  one-sided test
    known variance, 158
    unknown variance, 157
  point-null test
    known variance, 159

    unknown variance, 158
Detection limit
  multipler
    Eurachem, 236
    two options, 236
Detection probability, 107
  effect size, 108
  OC curves, 108
  power curves, 108
Detection regions, 107
  equivalence hypothesis, 163
  Power Approach, 163
Detection
  what has been detected?, 132
Dichotomous data, 17
Discrete data, 16
Discrete distributions, 33
Drinking water standards, 5, 52, 171

E

Effluent standards, 171
Error bars, 65
Error risk, 11
Errors-in-variables, 47
Exceedance probability, 2
Experimental design, 95
Exponential distribution, 32

F

F distribution
  mathematical details, 149
Face value stance, 5
Fail safe, 5
F-distribution, 27, 38
Fiducial probability, 12
Fisher (Sir Ronald Alymer), 38
Frequency distribution, 18
Frequency interpretation, 11
Frequency table, 17
Frequentist interpretation, 2, 58
Frequentist probability, 12

G

Gamma distribution, 26, 31
  and chi-square distribution, 142
  and exponential distribution, 142
  mathematical details, 141
Generalized Linear Models, 48
Geometric mean, 1, 21, 60
  versus median, 182

H

Histograms, 23
Hypergeometric distribution, 26, 36
  mathematical details, 147
Hypotheses

equivalence
  decision rules, 123
  detection probability, 123
Hypothesis test
  bioequivalence, 120
  equivalence, 81, 120
    Bayesian, 134
    burden of proof, 120, 128
    compared to point-null test, 125
    compared to Power Approach, 128
    detection probability, 120
    detection regions, 126
  inequivalence, 120
    two one-sided (TOST), 122
  one-sided
    burden of proof, 110
    detection probability, 108
    detection region, 113
    permissive, 80
    precautionary, 80
  point-null
    approximating an interval test, 130
    burden of proof, 81
    can be ultra-precautionary, 81, 120
    decision rules, 118
    detection probability, 115, 118
    detection regions, 119
    swinging burden of proof, 120, 125
    three-valued logic, 81
    versus confidence intervals, 132
Hypothesis
  alternative, 78
  interval, 81
  nil, 81
  null, 77
  one-sided, 79
    inferiority, 111
    superiority, 111
  point-null, 77, 80
    power and sample size, 116
  rejecting and accepting (maybe), 86
  research, 91

I

Impacts
  analysis of covariance, 231
  nominal data, 231
Independent events, 14
Inductive behavior, 79
Information needs
  of management, 4, 175
Interquartile range, 22
Interval scale, 16
Inverse probability, 12

J–K

Jeffreys' prior distribution, 33
Kendall's tau, 41

L

Language conventions, 1
Level of confidence, 4
Likelihood, 13, 102
Logistic regression, 48
Loglogistic distribution, 26
Lognormal distribution, 26, 30
  and geometric mean, 80
  mathematical details, 141

M

Mann–Kendall test, 42, 225
Mean, 1
Measurement errors, 7
Median, 60
  versus geometric mean, 182
Monotonic relationship, 49
Monte Carlo methods, 52
Most probable number (MPN), 16, 26
  binning method for distribution fitting, 220
  confidence intervals, 218
  credible intervals, 218
  decimal dilution series, 215
  exact calculation, 216, 222
    occupancy theory, 222
  occurrence probability, 217
  standard tables
    rounding, 216
  Thomas equation approximation, 216
Multinomial distribution, 26, 36
  and MPNs, 148, 224
  mathematical details, 148
Multiple comparisons, 95
  ANOVA, 95
  Bonferroni adjustment, 97
  false discovery rate, 97

N

Narrative standards, 171
Negative binomial distribution, 26, 36, 90
  and geometric distribution, 146
  mathematical details, 145
  overdispersed, 36, 146
Nominal scale, 17
Noncentral t-distribution
  algorithm, 155
Nondetects, 8
Nonexceedance probability, 19
Nonparametric analysis, 43
Nonparametric methods, 31
  Confidence of Compliance, 187
  percentile estimation, 180
Nonparametric regression, 49

Nonparametric tests, 98
  nondetects, 98
Normal deviate, 27
Normal distribution, 26–27
  mathematical details, 140

O

Objective probability, 11
Observational studies, 3
OC curve, 107–108
Occurrence probability, 21
One-sided intervals, 62, 69
One-tailed, 39
Ordinal scale, 16
Ordinary Least Squares, 47
Overdispersion, 26

P

Paired samples, 64
Pearson's correlation coefficient, 41
Percentile compliance rules
  Bayesian approach, 185
    calculating the posterior distribution, 189
    choosing a prior, 187
    Confidence of Compliance, 187
    Confidence of Failure, 187,
    conjugate priors, 187
    incomplete beta function ratio, 189
  Classical approach, 184
    proof of hazard, 184
    proof of safety, 185
  comparing classical and Bayesian approaches, 190
  permissible exceedances
    permissive approach, 190
    precautionary approach, 190
Percentile standards, 52, 5
  authorizing the status quo, 180
  burden of proof, 179
  drinking-water, 179
  effluents, 179
  receiving water, 179
  two forms, 179
Percentiles, 28, 51
  behavior
    compared to maximum, 182
  Blom method, 180
  calculation options, 180
  calculation
    minimum sample size, 181
  Excel method, 180
  Hazen method, 180
  Tukey method, 180
  Weibull method, 180
Permissive approach, 5
Poisson distribution, 26, 36
  mathematical details, 144
Poisson process, 145
Poisson series, 145
Polychotomous data, 17, 26, 36
Population, 6
Population parameters, 1
Posterior probability, 12
Power analysis, 94
  nonparametric model, 95
  of a performed test
    issues, 95
  point-null hypothesis, 95
  sample size, 95
Power curve, 107–108
Power
  retrospective, 95, 251
Precautionary approach, 5
Precision, 4, 7
Precision of estimate, 63
Prediction intervals, 66
Prior distribution, 26, 102
Prior probability, 12
Probability density function, 20
Probability mass function, 18
p-values, 2, 84
  behavior under different hypotheses, 131
  comparing, 87, 231
  culture, 104
  information criteria, 104
  intentions, 88
  peeking, 91
  point-null tests, 85
  posterior probability, 102
  rationale, 84
  what they are not, 87

Q

Quantitative risk assessment (QRA), 26, 145
  accounting for exposures, 196
  beta distribution, 199
  beta-Poisson model, 32
    derivation, 209
  bioaccumulation factors, 200
  conditional dose-response, 210
    beta-binomial curve, 211
  conditional versus average response, 212
  dichotomous virus data, 199
  dose-response, 201
    beta-Poisson, 201
    credible interval, 202
    exponential, 201
    Kummer hypergeometric curve, 202, 210
  exponential model
    derivation, 209
  exposure assessment, 198
  fitting MPN data, 199

four steps, 196
hazard identification, 196
hockey stick distribution, 199, 207
identifying the exposed population, 203
individual infection risk, 204
non-Poisson distribution, 211
risk profiling, 203
shellfish meal size
   loglogistic distribution, 199, 208
   truncate the distribution, 199
the infective dose
   meaningless, 201
uncertainties, 205
viruses
   added-zeroes distribution, 206
wastewater example, 196

R

Random samples, 6
Random sampling, 51, 83
Random variable, 41, 6
Randomization tests, 100
   bootstrap, 100
   jackknife, 100
   Monte Carlo, 100
Randomness, 6
Rank correlation, 41
Ratio scale, 16
Receiving water standards, 171
Regression, 47
Relative frequencies, 17
Representative, 83
Risk of error, 11

S

Sample size, 2
Sampling bias, 7
Sampling distributions, 27, 36, 83
Seasonal Kendall trend test, 42, 225
   anomalies, 227
Seasonality, 24, 6, 43
Sen slope estimator, 49, 225
Serial correlation, 23, 43
Significance level, 76
Skewness, 19
Spearman's rho, 41
Standard deviation, 22
Standard error, 2, 27, 50
   of estimate, 67
   of the mean, 60
Standardized variance, 37
Standards
   consistency

formulation versus implementation, 174
   enforceability, 174
   requirements, 175
   feasibility, 174
   key questions, 173
   measurability, 174
   statistically verifiable ideal standard, 175
Statistical inference, 7
Statistical model, 42
Statistical tables, 38
Stochastic process, 6
Stormwater standards, 171
Student's t-distribution, 27, 37
   mathematical details, 148
Subjective probability, 12
Summary statistics, 20
Systematic random sampling, 6, 83

T

Test statistic, 26, 76
Tolerance intervals, 70
   coverage, 70
Tolerance limits
   normal percentiles
      one-sided, 154
      two-sided, 154
Transformations, 30
Trend, 43
Trend analysis
   multi-site
      Bayesian approach, 228
Turing (Alan)
   Enigma code, 101
Two-sided intervals, 62
Two-tailed, 39
Type I error, 78, 236
Type I error risk, 82, 110
Type II error, 78, 236
Type II error risk, 70, 110

U

Unbiased estimate, 31
Unbiasedness, 7, 12
Uncertainty, 11
Underdispersion, 26
Uninformative prior distribution, 58
Unit process standards, 171

W

Wastewater discharges, 52
Wastewater effluent standards, 5
Water quality standards, 171

Appendix A
Statistical Tables

A.1 NORMAL DISTRIBUTION

A.1.1 Cumulative unit normal distribution

Z	..0	..1	..2	..3	..4	..5	..6	..7	..8	..9
0.0	0.5000	0.5040	0.5080	0.5120	0.5160	0.5199	0.5239	0.5279	0.5319	0.5359
0.1	0.5398	0.5438	0.5478	0.5517	0.5557	0.5596	0.5636	0.5675	0.5714	0.5753
0.2	0.5793	0.5832	0.5871	0.5910	0.5948	0.5987	0.6026	0.6064	0.6103	0.6141
0.3	0.6179	0.6217	0.6255	0.6293	0.6331	0.6368	0.6406	0.6443	0.6480	0.6517
0.4	0.6554	0.6591	0.6628	0.6664	0.6700	0.6736	0.6772	0.6808	0.6844	0.6879
0.5	0.6915	0.6950	0.6985	0.7019	0.7054	0.7088	0.7123	0.7157	0.7190	0.7224
0.6	0.7257	0.7291	0.7324	0.7357	0.7389	0.7422	0.7454	0.7486	0.7517	0.7549
0.7	0.7580	0.7611	0.7642	0.7673	0.7704	0.7734	0.7764	0.7794	0.7823	0.7852
0.8	0.7881	0.7910	0.7939	0.7967	0.7995	0.8023	0.8051	0.8078	0.8106	0.8133
0.9	0.8159	0.8186	0.8212	0.8238	0.8264	0.8289	0.8315	0.8340	0.8365	0.8389
1.0	0.8413	0.8438	0.8461	0.8485	0.8508	0.8531	0.8554	0.8577	0.8599	0.8621
1.1	0.8643	0.8665	0.8686	0.8708	0.8729	0.8749	0.8770	0.8790	0.8810	0.8830
1.2	0.8849	0.8869	0.8888	0.8907	0.8925	0.8944	0.8962	0.8980	0.8997	0.9015
1.3	0.9032	0.9049	0.9066	0.9082	0.9099	0.9115	0.9131	0.9147	0.9162	0.9177
1.4	0.9192	0.9207	0.9222	0.9236	0.9251	0.9265	0.9279	0.9292	0.9306	0.9319
1.5	0.9332	0.9345	0.9357	0.9370	0.9382	0.9394	0.9406	0.9418	0.9429	0.9441
1.6	0.9452	0.9463	0.9474	0.9484	0.9495	0.9505	0.9515	0.9525	0.9535	0.9545
1.7	0.9554	0.9564	0.9573	0.9582	0.9591	0.9599	0.9608	0.9616	0.9625	0.9633
1.8	0.9641	0.9649	0.9656	0.9664	0.9671	0.9678	0.9686	0.9693	0.9699	0.9706
1.9	0.9713	0.9719	0.9726	0.9732	0.9738	0.9744	0.9750	0.9756	0.9761	0.9767
2.0	0.9772	0.9778	0.9783	0.9788	0.9793	0.9798	0.9803	0.9808	0.9812	0.9817
2.1	0.9821	0.9826	0.9830	0.9834	0.9838	0.9842	0.9846	0.9850	0.9854	0.9857
2.2	0.9861	0.9864	0.9868	0.9871	0.9875	0.9878	0.9881	0.9884	0.9887	0.9890
2.3	0.9893	0.9896	0.9898	0.9901	0.9904	0.9906	0.9909	0.9911	0.9913	0.9916
2.4	0.9918	0.9920	0.9922	0.9925	0.9927	0.9929	0.9931	0.9932	0.9934	0.9936
2.5	0.9938	0.9940	0.9941	0.9943	0.9945	0.9946	0.9948	0.9949	0.9951	0.9952
2.6	0.9953	0.9955	0.9956	0.9957	0.9959	0.9960	0.9961	0.9962	0.9963	0.9964
2.7	0.9965	0.9966	0.9967	0.9968	0.9969	0.9970	0.9971	0.9972	0.9973	0.9974
2.8	0.9974	0.9975	0.9976	0.9977	0.9977	0.9978	0.9979	0.9979	0.9980	0.9981
2.9	0.9981	0.9982	0.9982	0.9983	0.9984	0.9984	0.9985	0.9985	0.9986	0.9986
3.0	0.9987	0.9987	0.9987	0.9988	0.9988	0.9989	0.9989	0.9989	0.9990	0.9990
3.1	0.9990	0.9991	0.9991	0.9991	0.9992	0.9992	0.9992	0.9992	0.9993	0.9993
3.2	0.9993	0.9993	0.9994	0.9994	0.9994	0.9994	0.9994	0.9995	0.9995	0.9995
3.3	0.9995	0.9995	0.9995	0.9996	0.9996	0.9996	0.9996	0.9996	0.9996	0.9997
3.4	0.9997	0.9997	0.9997	0.9997	0.9997	0.9997	0.9997	0.9997	0.9997	0.9998
3.5	0.9998	0.9998	0.9998	0.9998	0.9998	0.9998	0.9998	0.9998	0.9998	0.9998
3.6	0.9998	0.9998	0.9999	0.9999	0.9999	0.9999	0.9999	0.9999	0.9999	0.9999
3.7	0.9999	0.9999	0.9999	0.9999	0.9999	0.9999	0.9999	0.9999	0.9999	0.9999
3.8	0.9999	0.9999	0.9999	0.9999	0.9999	0.9999	0.9999	0.9999	0.9999	0.9999
3.9	1.0000	1.0000	1.0000	1.0000	1.0000	1.0000	1.0000	1.0000	1.0000	1.0000
4.0	1.0000	1.0000	1.0000	1.0000	1.0000	1.0000	1.0000	1.0000	1.0000	1.0000

The row labels give the first significant digit of the normal deviate (Z); the column labels give the second digit. For example, the area to the left of $Z = 2.57$ is $G(Z) = 0.9949$. The "1.0000" entries at the Table's bottom are values between 0.99995 and unity. For values of Z less than zero, subtract areas corresponding to $|Z|$ from one. For example, if $Z = -0.23$ the left-tail area is $1 - 0.5910 = 0.4090$. The area between $-Z$ and $+Z$ is given by $2G(Z) - 1$ [see Equation (2.18)]. For example, if $|Z| = 0.23$, the area between $-Z$ and $+Z$ is 0.1820. Areas for values of Z for more than two significant digits must be obtained by interpolation (see page 40), unless direct algorithms are used. This Table was computed using Equation (6.2).

A.1.2 Percentiles of the unit normal distribution

G	..0	..1	..2	..3	..4	..5	..6	..7	..8	..9
0.50	0.0000	0.0025	0.0050	0.0075	0.0100	0.0125	0.0150	0.0175	0.0201	0.0226
0.51	0.0251	0.0276	0.0301	0.0326	0.0351	0.0376	0.0401	0.0426	0.0451	0.0476
0.52	0.0502	0.0527	0.0552	0.0577	0.0602	0.0627	0.0652	0.0677	0.0702	0.0728
0.53	0.0753	0.0778	0.0803	0.0828	0.0853	0.0878	0.0904	0.0929	0.0954	0.0979
0.54	0.1004	0.1030	0.1055	0.1080	0.1105	0.1130	0.1156	0.1181	0.1206	0.1231
0.55	0.1257	0.1282	0.1307	0.1332	0.1358	0.1383	0.1408	0.1434	0.1459	0.1484
0.56	0.1510	0.1535	0.1560	0.1586	0.1611	0.1637	0.1662	0.1687	0.1713	0.1738
0.57	0.1764	0.1789	0.1815	0.1840	0.1866	0.1891	0.1917	0.1942	0.1968	0.1993
0.58	0.2019	0.2045	0.2070	0.2096	0.2121	0.2147	0.2173	0.2198	0.2224	0.2250
0.59	0.2275	0.2301	0.2327	0.2353	0.2378	0.2404	0.2430	0.2456	0.2482	0.2508
0.60	0.2533	0.2559	0.2585	0.2611	0.2637	0.2663	0.2689	0.2715	0.2741	0.2767
0.61	0.2793	0.2819	0.2845	0.2871	0.2898	0.2924	0.2950	0.2976	0.3002	0.3029
0.62	0.3055	0.3081	0.3107	0.3134	0.3160	0.3186	0.3213	0.3239	0.3266	0.3292
0.63	0.3319	0.3345	0.3372	0.3398	0.3425	0.3451	0.3478	0.3505	0.3531	0.3558
0.64	0.3585	0.3611	0.3638	0.3665	0.3692	0.3719	0.3745	0.3772	0.3799	0.3826
0.65	0.3853	0.3880	0.3907	0.3934	0.3961	0.3989	0.4016	0.4043	0.4070	0.4097
0.66	0.4125	0.4152	0.4179	0.4207	0.4234	0.4261	0.4289	0.4316	0.4344	0.4372
0.67	0.4399	0.4427	0.4454	0.4482	0.4510	0.4538	0.4565	0.4593	0.4621	0.4649
0.68	0.4677	0.4705	0.4733	0.4761	0.4789	0.4817	0.4845	0.4874	0.4902	0.4930
0.69	0.4959	0.4987	0.5015	0.5044	0.5072	0.5101	0.5129	0.5158	0.5187	0.5215
0.70	0.5244	0.5273	0.5302	0.5330	0.5359	0.5388	0.5417	0.5446	0.5476	0.5505
0.71	0.5534	0.5563	0.5592	0.5622	0.5651	0.5681	0.5710	0.5740	0.5769	0.5799
0.72	0.5828	0.5858	0.5888	0.5918	0.5948	0.5978	0.6008	0.6038	0.6068	0.6098
0.73	0.6128	0.6158	0.6189	0.6219	0.6250	0.6280	0.6311	0.6341	0.6372	0.6403
0.74	0.6433	0.6464	0.6495	0.6526	0.6557	0.6588	0.6620	0.6651	0.6682	0.6713
0.75	0.6745	0.6776	0.6808	0.6840	0.6871	0.6903	0.6935	0.6967	0.6999	0.7031
0.76	0.7063	0.7095	0.7128	0.7160	0.7192	0.7225	0.7257	0.7290	0.7323	0.7356
0.77	0.7388	0.7421	0.7454	0.7488	0.7521	0.7554	0.7588	0.7621	0.7655	0.7688
0.78	0.7722	0.7756	0.7790	0.7824	0.7858	0.7892	0.7926	0.7961	0.7995	0.8030
0.79	0.8064	0.8099	0.8134	0.8169	0.8204	0.8239	0.8274	0.8310	0.8345	0.8381
0.80	0.8416	0.8452	0.8488	0.8524	0.8560	0.8596	0.8633	0.8669	0.8705	0.8742
0.81	0.8779	0.8816	0.8853	0.8890	0.8927	0.8965	0.9002	0.9040	0.9078	0.9116
0.82	0.9154	0.9192	0.9230	0.9269	0.9307	0.9346	0.9385	0.9424	0.9463	0.9502
0.83	0.9542	0.9581	0.9621	0.9661	0.9701	0.9741	0.9782	0.9822	0.9863	0.9904
0.84	0.9945	0.9986	1.0027	1.0069	1.0110	1.0152	1.0194	1.0237	1.0279	1.0322
0.85	1.0364	1.0407	1.0450	1.0494	1.0537	1.0581	1.0625	1.0669	1.0714	1.0758
0.86	1.0803	1.0848	1.0893	1.0939	1.0985	1.1031	1.1077	1.1123	1.1170	1.1217
0.87	1.1264	1.1311	1.1359	1.1407	1.1455	1.1503	1.1552	1.1601	1.1650	1.1700
0.88	1.1750	1.1800	1.1850	1.1901	1.1952	1.2004	1.2055	1.2107	1.2160	1.2212
0.89	1.2265	1.2319	1.2372	1.2426	1.2481	1.2536	1.2591	1.2646	1.2702	1.2759
0.90	1.2816	1.2873	1.2930	1.2988	1.3047	1.3106	1.3165	1.3225	1.3285	1.3346
0.91	1.3408	1.3469	1.3532	1.3595	1.3658	1.3722	1.3787	1.3852	1.3917	1.3984
0.92	1.4051	1.4118	1.4187	1.4255	1.4325	1.4395	1.4466	1.4538	1.4611	1.4684
0.93	1.4758	1.4833	1.4909	1.4985	1.5063	1.5141	1.5220	1.5301	1.5382	1.5464
0.94	1.5548	1.5632	1.5718	1.5805	1.5893	1.5982	1.6072	1.6164	1.6258	1.6352
0.95	1.6449	1.6546	1.6646	1.6747	1.6849	1.6954	1.7060	1.7169	1.7279	1.7392
0.96	1.7507	1.7624	1.7744	1.7866	1.7991	1.8119	1.8250	1.8384	1.8522	1.8663
0.97	1.8808	1.8957	1.9110	1.9268	1.9431	1.9600	1.9774	1.9954	2.0141	2.0335
0.98	2.0537	2.0749	2.0969	2.1201	2.1444	2.1701	2.1973	2.2262	2.2571	2.2904
0.99	2.3263	2.3656	2.4089	2.4573	2.5121	2.5758	2.6521	2.7478	2.8782	3.0902

The row labels give the first two significant digits of the left tail area $[G(Z)]$ of the unit normal distribution; the column labels give the third digit. For example, the 97.5%ile is $Z = 1.9600$. For $G < 0.5$ the percentile is the negative of that for $1 - G$. For example, the 38%ile is $Z = -0.3055$. Percentiles for values of G with more than three significant digits must be obtained by interpolation (see page 40), unless direct algorithms are used. [This Table was computed by iterating Equation (6.2), from a starting value given by Equation (26.2.23) in Abramowitz & Stegun (1972 [1]).]

A.2 t-DISTRIBUTION

A.2.1 Cumulative Student's t-distribution

				abscissa (t)						
f	0.6	0.8	1.0	1.2	1.5	2.0	3.0	4.0	5.0	10.0
2	0.6953	0.7462	0.7887	0.8235	0.8638	0.9082	0.9523	0.9714	0.9811	0.9951
3	0.7046	0.7589	0.8045	0.8419	0.8847	0.9303	0.9712	0.9860	0.9923	0.9989
4	0.7096	0.7657	0.8130	0.8518	0.8960	0.9419	0.9800	0.9919	0.9963	0.9997
5	0.7127	0.7700	0.8184	0.8581	0.9030	0.9490	0.9850	0.9948	0.9979	0.9999
6	0.7148	0.7729	0.8220	0.8623	0.9079	0.9538	0.9880	0.9964	0.9988	1.0000
7	0.7163	0.7750	0.8247	0.8654	0.9114	0.9572	0.9900	0.9974	0.9992	1.0000
8	0.7174	0.7766	0.8267	0.8678	0.9140	0.9597	0.9915	0.9980	0.9995	1.0000
9	0.7183	0.7778	0.8283	0.8696	0.9161	0.9617	0.9925	0.9984	0.9996	1.0000
10	0.7191	0.7788	0.8296	0.8711	0.9177	0.9633	0.9933	0.9987	0.9997	1.0000
11	0.7197	0.7797	0.8306	0.8723	0.9191	0.9646	0.9940	0.9990	0.9998	1.0000
12	0.7202	0.7804	0.8315	0.8734	0.9203	0.9657	0.9945	0.9991	0.9998	1.0000
13	0.7206	0.7810	0.8322	0.8742	0.9212	0.9666	0.9949	0.9992	0.9999	1.0000
14	0.7210	0.7815	0.8329	0.8750	0.9221	0.9674	0.9952	0.9993	0.9999	1.0000
15	0.7213	0.7819	0.8334	0.8756	0.9228	0.9680	0.9955	0.9994	0.9999	1.0000
16	0.7215	0.7823	0.8339	0.8762	0.9235	0.9686	0.9958	0.9995	0.9999	1.0000
17	0.7218	0.7826	0.8343	0.8767	0.9240	0.9691	0.9960	0.9995	0.9999	1.0000
18	0.7220	0.7829	0.8347	0.8772	0.9245	0.9696	0.9962	0.9996	1.0000	1.0000
19	0.7222	0.7832	0.8351	0.8776	0.9250	0.9700	0.9963	0.9996	1.0000	1.0000
20	0.7224	0.7834	0.8354	0.8779	0.9254	0.9704	0.9965	0.9996	1.0000	1.0000
21	0.7225	0.7837	0.8357	0.8782	0.9258	0.9707	0.9966	0.9997	1.0000	1.0000
22	0.7227	0.7839	0.8359	0.8785	0.9261	0.9710	0.9967	0.9997	1.0000	1.0000
23	0.7228	0.7841	0.8361	0.8788	0.9264	0.9713	0.9968	0.9997	1.0000	1.0000
24	0.7229	0.7842	0.8364	0.8791	0.9267	0.9715	0.9969	0.9997	1.0000	1.0000
25	0.7230	0.7844	0.8366	0.8793	0.9269	0.9718	0.9970	0.9998	1.0000	1.0000
26	0.7231	0.7845	0.8367	0.8795	0.9272	0.9720	0.9971	0.9998	1.0000	1.0000
27	0.7232	0.7847	0.8369	0.8797	0.9274	0.9722	0.9971	0.9998	1.0000	1.0000
28	0.7233	0.7848	0.8371	0.8799	0.9276	0.9724	0.9972	0.9998	1.0000	1.0000
29	0.7234	0.7849	0.8372	0.8801	0.9278	0.9725	0.9973	0.9998	1.0000	1.0000
30	0.7235	0.7850	0.8373	0.8802	0.9280	0.9727	0.9973	0.9998	1.0000	1.0000
31	0.7236	0.7851	0.8375	0.8804	0.9281	0.9728	0.9974	0.9998	1.0000	1.0000
32	0.7236	0.7852	0.8376	0.8805	0.9283	0.9730	0.9974	0.9998	1.0000	1.0000
33	0.7237	0.7853	0.8377	0.8807	0.9284	0.9731	0.9974	0.9998	1.0000	1.0000
34	0.7238	0.7854	0.8378	0.8808	0.9286	0.9732	0.9975	0.9998	1.0000	1.0000
35	0.7238	0.7854	0.8379	0.8809	0.9287	0.9733	0.9975	0.9998	1.0000	1.0000
36	0.7239	0.7855	0.8380	0.8810	0.9288	0.9735	0.9976	0.9998	1.0000	1.0000
37	0.7239	0.7856	0.8381	0.8811	0.9290	0.9736	0.9976	0.9999	1.0000	1.0000
38	0.7240	0.7857	0.8382	0.8812	0.9291	0.9737	0.9976	0.9999	1.0000	1.0000
39	0.7240	0.7857	0.8383	0.8813	0.9292	0.9738	0.9977	0.9999	1.0000	1.0000
39	0.7240	0.7857	0.8383	0.8813	0.9292	0.9738	0.9977	0.9999	1.0000	1.0000
45	0.7242	0.7860	0.8387	0.8818	0.9297	0.9742	0.9978	0.9999	1.0000	1.0000
50	0.7244	0.7863	0.8389	0.8821	0.9300	0.9745	0.9979	0.9999	1.0000	1.0000
55	0.7245	0.7864	0.8392	0.8824	0.9303	0.9748	0.9980	0.9999	1.0000	1.0000
60	0.7246	0.7866	0.8393	0.8826	0.9306	0.9750	0.9980	0.9999	1.0000	1.0000
70	0.7248	0.7868	0.8396	0.8829	0.9309	0.9753	0.9981	0.9999	1.0000	1.0000
80	0.7249	0.7870	0.8398	0.8832	0.9312	0.9756	0.9982	0.9999	1.0000	1.0000
90	0.7250	0.7871	0.8400	0.8834	0.9314	0.9757	0.9983	0.9999	1.0000	1.0000
100	0.7251	0.7872	0.8401	0.8835	0.9316	0.9759	0.9983	0.9999	1.0000	1.0000
200	0.7254	0.7877	0.8407	0.8842	0.9324	0.9766	0.9985	1.0000	1.0000	1.0000
500	0.7256	0.7880	0.8411	0.8846	0.9329	0.9770	0.9986	1.0000	1.0000	1.0000
1000	0.7257	0.7880	0.8412	0.8848	0.9330	0.9771	0.9986	1.0000	1.0000	1.0000
10000	0.7257	0.7881	0.8413	0.8849	0.9332	0.9772	0.9986	1.0000	1.0000	1.0000
Infty	0.7257	0.7881	0.8413	0.8849	0.9332	0.9772	0.9987	1.0000	1.0000	1.0000

Cumulative areas for values of f or t not appearing on this table can be obtained by interpolation (see page 40). This table was prepared using the algorithm given in Section 6.2.10.

A.2.2 Percentiles of Student's t-distribution

f	|	0.75	0.80	0.90	0.95	0.975	0.99	0.995	0.9975	0.999	0.9995
	|	\multicolumn{10}{c}{cumulative area}

f	0.75	0.80	0.90	0.95	0.975	0.99	0.995	0.9975	0.999	0.9995
2	0.8165	1.0607	1.8856	2.9200	4.3027	6.9646	9.9248	14.0890	22.3271	31.5991
3	0.7649	0.9785	1.6377	2.3534	3.1824	4.5407	5.8409	7.4533	10.2145	12.9240
4	0.7407	0.9410	1.5332	2.1318	2.7764	3.7469	4.6041	5.5976	7.1732	8.6103
5	0.7267	0.9195	1.4759	2.0150	2.5706	3.3649	4.0321	4.7733	5.8934	6.8688
6	0.7176	0.9057	1.4398	1.9432	2.4469	3.1427	3.7074	4.3168	5.2076	5.9588
7	0.7111	0.8960	1.4149	1.8946	2.3646	2.9980	3.4995	4.0293	4.7853	5.4079
8	0.7064	0.8889	1.3968	1.8595	2.3060	2.8965	3.3554	3.8325	4.5008	5.0413
9	0.7027	0.8834	1.3830	1.8331	2.2622	2.8214	3.2498	3.6897	4.2968	4.7809
10	0.6998	0.8791	1.3722	1.8125	2.2281	2.7638	3.1693	3.5814	4.1437	4.5869
11	0.6974	0.8755	1.3634	1.7959	2.2010	2.7181	3.1058	3.4966	4.0247	4.4370
12	0.6955	0.8726	1.3562	1.7823	2.1788	2.6810	3.0545	3.4284	3.9296	4.3178
13	0.6938	0.8702	1.3502	1.7709	2.1604	2.6503	3.0123	3.3725	3.8520	4.2208
14	0.6924	0.8681	1.3450	1.7613	2.1448	2.6245	2.9768	3.3257	3.7874	4.1405
15	0.6912	0.8662	1.3406	1.7531	2.1314	2.6025	2.9467	3.2860	3.7328	4.0728
16	0.6901	0.8647	1.3368	1.7459	2.1199	2.5835	2.9208	3.2520	3.6862	4.0150
17	0.6892	0.8633	1.3334	1.7396	2.1098	2.5669	2.8982	3.2224	3.6458	3.9651
18	0.6884	0.8620	1.3304	1.7341	2.1009	2.5524	2.8784	3.1966	3.6105	3.9216
19	0.6876	0.8610	1.3277	1.7291	2.0930	2.5395	2.8609	3.1737	3.5794	3.8834
20	0.6870	0.8600	1.3253	1.7247	2.0860	2.5280	2.8453	3.1534	3.5518	3.8495
21	0.6864	0.8591	1.3232	1.7207	2.0796	2.5176	2.8314	3.1352	3.5272	3.8193
22	0.6858	0.8583	1.3212	1.7171	2.0739	2.5083	2.8188	3.1188	3.5050	3.7921
23	0.6853	0.8575	1.3195	1.7139	2.0687	2.4999	2.8073	3.1040	3.4850	3.7676
24	0.6848	0.8569	1.3178	1.7109	2.0639	2.4922	2.7969	3.0905	3.4668	3.7454
25	0.6844	0.8562	1.3163	1.7081	2.0595	2.4851	2.7874	3.0782	3.4502	3.7251
26	0.6840	0.8557	1.3150	1.7056	2.0555	2.4786	2.7787	3.0669	3.4350	3.7066
27	0.6837	0.8551	1.3137	1.7033	2.0518	2.4727	2.7707	3.0565	3.4210	3.6896
28	0.6834	0.8546	1.3125	1.7011	2.0484	2.4671	2.7633	3.0469	3.4082	3.6739
29	0.6830	0.8542	1.3114	1.6991	2.0452	2.4620	2.7564	3.0380	3.3962	3.6594
30	0.6828	0.8538	1.3104	1.6973	2.0423	2.4573	2.7500	3.0298	3.3852	3.6460
31	0.6825	0.8534	1.3095	1.6955	2.0395	2.4528	2.7440	3.0221	3.3749	3.6335
32	0.6822	0.8530	1.3086	1.6939	2.0369	2.4487	2.7385	3.0149	3.3653	3.6218
33	0.6820	0.8526	1.3077	1.6924	2.0345	2.4448	2.7333	3.0082	3.3563	3.6109
34	0.6818	0.8523	1.3070	1.6909	2.0322	2.4411	2.7284	3.0020	3.3479	3.6007
35	0.6816	0.8520	1.3062	1.6896	2.0301	2.4377	2.7238	2.9960	3.3400	3.5911
36	0.6814	0.8517	1.3055	1.6883	2.0281	2.4345	2.7195	2.9905	3.3326	3.5821
37	0.6812	0.8514	1.3049	1.6871	2.0262	2.4314	2.7154	2.9852	3.3256	3.5737
38	0.6810	0.8512	1.3042	1.6860	2.0244	2.4286	2.7116	2.9803	3.3190	3.5657
39	0.6808	0.8509	1.3036	1.6849	2.0227	2.4258	2.7079	2.9756	3.3128	3.5581
39	0.6808	0.8509	1.3036	1.6849	2.0227	2.4258	2.7079	2.9756	3.3128	3.5581
45	0.6800	0.8497	1.3006	1.6794	2.0141	2.4121	2.6896	2.9521	3.2815	3.5203
50	0.6794	0.8489	1.2987	1.6759	2.0086	2.4033	2.6778	2.9370	3.2614	3.4960
55	0.6790	0.8482	1.2971	1.6730	2.0040	2.3961	2.6682	2.9247	3.2451	3.4764
60	0.6786	0.8477	1.2958	1.6706	2.0003	2.3901	2.6603	2.9146	3.2317	3.4602
70	0.6780	0.8468	1.2938	1.6669	1.9944	2.3808	2.6479	2.8987	3.2108	3.4350
80	0.6776	0.8461	1.2922	1.6641	1.9901	2.3739	2.6387	2.8870	3.1953	3.4163
90	0.6772	0.8456	1.2910	1.6620	1.9867	2.3685	2.6316	2.8779	3.1833	3.4019
100	0.6770	0.8452	1.2901	1.6602	1.9840	2.3642	2.6259	2.8707	3.1737	3.3905
200	0.6757	0.8434	1.2858	1.6525	1.9719	2.3451	2.6006	2.8385	3.1315	3.3398
500	0.6750	0.8423	1.2832	1.6479	1.9647	2.3338	2.5857	2.8195	3.1066	3.3101
1000	0.6747	0.8420	1.2824	1.6464	1.9623	2.3301	2.5808	2.8133	3.0984	3.3003
10000	0.6745	0.8417	1.2816	1.6450	1.9602	2.3267	2.5763	2.8077	3.0910	3.2915
Infty	0.6745	0.8416	1.2816	1.6449	1.9600	2.3263	2.5758	2.8070	3.0902	3.2905

Percentiles for values of f or cumulative areas not appearing on this table can be obtained by interpolation (see page 40). This table was prepared using an iterative interval-halving technique, using the same probability calculator as used for Table A.2.1. Note that the least line of the table (for infinite degrees of freedom) corresponds to the appropriate percentiles in Table A.1.2.

A.3 BINOMIAL DISTRIBUTION

A.3.1 Cumulative binomial probabilities

Probability that a single random sample will be an exceedance (θ)

n	e	0.01	0.02	0.05	0.10	0.20	0.50	0.80	0.90	0.95	0.98	0.99
7	0	0.9321	0.8681	0.6983	0.4783	0.2097	0.0078	0.0000	0.0000	0.0000	0.0000	0.0000
	1	0.9980	0.9921	0.9556	0.8503	0.5767	0.0625	0.0004	0.0000	0.0000	0.0000	0.0000
	2	1.0000	0.9997	0.9962	0.9743	0.8520	0.2266	0.0047	0.0002	0.0000	0.0000	0.0000
	3	1.0000	1.0000	0.9998	0.9973	0.9667	0.5000	0.0333	0.0027	0.0002	0.0000	0.0000
	4	1.0000	1.0000	1.0000	0.9998	0.9953	0.7734	0.1480	0.0257	0.0038	0.0003	0.0000
	5	1.0000	1.0000	1.0000	1.0000	0.9996	0.9375	0.4233	0.1497	0.0444	0.0079	0.0020
	6	1.0000	1.0000	1.0000	1.0000	1.0000	0.9922	0.7903	0.5217	0.3017	0.1319	0.0679
	7	1.0000	1.0000	1.0000	1.0000	1.0000	1.0000	1.0000	1.0000	1.0000	1.0000	1.0000
14	0	0.8687	0.7536	0.4877	0.2288	0.0440	0.0001	0.0000	0.0000	0.0000	0.0000	0.0000
	1	0.9916	0.9690	0.8470	0.5846	0.1979	0.0009	0.0000	0.0000	0.0000	0.0000	0.0000
	2	0.9997	0.9975	0.9699	0.8416	0.4481	0.0065	0.0000	0.0000	0.0000	0.0000	0.0000
	3	1.0000	0.9999	0.9958	0.9559	0.6982	0.0287	0.0000	0.0000	0.0000	0.0000	0.0000
	4	1.0000	1.0000	0.9996	0.9908	0.8702	0.0898	0.0000	0.0000	0.0000	0.0000	0.0000
	5	1.0000	1.0000	1.0000	0.9985	0.9561	0.2120	0.0004	0.0000	0.0000	0.0000	0.0000
	6	1.0000	1.0000	1.0000	0.9998	0.9884	0.3953	0.0024	0.0000	0.0000	0.0000	0.0000
	7	1.0000	1.0000	1.0000	1.0000	0.9976	0.6047	0.0116	0.0002	0.0000	0.0000	0.0000
	8	1.0000	1.0000	1.0000	1.0000	0.9996	0.7880	0.0439	0.0015	0.0000	0.0000	0.0000
	9	1.0000	1.0000	1.0000	1.0000	1.0000	0.9102	0.1298	0.0092	0.0004	0.0000	0.0000
	10	1.0000	1.0000	1.0000	1.0000	1.0000	0.9713	0.3018	0.0441	0.0042	0.0001	0.0000
	11	1.0000	1.0000	1.0000	1.0000	1.0000	0.9935	0.5519	0.1584	0.0301	0.0025	0.0003
	12	1.0000	1.0000	1.0000	1.0000	1.0000	0.9991	0.8021	0.4154	0.1530	0.0310	0.0084
	13	1.0000	1.0000	1.0000	1.0000	1.0000	0.9999	0.9560	0.7712	0.5123	0.2464	0.1313
	14	1.0000	1.0000	1.0000	1.0000	1.0000	1.0000	1.0000	1.0000	1.0000	1.0000	1.0000
28	0	0.7547	0.5680	0.2378	0.0523	0.0019	0.0000	0.0000	0.0000	0.0000	0.0000	0.0000
	1	0.9682	0.8925	0.5883	0.2152	0.0155	0.0000	0.0000	0.0000	0.0000	0.0000	0.0000
	2	0.9973	0.9820	0.8373	0.4594	0.0612	0.0000	0.0000	0.0000	0.0000	0.0000	0.0000
	3	0.9998	0.9978	0.9509	0.6946	0.1602	0.0000	0.0000	0.0000	0.0000	0.0000	0.0000
	4	1.0000	0.9998	0.9883	0.8579	0.3149	0.0001	0.0000	0.0000	0.0000	0.0000	0.0000
	5	1.0000	1.0000	0.9977	0.9450	0.5005	0.0005	0.0000	0.0000	0.0000	0.0000	0.0000
	6	1.0000	1.0000	0.9996	0.9821	0.6784	0.0019	0.0000	0.0000	0.0000	0.0000	0.0000
	7	1.0000	1.0000	1.0000	0.9950	0.8182	0.0063	0.0000	0.0000	0.0000	0.0000	0.0000
	8	1.0000	1.0000	1.0000	0.9988	0.9100	0.0178	0.0000	0.0000	0.0000	0.0000	0.0000
	9	1.0000	1.0000	1.0000	0.9998	0.9609	0.0436	0.0000	0.0000	0.0000	0.0000	0.0000
	10	1.0000	1.0000	1.0000	1.0000	0.9851	0.0925	0.0000	0.0000	0.0000	0.0000	0.0000
	11	1.0000	1.0000	1.0000	1.0000	0.9950	0.1725	0.0000	0.0000	0.0000	0.0000	0.0000
	12	1.0000	1.0000	1.0000	1.0000	0.9985	0.2858	0.0000	0.0000	0.0000	0.0000	0.0000
	13	1.0000	1.0000	1.0000	1.0000	0.9996	0.4253	0.0001	0.0000	0.0000	0.0000	0.0000
	14	1.0000	1.0000	1.0000	1.0000	0.9999	0.5747	0.0004	0.0000	0.0000	0.0000	0.0000
	15	1.0000	1.0000	1.0000	1.0000	1.0000	0.7142	0.0015	0.0000	0.0000	0.0000	0.0000
	16	1.0000	1.0000	1.0000	1.0000	1.0000	0.8275	0.0050	0.0000	0.0000	0.0000	0.0000
	17	1.0000	1.0000	1.0000	1.0000	1.0000	0.9075	0.0149	0.0000	0.0000	0.0000	0.0000
	18	1.0000	1.0000	1.0000	1.0000	1.0000	0.9564	0.0391	0.0002	0.0000	0.0000	0.0000
	19	1.0000	1.0000	1.0000	1.0000	1.0000	0.9822	0.0900	0.0012	0.0000	0.0000	0.0000
	20	1.0000	1.0000	1.0000	1.0000	1.0000	0.9937	0.1818	0.0050	0.0000	0.0000	0.0000
	21	1.0000	1.0000	1.0000	1.0000	1.0000	0.9981	0.3216	0.0179	0.0004	0.0000	0.0000
	22	1.0000	1.0000	1.0000	1.0000	1.0000	0.9995	0.4995	0.0550	0.0023	0.0000	0.0000
	23	1.0000	1.0000	1.0000	1.0000	1.0000	0.9999	0.6851	0.1421	0.0117	0.0002	0.0000
	24	1.0000	1.0000	1.0000	1.0000	1.0000	1.0000	0.8398	0.3054	0.0491	0.0022	0.0002
	25	1.0000	1.0000	1.0000	1.0000	1.0000	1.0000	0.9388	0.5406	0.1627	0.0180	0.0027
	26	1.0000	1.0000	1.0000	1.0000	1.0000	1.0000	0.9845	0.7848	0.4117	0.1075	0.0318
	27	1.0000	1.0000	1.0000	1.0000	1.0000	1.0000	0.9981	0.9477	0.7622	0.4320	0.2453
	28	1.0000	1.0000	1.0000	1.0000	1.0000	1.0000	1.0000	1.0000	1.0000	1.0000	1.0000

Cumulative binomial probabilities of getting up to e exceedances in n random trials when each trial has a probability θ of being an exceedance. The probabilities were calculated using the recursive algorithm given in Section 6.2.5.

A.4 NORMAL CONFIDENCE AND TOLERANCE INTERVALS

A.4.1 k factor for two-sided 95% confidence intervals on a normal percentile

n	80%ile LCL	80%ile UCL	90%ile LCL	90%ile UCL	95%ile LCL	95%ile UCL	99%ile LCL	99%ile UCL
2	-1.2285	28.1397	-0.1429	41.2008	0.2730	52.5593	0.7607	74.2338
3	-0.3803	6.3433	0.1585	8.7969	0.4785	10.9268	0.9580	15.0425
4	-0.1579	3.9145	0.2977	5.3541	0.6007	6.6015	1.0878	9.0176
5	-0.0375	3.0584	0.3886	4.1662	0.6868	5.1242	1.1823	6.9795
6	0.0431	2.6209	0.4552	3.5681	0.7522	4.3854	1.2556	5.9675
7	0.1028	2.3527	0.5073	3.2055	0.8044	3.9401	1.3148	5.3607
8	0.1496	2.1697	0.5497	2.9605	0.8475	3.6404	1.3640	4.9543
9	0.1877	2.0359	0.5851	2.7826	0.8839	3.4238	1.4059	4.6616
10	0.2197	1.9331	0.6153	2.6469	0.9152	3.2590	1.4422	4.4398
11	0.2470	1.8513	0.6416	2.5395	0.9427	3.1290	1.4741	4.2652
12	0.2707	1.7843	0.6648	2.4520	0.9669	3.0233	1.5024	4.1237
13	0.2917	1.7283	0.6854	2.3792	0.9886	2.9355	1.5278	4.0064
14	0.3103	1.6806	0.7039	2.3174	1.0082	2.8612	1.5507	3.9074
15	0.3270	1.6394	0.7206	2.2643	1.0259	2.7974	1.5716	3.8224
16	0.3421	1.6034	0.7359	2.2180	1.0421	2.7419	1.5908	3.7487
17	0.3559	1.5716	0.7499	2.1772	1.0571	2.6930	1.6084	3.6839
18	0.3686	1.5433	0.7627	2.1409	1.0708	2.6497	1.6247	3.6265
19	0.3803	1.5178	0.7747	2.1084	1.0836	2.6109	1.6398	3.5752
20	0.3910	1.4948	0.7858	2.0791	1.0955	2.5760	1.6540	3.5291
21	0.4011	1.4738	0.7961	2.0525	1.1066	2.5443	1.6672	3.4873
22	0.4104	1.4547	0.8058	2.0282	1.1170	2.5154	1.6796	3.4492
23	0.4191	1.4371	0.8149	2.0060	1.1268	2.4890	1.6912	3.4144
24	0.4274	1.4209	0.8234	1.9855	1.1360	2.4646	1.7023	3.3823
25	0.4351	1.4058	0.8315	1.9665	1.1447	2.4421	1.7127	3.3528
26	0.4424	1.3918	0.8391	1.9489	1.1530	2.4212	1.7225	3.3253
27	0.4492	1.3788	0.8463	1.9325	1.1608	2.4018	1.7319	3.2999
28	0.4558	1.3666	0.8532	1.9172	1.1683	2.3837	1.7408	3.2761
29	0.4620	1.3552	0.8597	1.9028	1.1754	2.3667	1.7493	3.2539
30	0.4679	1.3444	0.8659	1.8894	1.1821	2.3508	1.7575	3.2330
31	0.4735	1.3343	0.8719	1.8767	1.1886	2.3358	1.7652	3.2134
32	0.4788	1.3247	0.8776	1.8647	1.1948	2.3217	1.7726	3.1950
33	0.4840	1.3156	0.8830	1.8534	1.2007	2.3083	1.7798	3.1775
34	0.4889	1.3070	0.8882	1.8427	1.2064	2.2957	1.7866	3.1610
35	0.4935	1.2989	0.8932	1.8326	1.2118	2.2837	1.7932	3.1454
40	0.5144	1.2635	0.9155	1.7886	1.2362	2.2319	1.8226	3.0779
45	0.5318	1.2350	0.9342	1.7533	1.2566	2.1905	1.8473	3.0240
50	0.5466	1.2114	0.9502	1.7242	1.2742	2.1564	1.8686	2.9797
55	0.5594	1.1915	0.9641	1.6997	1.2895	2.1277	1.8871	2.9426
60	0.5706	1.1744	0.9763	1.6788	1.3029	2.1032	1.9035	2.9109
70	0.5895	1.1464	0.9969	1.6446	1.3256	2.0632	1.9311	2.8592
80	0.6047	1.1243	1.0137	1.6177	1.3442	2.0319	1.9538	2.8188
90	0.6175	1.1063	1.0277	1.5958	1.3597	2.0064	1.9728	2.7861
100	0.6284	1.0913	1.0397	1.5776	1.3730	1.9853	1.9891	2.7589
200	0.6881	1.0131	1.1064	1.4835	1.4473	1.8763	2.0804	2.6193
500	0.7429	0.9475	1.1682	1.4055	1.5167	1.7865	2.1669	2.5061
1000	0.7712	0.9156	1.2004	1.3680	1.5532	1.7437	2.2118	2.4510
2000	0.7915	0.8935	1.2237	1.3421	1.5793	1.7139	2.2444	2.4133
5000	0.8097	0.8743	1.2447	1.3195	1.6030	1.6881	2.2740	2.3807
10000	0.8190	0.8646	1.2554	1.3083	1.6151	1.6753	2.2891	2.3646
Infty	0.8416	0.8416	1.2816	1.2816	1.6449	1.6449	2.3263	2.3263

Factors k_{LCL} and k_{UCL} for lower and upper two-sided 95% confidence limits on four percentiles of a normal distribution. The lower limit is $\bar{X} + k_{\text{LCL}} S$ where $\bar{X}$ and S are the mean and standard deviation of n samples. Similarly the upper limit is $\bar{X} + k_{\text{UCL}} S$. The method of calculation is described in Section 6.3.1.

A.4.2 k factor for one-sided confidence (or tolerance) intervals on a normal percentile

	50%ile Level of confidence			80%ile Level of confidence			90%ile Level of confidence		
n	5%	50%	95%	5%	50%	95%	5%	50%	95%
2	-4.4645	0.0000	4.4645	-0.5214	1.1270	14.0509	0.1380	1.7842	20.5815
3	-1.6859	0.0000	1.6859	-0.1274	0.9707	4.4241	0.3345	1.4985	6.1553
4	-1.1767	0.0000	1.1767	0.0207	0.9246	3.0260	0.4439	1.4189	4.1619
5	-0.9534	0.0000	0.9534	0.1100	0.9026	2.4833	0.5188	1.3818	3.4066
6	-0.8226	0.0000	0.8226	0.1727	0.8898	2.1907	0.5748	1.3605	3.0063
7	-0.7345	0.0000	0.7345	0.2204	0.8815	2.0051	0.6190	1.3466	2.7554
8	-0.6698	0.0000	0.6698	0.2583	0.8756	1.8754	0.6552	1.3369	2.5819
9	-0.6198	0.0000	0.6198	0.2896	0.8712	1.7787	0.6856	1.3296	2.4538
10	-0.5797	0.0000	0.5797	0.3160	0.8678	1.7035	0.7116	1.3241	2.3546
11	-0.5465	0.0000	0.5465	0.3387	0.8651	1.6429	0.7342	1.3197	2.2753
12	-0.5184	0.0000	0.5184	0.3585	0.8629	1.5928	0.7541	1.3161	2.2101
13	-0.4943	0.0000	0.4943	0.3761	0.8611	1.5506	0.7719	1.3132	2.1554
14	-0.4733	0.0000	0.4733	0.3917	0.8596	1.5144	0.7878	1.3107	2.1088
15	-0.4548	0.0000	0.4548	0.4058	0.8583	1.4830	0.8023	1.3085	2.0684
20	-0.3866	0.0000	0.3866	0.4598	0.8538	1.3714	0.8585	1.3013	1.9260
30	-0.3102	0.0000	0.3102	0.5250	0.8496	1.2530	0.9276	1.2944	1.7773
40	-0.2664	0.0000	0.2664	0.5645	0.8475	1.1884	0.9702	1.2911	1.6972
50	-0.2371	0.0000	0.2371	0.5919	0.8463	1.1465	1.0000	1.2891	1.6456
52	-0.2323	0.0000	0.2323	0.5965	0.8461	1.1397	1.0050	1.2888	1.6373
60	-0.2157	0.0000	0.2157	0.6123	0.8455	1.1165	1.0224	1.2878	1.6089
80	-0.1861	0.0000	0.1861	0.6414	0.8445	1.0758	1.0544	1.2862	1.5594
100	-0.1660	0.0000	0.1660	0.6614	0.8439	1.0488	1.0767	1.2853	1.5267
200	-0.1169	0.0000	0.1169	0.7121	0.8428	0.9844	1.1335	1.2834	1.4496
500	-0.0737	0.0000	0.0737	0.7585	0.8421	0.9301	1.1860	1.2823	1.3851
1000	-0.0521	0.0000	0.0521	0.7823	0.8419	0.9035	1.2132	1.2819	1.3538
10000	-0.0165	0.0000	0.0165	0.8226	0.8416	0.8609	1.2596	1.2816	1.3040
Infty	0.0000	0.0000	0.0000	0.8416	0.8416	0.8416	1.2816	1.2816	1.2816

	95%ile			98%ile			99%ile		
2	0.4748	2.3387	26.2597	0.7756	2.9624	32.7500	0.9538	3.3760	37.0936
3	0.6391	1.9384	7.6559	0.9416	2.4342	9.3855	1.1297	2.7645	10.5527
4	0.7433	1.8295	5.1439	1.0512	2.2923	6.2767	1.2462	2.6008	7.0424
5	0.8178	1.7793	4.2027	1.1308	2.2272	5.1206	1.3309	2.5258	5.7411
6	0.8748	1.7505	3.7077	1.1923	2.1899	4.5158	1.3964	2.4828	5.0620
7	0.9204	1.7318	3.3995	1.2417	2.1657	4.1409	1.4492	2.4551	4.6417
8	0.9580	1.7187	3.1873	1.2828	2.1489	3.8836	1.4931	2.4357	4.3539
9	0.9899	1.7091	3.0312	1.3176	2.1364	3.6950	1.5303	2.4213	4.1430
10	1.0173	1.7016	2.9110	1.3476	2.1268	3.5499	1.5625	2.4103	3.9811
11	1.0413	1.6957	2.8150	1.3740	2.1192	3.4345	1.5908	2.4016	3.8523
12	1.0625	1.6910	2.7363	1.3973	2.1131	3.3400	1.6158	2.3945	3.7471
13	1.0814	1.6870	2.6705	1.4182	2.1080	3.2611	1.6382	2.3886	3.6592
14	1.0985	1.6837	2.6144	1.4371	2.1037	3.1939	1.6585	2.3837	3.5845
15	1.1140	1.6808	2.5660	1.4542	2.1000	3.1360	1.6769	2.3795	3.5201
20	1.1746	1.6712	2.3960	1.5215	2.0876	2.9334	1.7492	2.3652	3.2952
30	1.2498	1.6620	2.2198	1.6054	2.0758	2.7246	1.8397	2.3516	3.0639
40	1.2966	1.6575	2.1255	1.6578	2.0701	2.6133	1.8963	2.3451	2.9409
50	1.3294	1.6549	2.0650	1.6947	2.0667	2.5422	1.9362	2.3412	2.8624
52	1.3349	1.6545	2.0553	1.7009	2.0662	2.5308	1.9428	2.3406	2.8499
60	1.3541	1.6532	2.0222	1.7226	2.0645	2.4919	1.9663	2.3387	2.8071
80	1.3896	1.6511	1.9644	1.7627	2.0618	2.4244	2.0097	2.3355	2.7326
100	1.4143	1.6498	1.9265	1.7907	2.0601	2.3801	2.0401	2.3337	2.6840
200	1.4778	1.6473	1.8372	1.8627	2.0569	2.2761	2.1182	2.3300	2.5697
500	1.5367	1.6458	1.7630	1.9298	2.0550	2.1901	2.1917	2.3275	2.4760
1000	1.5676	1.6453	1.7274	1.9652	2.0543	2.1489	2.2298	2.3269	2.4304
10000	1.6199	1.6449	1.6704	2.0251	2.0538	2.0831	2.2950	2.3264	2.3584
Infty	1.6449	1.6449	1.6449	2.0537	2.0537	2.0537	2.3263	2.3263	2.3263

Factors k for lower (5%) and upper (95%) one-sided confidence limits on six percentiles of a normal distribution. The limits are given by $\bar{X} + kS$, where $\bar{X}$ and S are the mean and standard deviation of n samples. The method of calculation is described in Section 6.3.2.

A.4.3 k factor for two-sided tolerance intervals on a normal percentile

	90% confidence				95% confidence				99% confidence			
	Coverage				Coverage				Coverage			
n	80%	90%	95%	99%	80%	90%	95%	99%	80%	90%	95%	99%
3	4.577	5.788	6.823	8.819	6.572	8.306	9.789	12.647	14.867	18.782	22.131	28.586
4	3.276	4.157	4.913	6.372	4.233	5.368	6.341	8.221	7.431	9.416	11.118	14.405
5	2.750	3.499	4.142	5.387	3.375	4.291	5.077	6.598	5.240	6.655	7.870	10.220
6	2.464	3.141	3.723	4.850	2.930	3.733	4.422	5.758	4.231	5.383	6.373	8.292
7	2.282	2.913	3.456	4.508	2.657	3.390	4.020	5.241	3.656	4.658	5.520	7.191
8	2.156	2.754	3.270	4.271	2.472	3.156	3.746	4.889	3.284	4.189	4.968	6.479
9	2.062	2.637	3.132	4.094	2.337	2.986	3.546	4.633	3.024	3.860	4.581	5.980
10	1.990	2.546	3.026	3.958	2.234	2.856	3.393	4.437	2.831	3.617	4.294	5.610
11	1.932	2.473	2.941	3.849	2.152	2.754	3.273	4.282	2.682	3.429	4.073	5.324
12	1.885	2.414	2.871	3.759	2.086	2.670	3.175	4.156	2.563	3.279	3.896	5.096
13	1.846	2.364	2.812	3.684	2.031	2.601	3.093	4.051	2.466	3.156	3.751	4.909
14	1.812	2.322	2.762	3.620	1.985	2.542	3.024	3.962	2.386	3.054	3.631	4.753
15	1.783	2.285	2.720	3.565	1.945	2.492	2.965	3.885	2.317	2.967	3.529	4.621
16	1.758	2.254	2.682	3.517	1.911	2.449	2.913	3.819	2.259	2.893	3.441	4.507
17	1.736	2.226	2.649	3.474	1.881	2.410	2.868	3.761	2.207	2.828	3.364	4.408
18	1.717	2.201	2.620	3.436	1.854	2.376	2.828	3.709	2.163	2.771	3.297	4.321
19	1.699	2.178	2.593	3.402	1.830	2.346	2.793	3.663	2.123	2.720	3.237	4.244
20	1.683	2.158	2.570	3.372	1.809	2.319	2.760	3.621	2.087	2.675	3.184	4.175
21	1.669	2.140	2.548	3.344	1.789	2.294	2.731	3.583	2.055	2.635	3.136	4.113
22	1.656	2.123	2.528	3.318	1.772	2.272	2.705	3.549	2.026	2.598	3.092	4.056
23	1.644	2.108	2.510	3.295	1.755	2.251	2.681	3.518	2.000	2.564	3.053	4.005
24	1.633	2.094	2.494	3.274	1.741	2.232	2.658	3.489	1.976	2.534	3.017	3.958
25	1.623	2.081	2.479	3.254	1.727	2.215	2.638	3.462	1.954	2.506	2.984	3.915
26	1.613	2.069	2.464	3.235	1.714	2.199	2.619	3.437	1.934	2.480	2.953	3.875
27	1.604	2.058	2.451	3.218	1.703	2.184	2.601	3.415	1.915	2.456	2.925	3.838
28	1.596	2.048	2.439	3.202	1.692	2.170	2.585	3.393	1.898	2.434	2.898	3.804
29	1.589	2.038	2.427	3.187	1.682	2.157	2.569	3.373	1.882	2.413	2.874	3.772
30	1.581	2.029	2.417	3.173	1.672	2.145	2.555	3.355	1.866	2.394	2.851	3.742
31	1.575	2.020	2.406	3.160	1.663	2.134	2.541	3.337	1.852	2.376	2.829	3.715
32	1.568	2.012	2.397	3.148	1.655	2.123	2.529	3.320	1.839	2.359	2.809	3.688
33	1.562	2.005	2.388	3.136	1.647	2.113	2.517	3.305	1.826	2.343	2.790	3.664
34	1.557	1.997	2.379	3.125	1.639	2.103	2.505	3.290	1.815	2.328	2.773	3.640
35	1.551	1.991	2.371	3.114	1.632	2.094	2.495	3.276	1.803	2.314	2.756	3.618
36	1.546	1.984	2.363	3.104	1.626	2.086	2.484	3.263	1.793	2.300	2.740	3.598
37	1.541	1.978	2.356	3.095	1.619	2.077	2.475	3.250	1.783	2.287	2.725	3.578
38	1.537	1.972	2.349	3.086	1.613	2.070	2.466	3.238	1.773	2.275	2.710	3.559
39	1.532	1.966	2.343	3.077	1.607	2.062	2.457	3.227	1.764	2.264	2.697	3.541
40	1.528	1.961	2.336	3.069	1.602	2.055	2.448	3.216	1.756	2.253	2.684	3.524
45	1.510	1.938	2.308	3.032	1.578	2.024	2.412	3.168	1.718	2.205	2.627	3.450
50	1.495	1.918	2.285	3.003	1.558	1.999	2.382	3.129	1.688	2.166	2.580	3.390
55	1.482	1.902	2.266	2.978	1.541	1.978	2.356	3.096	1.663	2.134	2.542	3.339
60	1.471	1.888	2.250	2.956	1.527	1.960	2.335	3.068	1.641	2.106	2.509	3.297
70	1.454	1.866	2.224	2.922	1.504	1.931	2.300	3.023	1.607	2.062	2.457	3.228
80	1.441	1.849	2.203	2.895	1.487	1.908	2.274	2.988	1.580	2.028	2.416	3.175
90	1.430	1.835	2.186	2.873	1.473	1.890	2.252	2.959	1.559	2.001	2.384	3.133
100	1.421	1.823	2.172	2.855	1.461	1.875	2.234	2.936	1.541	1.978	2.357	3.098
200	1.375	1.764	2.102	2.763	1.401	1.798	2.143	2.816	1.454	1.866	2.223	2.921
500	1.338	1.717	2.046	2.689	1.354	1.737	2.070	2.721	1.385	1.777	2.117	2.783
1000	1.320	1.695	2.019	2.654	1.331	1.709	2.036	2.676	1.352	1.736	2.068	2.718
2000	1.309	1.679	2.001	2.630	1.316	1.689	2.013	2.645	1.331	1.708	2.035	2.675
5000	1.298	1.666	1.986	2.610	1.303	1.673	1.993	2.619	1.312	1.684	2.007	2.637
10000	1.293	1.660	1.978	2.600	1.297	1.664	1.983	2.606	1.303	1.672	1.993	2.619
Infty	1.282	1.645	1.960	2.576	1.282	1.645	1.960	2.576	1.282	1.645	1.960	2.576

Factors k for two-sided tolerance limits (control center) for four coverages of a normal distribution. The limits are given by $\bar{X} \pm kS$, where $\bar{X}$ and S are the mean and standard deviation of n samples. The method of calculation is described in Section 6.3.3.

WILEY SERIES IN STATISTICS IN PRACTICE

Advisory Editor, MARIAN SCOTT, *University of Glasgow, Scotland, UK*

Founding Editor, VIC BARNETT, *Nottingham Trent University, UK*

Human and Biological Sciences

    Brown and Prescott · Applied Mixed Models in Medicine
    Ellenberg, Fleming and DeMets · Data Monitoring Committees in Clinical Trials:
      A Practical Perspective
    Lawson, Browne and Vidal Rodeiro · Disease Mapping With WinBUGS and MLwiN
    Lui · Statistical Estimation of Epidemiological Risk
  *Marubini and Valsecchi · Analysing Survival Data from Clinical Trials and
      Observation Studies
    Parmigiani · Modeling in Medical Decision Making: A Bayesian Approach
    Senn · Cross-over Trials in Clinical Research, *Second Edition*
    Senn · Statistical Issues in Drug Development
    Spiegelhalter, Abrams and Myles · Bayesian Approaches to Clinical Trials and Health-
      Care Evaluation
    Whitehead · Design and Analysis of Sequential Clinical Trials, *Revised Second Edition*
    Whitehead · Meta-Analysis of Controlled Clinical Trials

Earth and Environmental Sciences

    Buck, Cavanagh and Litton · Bayesian Approach to Interpreting Archaeological Data
    Glasbey and Horgan · Image Analysis in the Biological Sciences
    McBride · Using Statistical Methods for Water Quality Management: Issues, Problems
      and Solutions
    Helsel · Nondetects And Data Analysis: Statistics for Censored Environmental Data
    Webster and Oliver · Geostatistics for Environmental Scientists

Industry, Commerce and Finance

    Aitken and Taroni · Statistics and the Evaluation of Evidence for Forensic Scientists,
      Second Edition
    Lehtonen and Pahkinen · Practical Methods for Design and Analysis of Complex Surveys,
      Second Edition
    Ohser and Mücklich · Statistical Analysis of Microstructures in Materials Science

*Now available in paperback.